ADVANCES IN BIOCHEMICAL ENGINEERING

Volume 12

Editors: T. K. Ghose, A. Fiechter,
N. Blakebrough

Managing Editor: A. Fiechter

With 99 Figures

Springer-Verlag
Berlin Heidelberg New York 1979

ISBN 3-540-09262-5 Springer-Verlag Berlin Heidelberg New York
ISBN 0-387-09262-5 Springer-Verlag New York Heidelberg Berlin

Library of Congress Catalog Card Number 72-152360
Printed in Germany

Typesetting, printing, and bookbinding: Brühlsche Universitätsdruckerei Lahn-Gießen.
2152/3140-543210

Contents

Enzyme Production During Transient Growth

The Reorganization of Protein Synthesis

H. Michael Koplove[a] and Charles L. Cooney[b]
Department of Nutrition and Food Science
Massachusetts Institute of Technology[c]
Cambridge, MA 02139, USA

The problems of enzyme production are examined in the light of molecular biological events. An understanding of the dynamics of protein synthesis that occur during transient growth are reviewed to provide insight into the potentials and the limitations of differential enzyme synthesis.

---

[a] Present address: Union Carbide Corp., South Charleston, West Virginia.
[b] Author to whom correspondence should be addressed.
[c] Publication Number 3359 from the Department of Nutrition and Food Science, Massachusetts Institute of Technology, Cambridge, Massachusetts 02139.

1 Preface

The well-controlled environmental conditions of a laboratory fermentor are, more often than not, very different from the ecological niche from which the microorganisms in the reactor have evolved. The "natural" environment of many microorganisms is, in fact, one in which the nutrient supply may rapidly change. Therefore, a discussion of the adaptive response of a microorganism to its environment must include not only its response to a wide variety of constant environmental factors, such as medium formulation, pH, and temperature, but also its manner of reorganizing macro-molecular synthesis to accommodate environmental fluctuations. Koch[1)] has called this form of ecology a "feast or famine existence" and, in discussing enteric microorganisms, writes:

The [enteric] microorganisms have not only been selected for ability to grow under chronic starvation, but also for ability to respond quickly to unannounced and irregular windfalls of food. Selection is still directed toward growth. However, at one time, the emphasis is on ability to accelerate growth and, at a later time, the emphasis is on coping with a deceleration of growth.

In balanced growth, microorganisms have supplies of ribosomal RNA, ribosomal proteins, RNA polymerase and perhaps transfer RNA which increase as a function of growth rate. This pattern of response is referred to as "growth-associated". DNA, protein, and messenger RNA concentrations are growth rate independent. Following a change in the microbial environment, which allows the microorganism to grow more rapidly, the cell will preferentially synthesize growth-associated components in a short period of time. An increase in "general" protein synthesis, however, often lags the initial burst of ribosome synthesis by 15–45 min and the DNA synthesis rate may remain at the pre-shift rate for 60 min or longer.

The purpose of this article is to present to the biochemical engineering audience selected investigations which relate to the phenomenon of temporal organization. Studies will be cited from both fundamental molecular biology and from investigations of a more general scope, with a focus upon enzyme synthesis during unbalanced growth. It is hoped that this presentation will help the reader to establish a background of relevant physiological and molecular biological studies as he reads this Literature Review and enable him/her to conceptualize the events which occur during microbial adaptations to rapidly changing environments. We believe that the regulatory mechanisms which are involved in temporal organization have been relatively untapped with regards to their process development potential and we hope that this review will stimulate thinking along these lines. With an improved understanding of the molecular biology and the regulatory mechanisms underlying enzyme synthesis it should be possible to better direct via process control the overproduction of desired enzymes. In other words, it is time to now capitalize on our understanding of the molecular biology of the gene and its expression.

2 Introduction

The terms "balanced growth" and "steady-state" growth are often used interchangeably although they really refer to somewhat different phenomena. Steady state, in the engineering sense, is defined as the condition in which derivatives of all system parameters with respect to time are now, and in principle, will remain zero. This definition can apply in a rigorous sense only to continuous cultures. Nagai et al.[2)] have incorporated this concept of steady state into a definition of balanced growth by stating that, for balanced growth,

$$\frac{d(Ci)_f}{dt} = 0 \tag{1}$$

in which $(Ci)_f$ is the fraction of Ci, the concentration of the ith component of cell material, and t is time.

The "steady state" discussed by Maaloe and Kjeldgaard[3)] and many other investigators who do not utilize chemostats is actually the balanced growth of microorganisms in a batch culture carried out in a particular medium at a time at which all nutrients are present in excess. During certain periods of time which may last several cell doubling times, the steady-state criterion of Eq. (1) may be satisfied while the chemical environment is changing. In the case of bacteria, both the cell mass and number increase exponentially. In this situation, which is often called "unrestricted growth", the cells are growing at their maximum specific growth rate permitted by the physical-chemical environment. In order to change growth rates in a batch culture, investigators have changed a nutrient or other environmental parameter in the medium. Typically, investigators of exponential growth have formulated media with a constant amount of the necessary nitrogen, sulfur, phosphate, magnesium, and trace salts and have varied the carbon source. For instance, in a culture of *Escherichia coli*, a change from alanine to glucose will change the growth rate from 0.014 to 1.0 h^{-1}. In some cases, amino acids are added to the growth medium; this type of addition is called a nutritional enrichment. Throughout this review, to avoid confusion, a medium which is referred to as "succinate medium" or "glucose medium" will mean a medium whose sole carbon source is succinate or glucose and the remainder of the medium is a mineral salts solution which supplies the necessary quantity of non-organic nutrients for growth.

A change in the carbon source to alter the specific growth rate probably causes derepression or induction of a region of the genome not previously transcribed and translated. On the other hand, alterations in the specific growth rate in continuous culture may be achieved by altering the flow rate of medium through the system[4)]. This change in growth rate may be mediated through physiological regulation, i.e. by the control of enzyme activity or concentration, as opposed to molecular biological events such as enzyme synthesis[5, 6)].

Throughout the following discussion, a persistant question may be raised by the reader: "Are the authors quoted here truly observing what they claim to be key regulatory events or are they seeing the end molecular biological result of very complex physiological interactions?" This is an important and complex query which focuses on the relationship of the regulation of protein synthesis and the physiological effects of the regulation.

3 Balanced Growth

3.1 Batch Culture

Maaloe and Kjeldgaard[3] are considered the founding fathers of the branch of molecular biology which focuses upon the regulation of macromolecular synthesis in response to growth rate changes, a set of regulatory events which has been also called temporal organization[7] or metabolic regulation[8]. In 1966, Maaloe and Kjeldgaard compiled the primer of the field, a monograph entitled *The Control of Macromolecular Synthesis,* in which they summarized and discussed the literature concerned with balanced and unbalanced growth[3].

During balanced exponential growth of *Salmonella typhimurium,* Maaloe and Kjeldgaard[3] reported that the amount of total stable RNA per unit of cell mass was a linear function of specific growth rate, whereas the DNA content per unit of cell mass decreased slightly with increasing growth rate. Of the total stable RNA, the ribosomal RNA (rRNA) concentration increased linearly with growth rate, while the transfer RNA (tRNA) concentration was essentially independent of growth rate. Total protein per unit of cell mass decreased slightly with increasing growth rate and the number of genomes or nuclei per cell was shown to increase linearly at growth rates exceeding 0.85 h^{-1}, up to a maximum of 4.5 genomes/cell. (This number was later confirmed by Chai and Lark[9] in an elegant autoradiography experiment.) A summary of results of this key experiment is shown in Table 1. Maaloe and Kjeldgaard[3], using a variety of assumptions, proceeded to calculate the peptide chain elongation rate per ribosome, often termed the ribosomal efficiency, and determined that the elongation rate was constant at 14 amino acids per ribosome per second at growth rates below 0.85 h^{-1} and increased slightly to 16 amino acids per ribosome-second at a growth rate of 1.6 h^{-1}.

The data which they presented has been corroborated for the most part by more recent investigations. For instance, Rosset et al.[10] separated the RNA constituents of *Escherichia coli, Aerobacter aerogenes,* and *S. typhimurium* on methylated serum albumin columns and found that the tRNA content is indeed constant as a function of growth rate and comprises a larger fraction of the total RNA at lower growth rates. On the basis of these results, Rosset et al.[10] postulated that tRNA and rRNA are regulated by different control mechanisms.

Dennis and Bremer[11] compiled a large number of facts regarding the macromolecular constituents of *E. coli* during balanced, exponential growth. Some of their more interesting observations were that both the RNA/DNA and the RNA/protein

Table 1 (see ref. 3 for further details)
A

μ (h^{-1})	Cells ($X10^{-12}$) per g dry weight	Total[a]				
		DNA	RNA	rRNA	tRNA	Protein[b]
		($mg \cdot g^{-1}$)	($mg \cdot g^{-1}$)	($mg \cdot g^{-1}$)	($mg \cdot g^{-1}$)	($mg \cdot g^{-1}$)
2.4	1.3	30	310	250	60	670
1.2	3.1	35	220	135	85	740
0.6	4.8	37	180	90	90	780
0.2	6.3	40	120	35	85	830

[a] Per g dry weight
[b] To a certain extent, membranes and cell-wall material are included in this figure. We have not attempted to estimate these quantities separately in *S. typhimurium*

B

μ (h^{-1})	Genome equivalents[a] (per cell)	70 S ribosomes (per genome)	tRNA molecules (per genome $X 10^{-5}$)	Amino acids[b] (per genome $X 10^{-8}$)	Amino acids[c] (per 70 S ribosome per second)
2.4	4.5	15,500	2.4	5.4	16
1.2	2.4	6,800	2.8	4.9	17
0.6	1.7	4,200	2.7	4.7	14
0.2	1.4	1,450	2.3	4.5	14

[a] Calculated from Eq. (3-4a)
[b] Based on an average molecular weight of 125 for the amino acids
[c] Calculated on the basis of the average number of ribosomes present in the cell during the cycle, i.e., the figures of Column 3 multiplied by 1.5

fractions were linearly growth-associated, but that the protein/DNA ratio, paradoxically, was also growth-associated although only in certain regions at growth rates less than 0.7 h^{-1}. "Growth associated" in this case, and in all future references, will be used to indicate a positive correlation between growth-rate and the value of the parameters. A linearly growth-associated RNA/DNA ratio thus means that the ratio increases linearly with growth rate.

Dennis and Bremer[11)] also observed that the rRNA/total RNA fraction was constant at 0.85; this observation implies that tRNA must also be a function of growth rate, a result which appears to contradict Maaloe and Kjeldgaard[3)]. Thus, there appears to be a discrepancy in the literature regarding the correlation between tRNA and growth rate. More discussion will be presented regarding tRNA and this discrepancy in Sect. 4.5 dealing with the response of tRNA to unbalanced growth.

Based upon their data, Dennis and Bremer calculated a constant ribosome efficiency in the growth rate range of 0.5–0.85 h^{-1} of 13.5 amino acids per ribosome-second. At lower growth rates, they claimed that the ribosome efficiency, which is also called the peptide chain elongation rate, decreased. Perhaps the most interesting statement they make, which is a concept overlooked or at least rarely addressed by many other molecular biologists, is that the difference in observations between high and low growth rates may reflect a change in the nature of the metabolic limitation: at low growth rates (e.g. $<0.7\ h^{-1}$), the limitation may result from an enzyme deficiency (a physiological deficiency), and at higher growth rates, the limitation may result from ribosomal efficiency (a molecular biological limitation).

Waldron and LaCroute[12)] extended balanced growth observations to eucaryotes by studying *Saccharomyces cerevisiae.* The RNA/DNA fraction of the cell was found to increase linearly from 0.02 to 0.12 in the growth rate range of 0.14 to 0.5 h^{-1}. At 0.5 h^{-1}, rRNA comprised 85% of the total stable RNA. Waldron and LaCroute determined that the ribosomal efficiency varied linearly from 0.5 to 6.0 amino acids per ribosome-second in the growth rate range of 0.035–0.42 h^{-1}. They concluded from this result that in eucaryotes, unlike bacteria, ribosomal efficiency is a function of growth rate.

3.2 Continuous Culture

Most of the research concerned with the dependence of macromolecular composition and regulatory mechanisms on growth rates has been done with balanced growth in batch cultures. Several studies in chemostats, however, have explored the same phenomena and are highlighted in this section. The primary objective of this section and the previous one is to present a picture of macromolecular synthesis as a function of growth rate during balanced growth. In later sections, the data presented here will be used to develop a conceptual model of unbalanced growth.

Herbert[13)] studied the effect of dilution rate on the cellular RNA, DNA and mean cell mass of *A. aerogenes* and *Bacillus megaterium.* His results, shown in Fig. 1, were qualitatively similar to those reported by Maaloe and Kjeldgaard[3)]: the RNA fraction per dry cell weight and the mean cell mass increased with increasing growth rate (or dilution rate) while both the protein and DNA weight fraction decreased slightly as the growth rate increased. Similar results were obtained with *A. aerogenes* with both glycerol and ammonia limitations from which evidence Herbert concluded that growth rate, rather than the type of growth limitation, was the prime controlling factor of cellular composition.

Dean and Rogers[14)] studied cell size and macromolecular composition of *A. aerogenes* under a variety of growth limiting conditions in chemostats and found that the RNA/DNA fraction increased linearly with growth rate, the protein/RNA decreased with growth rate, and the DNA/cell mass was independent of growth rate for glucose, nitrogen, and phosphate limitations. The data obtained from sulfate limitation varied significantly from the above results and the authors claimed that anomalous RNA

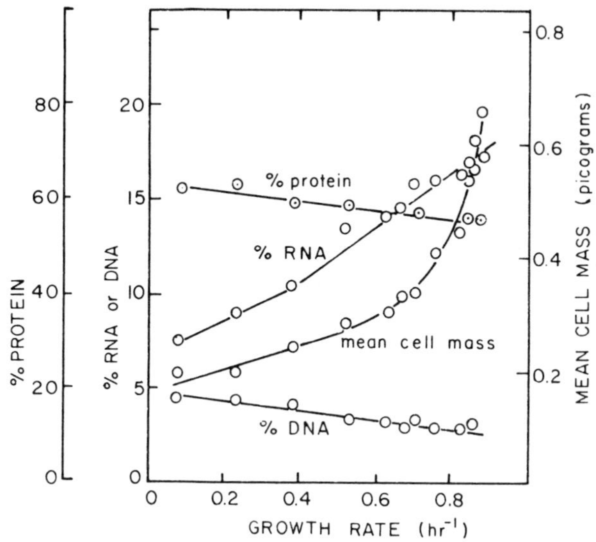

Fig. 1. Changes in cell composition with dilution on specific growth rate of *Aerobacter aerogenes* grown in a nitrogen-limited chemostat[13]

was produced during sulfate limitation. Further support of this conclusion has been provided by other investigators and will be referred to at a later point in the Literature Review.

Cooney et al.[15] studied macromolecular composition under a variety of single or dual nutrient limitation (carbon and/or nitrogen limitation; nitrogen and/or phosphate limitation) in aerobic continuous cultures of *Enterobacter (Aerobacter) aerogenes.* In all cases, the RNA concentrations were observed to be linearly growth associated. The authors report that the absolute concentrations of protein, RNA, and total cellular carbohydrates were functions of the type of growth-limitation; however, the ratios of protein to RNA were equivalent in all cases with the exception of growth limitation by phosphate alone (Fig. 2). Cooney et al.[15] also concluded that the protein to RNA ratio was solely a function of the growth rate and not a function of the input carbon to nitrogen ratio with the exception of phosphate limitation. In phosphate-limited growth, the authors speculated that a limitation in a nucleic acid biosynthetic pathway restricts the quantity of nucleic acids synthesized.

Sykes and Young[16] also studied *A. aerogenes* in chemostats and found that, for growth rates of less than 0.4 h^{-1}, bulk rRNA and the 4S rRNA subunit increased linearly with growth rate. At growth rates exceeding 0.5 h^{-1}, the 4S rRNA continued to increase linearly while the bulk rRNA increased more rapidly as a function of growth rate. The ratio of 50S to 30S subunits remained at 1.0 independent of growth rate. The authors claimed that ribosomal efficiency varied with growth rate.

Norris and Koch[17] conducted a very detailed chemostat study of an *E. coli* uracil auxotroph during glucose-limited growth in which they measured the relative quantities of mRNA, 16S plus 23S rRNA, and tRNA over a range of dilution rates. mRNA was found to compose 3–4.5% of the total RNA at all growth rates. The results in Table 2 summarize some of their other data. It is obvious that, even though the

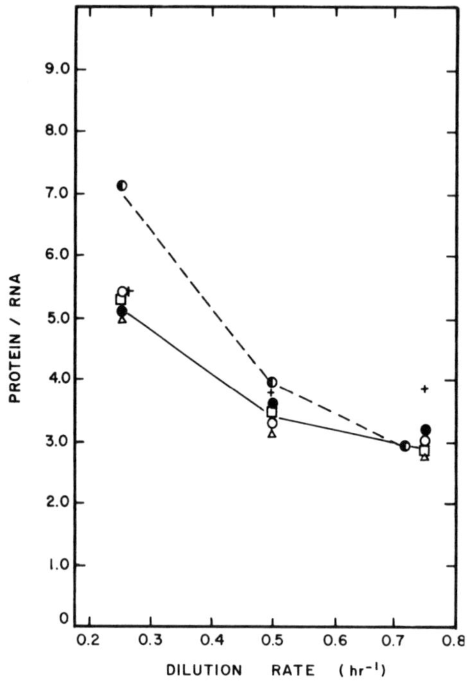

Fig. 2. Protein to RNA ratio in *Enterobacter (Aerobacter) aerogenes* as a function of dilution rate in single and dual nutrient-limited chemostats: ●, Phosphate-limited; △, Nitrogen-limited; □, carbon-limited; ○, Nitrogen- and Phosphate-limited; t, Nitrogen- and Phosphate-limited, ◑, Nitrogen- and Carbon-limited[15)]

Table 2. Results of Norris and Koch[17)] for growth of *E. coli* in a chemostat

	Growth rate (h^{-1})		
	0.07	0.12	0.70
16S + 23S rRNA fraction[a]	73	77	80
tRNA fraction[a]	24	20	15
Relative synthesis rate: mRNA/16S + 23S rRNA	3.2	–	1.6
Relative synthesis rate: 16S + 23S rRNA/tRNA	1.5	–	5.0

[a] % of total RNA

%mRNA is constant at a small percentage of the total RNA, it is synthesized at the highest rate of all classes of RNA and, therefore, engages the greatest amount of RNA polymerase; this is an important point that will be recalled later. Norris and Koch[17)] concluded from their work that the rates of synthesis of m-, r-, and tRNA are controlled independently.

From the studies of macromolecular synthesis in batch and continuous cultures, several generalizations can be made. The rRNA concentration has been found by all investigators to be a linear function of growth rate. The number of ribosomes and, consequently, the r-protein concentration has also been observed to be growth-associated. The mRNA, DNA, and protein concentrations are essentially constant and independent of growth rate, although several investigators have reported that these variables are weakly dependent upon growth rate. There are some disagreements regarding the relationship between tRNA content and growth rate. The synthesis rate of mRNA was shown to be greater than the synthesis rates of the other classes of RNA. Its low cell concentration is a result of its rapid degradation.

3.3 Enzyme Production During Balanced Growth

The literature on enzyme composition of cells during balanced growth is vast; this review is selective with the objective of providing a relevant introduction to the topic. The main focus here is upon the enzymes of central metabolism, or the Group I enzymes form the pathways which provide,

... for most bacteria, the main route for the formation of carbon skeletons for biosynthesis by degradation of carbohydrates and related compounds and for the production of energy and reducing power needed in biosynthetic reactions ...[18].

The main glucose catabolic pathways are (1) Embden-Meyerhof-Parnas glycolysis, (2) pentose phosphate cycle, (3) Entner-Doudoroff pathway, and (4) phosphoketolase pathway[18]. The ATP synthesis resulting from glucose catabolism occurs either through oxidative phosphorylations via a cytochrome system or through substrate-level phosphorylations. The critical factors in determining which of the phosphorylating pathways is utilized appear to be both oxygen availability and growth rate[19, 20].

The enzymes of central metabolism also function in anaplerotic reactions and gluconeogenesis as well as in carbohydrate catabolism. Hence, many are amphibolic[21].

As Clarke and Lilly[22] stated:

Most variation is observed for catabolic enzymes which may vary several thousand-fold under different growth conditions. On the whole, biosynthetic enzymes tend to vary less, but differences of a hundred-fold have been encountered between repressed states. Much less information is available about regulation of the enzymes of the central metabolic pathways ...

The regulation of these enzymes in the enterobacteria is very complex. Most enzymes of central metabolism are constituitive and the control of enzyme activity appears to be of more regulatory significance than the control of enzyme synthesis[18]. The main mechanisms of the control of enzyme activity include substrate and product activations and inhibitions; energy charge interactions; ATP, ADP, and/or AMP activation and inhibition; and, for the amphibolic enzymes, precursor activation[18, 21, 23]. In many cases, these enzyme activity controls have been studied in detail in vitro and may be considered "textbook examples". Since a thorough review of these controls is beyond the scope of this literatur review, Mandelstam and McQuillen[18] is highly recommended to the curious reader.

Unfortunately, even though many investigators have studied enzyme activity as a function of growth rate for the enzymes of central metabolism, few generalizations can be made. The response of microorganisms appears to be strain-specific as can be seen from the review of Dean[24], which provides an excellent compilation of data generated from studies on enzyme production in steady-state continuous cultures. The specific activity of various enzymes show several common forms of growth rate dependency. These are: an increase or decrease as a function of dilution rate, a maximum or minimum at a specific dilution rate, or an independence of growth rate. In fact, a particular enzyme may show different kinetic patterns in different organisms; for instance, a glucose-6-phosphate dehydrogenase increases with dilution rate in *Aspergillus niger*, decreases with dilution rate in *Candida utilis*, and is independent of dilution rates exceeding 0.3 h^{-1} in *A. aerogenes*. Beck and von Meyenburg[20] found that, in glucose-limited chemostats of *Saccharomyces cerevisiae*, the Group I (oxidative and related) enzymes exhibited various patterns of repression as a function of dilution rates: succinate-cytochrome c oxidoreductase activity decreased five-fold with increasing dilution rate, malate and NAD^+-linked glutamate dehydrogenase activities decreased sharply between dilution rates of 0.18 and 0.25 h^{-1} producing a 10-fold repression, and isocitrate lyase and malate synthetase, glyoxylate cycle enzymes, showed maximum activities at approximately 0.1 h^{-1}. On the other hand, aldolase and $NADP^+$-linked glutamate dehydrogenase, both Group II (biosynthetic) enzymes, increased activity over five-fold with increasing growth rates ($\mu = 0.05$ to 0.65 h^{-1}). Matin et al.[25] working with an aquatic *Pseudomonas sp.*, found that lactate dehydrogenase, aconitase, and glucose-6-phosphate dehydrogenase, all Group I enzymes, increased 10-fold with decreasing dilution rate (0.5 h^{-1} to 0.02 h^{-1}) while NADH oxidase activity increased linearly with increasing dilution rate.

Dean[24] speculated upon the control mechanisms which might produce the observed enzyme patterns: a decreasing concentration of enzyme with increasing growth rate may indicate that catabolite repression, which would result from increased substrate levels at higher growth rates, may cause the decrease in activity. An increase of enzyme concentration with growth rate (growth association), as reported by Harvey[6] for glutamic dehydrogenase, glutamic-oxalacetic transaminase, and aspartokinase, might result from the induction of enzyme synthesis by either an increase in a substrate concentration or as a consequence of the "Network Theorem". If a cell requires or is limited by a particular enzymatic product at a specific growth rate, i.e. a bottleneck, it would be necessary to increase the specific bottleneck enzyme level in order to grow at higher growth rates[5,6]. Presumably, the control of enzyme level results from complicated interactions of the metabolic reaction network, with the ultimate control resulting from a particular metabolic pool size.

The explanation of a maximum or minimum enzyme concentration as a function of dilution rate follows directly from the above concepts: over one growth rate region, a form of induction (derepression) is observed, over another region, catabolic repression is dominant. Thus, a spike in the curve is observed. This explanation was shown by Clarke and Lilly[22] to be correct for the aliphatic amidase of *Pseudomonas aeruginosa*. The microorganism was grown in a chemostat containing an inducer (acetamide) and a catabolite repressing carbon source (succinate); a maximum in activity was observed at a dilution rate of 0.3 h^{-1}. Amidase is considered to be a peripheral enzyme of central metabolism. Clarke and Lilly[22] report other examples of central metabolic enzymes which can show a maximum or minimum as a function of growth rate: glucose dehydrogenase, glucose-6-phosphate dehydrogenase and 6-phosphogluconate dehydrogenase, all in *P. aeruginosa.*

There can be no simple summary of this section other than to say that the enzymes of central metabolism are very tightly regulated with respect to both synthesis and activity. Most studies of these enzymes in balanced growth produce varying forms of growth rate dependecy. The same enzyme, such as glucose-6-phosphate dehydrogenase may be growth-associated, inversely growth-associated, or growth-independent depending on the microorganism studied. A summary of the data is presented in Table 3. The effects of environmental factors on the enzymes of central metabolism, such as pH temperature, and dissolved oxygen concentration, have been intentionally excluded from this discussion.

Table 3. Growth rate dependency of selected enzymes of central metabolism

Enzyme	Microorganism	Growth rate dependency	Ref.
Glucose-6-phosphate dehydrogenase	*A. niger*	Positive growth association	(See[24])
Glucose-6-phosphate dehydrogenase	*C. utilis*	Negative growth association	(See[24])
Glucose-6-phosphate dehydrogenase	*A. aerogenes*	Independent of growth rate	(See[24])
Glucose-6-phosphate dehydrogenase	*Pseudomonas sp.*	Negative growth association	Matin et al.[25]
Glucose-6-phosphate dehydrogenase	*P. aeruginosa*	Minimum observed	Clark and Lilly[22]
Succinate-cytochrome c oxidoreductase	*S. cerevisiae*	Negative growth association	Beck and von Meyenburg[20]
Malate dehydrogenase	*S. cerevisiae*	Negative growth association	Beck and von Meyenburg[20]
NAD^+-linked glutamate dehydrogenase	*S. cerevisiae*	Negative growth association	Beck and von Meyenburg[20]
Isocitrate lyase	*S. cerevisiae*	Maximum observed	Beck and von Meyenburg[20]
Malate synthetase	*S. cerevisiae*	Maximum observed	Beck and von Meyenburg[20]
Lactate dehydrogenase	*Pseudomonas sp.*	Negative growth associated	Matin et al.[25]
Aconitase	*Pseudomonas sp.*	Negative growth associated	Matin et al.[25]
NADH oxidase	*Pseudomonas sp.*	Possitive growth association	Matin et al.[25]
Aliphatic amidase	*P. aeruginosa*	Maximum observed	Clark and Lilly[22]
Glucose dehydrogenase	*P. aeruginosa*	Minimum observed	Clark and Lilly[22]
6-phosphogluconate	*P. aeruginosa*	Minimum observed	Clark and Lilly[22]

3.4 Synchronous Culture

Synchronous cultures are examples of growth which violate the criterion of balanced growth given in Eq. (1). This violation results from the fact that the division cycles of the cells in a synchronous culture occur simultaneously. In non-synchronous cultures, the division cycles are randomly distributed in the population. A particular advantage of a synchronous culture comes from its ability to magnify those regulatory events which are correlated with the division cycle of the cell.

One of the most interesting observations derived from synchronous culture studies is related to the potential for differential enzyme synthesis and the actual expression of this potential. Kuempel et al.[26)] and Donachie and Masters[27)] define the potential for enzyme synthesis as the ability of the cell to produce enzyme in the presence of an inducer. If, *E. coli* or *Bacillus subtilis* growing in synchronous cultures without inducers are inoculated into media containing inducers, all the enzymes which were assayed for could be synthesized immediately regardless of what point in the cells' division cycle the induction took place. If the inducer is included in the synchronous culture medium, many enzymes, such as tryptophanase and aspartyl transcarbamylase, show step-wise increases in enzyme activity at discrete times in the cell cycle. Apparently, this step-wise increase in activity results from the synchronous replication of the genes which code for the enzymes. The authors postulated that the number of genes in a cell which code for a particular enzyme determines the potential of the cell for synthesizing an enzyme in an environment which derepresses the operon.

Experimental evidence which supported this postulate regarding the correlation of gene dosage and enzyme potential had already been provided by Gorman et al.[28)]. In this experiment, shown in Fig. 3, they were able to produce yeast cells which were diploid in the β-galactosidase gene. However, the diploid mutant produced two periods of increase in synthesis potential correlating to the replication of each of the two genes.

Kuempel et al.[26)] discussed an example of an enzyme, alkaline phosphatase, in which a step-wise exponential increase in activity occurred in the presence of an inducer rather than the step increases observed in tryptophanase and aspartyl transcarbamylase. With a simple but convincing model, they were able to demonstrate that this step-wise exponential increase may result from the feedback control of an oscillating repressor pool. This form of internal control of enzyme synthesis was called "autogenous control" by Kuempel et al.[26)] and Donachie and Masters[27)] and indicates that, even if the number of genes is increased, the synthesis rate of the resulting enzyme may not increase accordingly.

In general, the authors claim that the potential of a microorganism to synthesize an enzyme is a function of the number of genes in the cell which code for the enzyme. Procaryotic cells can attain the maximum potential enzyme synthesis rate for inducible enzymes at any point in the cell cycle, given optimal environmental conditions for enzyme synthesis (e.g. inducer). However, for enzymes under internal feedback control, complex interactions occur between repressors, inducers, and operators; and various patterns of synthesis may be observed for these enzymes.

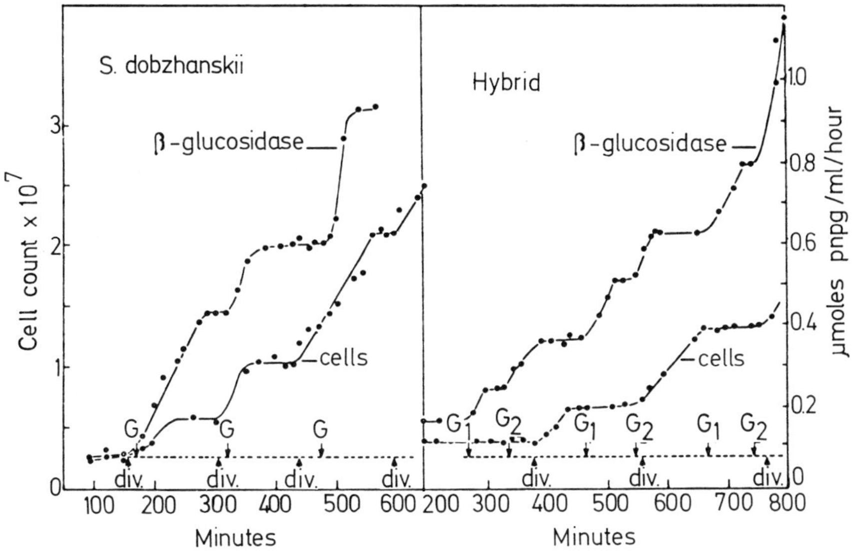

Fig. 3. Synthesis of β-glucosidase in synchronous cultures of *Saccharomyces dobzhanskii* and a hybrid *(S. dobzhanskii S. fragilis)*. At intervals, 6 ml aliquots were removed, washed twice by centrifugation and prepared and assayed for β-glucosidase. In the lower bar graph, the time of initiation of cell division (div.) and of β-glucosidase synthesis (G) are indicated by arrows[28)]

Therefore, from these studies one can conclude that an increased synthesis rate of an enzyme may result from both increasing the number of genes coding for the enzyme and removing internal repression. However, an increased gene dosage itself will not necessarily be reflected in increased enzyme synthesis if the internal repression is not overcome.

The concept of gene dosage has been postulated to be of importance in the regulation of "passively" controlled genes such as r-protein genes by Maaloe[29)]. This subject will be addressed more thoroughly in Sect. 4.9.

4 Unbalanced Growth

This section is an introduction to unbalanced growth, followed by more detailed information on the response of protein synthesizing components, such as rRNA and RNA polymerase. The introductory section includes selective information from batch and continuous cultures.

4.1 Cell Response to Unbalanced Growth

Relying heavily on earlier publication[30–32], Maaloe and Kjeldgaard[3] stated that when *E. coli* and *S. typhimurium* are transferred to a relatively richer medium which supports a higher maximum growth rate (a "nutritional shift-up") an increase in RNA, protein, and DNA synthesis rate occurs. As shown in Fig. 4, by following the uptake of either radioactively labelled phosphate or uracil and radioactively labelled amino acids, Maaloe and Kjeldgaard[3] were able to establish that, in *E. coli,* the RNA synthesis rate increased almost immediately after the shift-up and exceeded the preshift rate four to six-fold. An increase in protein synthesis was delayed by approximately 5–20 min depending upon the type of shift-up, and the DNA synthesis rate was constant at the pre-shift rate for as long as 60–70 min.

The observations of Maaloe and Kjeldgaard have been carefully verified by Koch[33, 34]: if a nutritional enrichment consisting of glucose and amino acids is made to *E. coli* growing either in batch culture on alanine ($\mu = 0.14\ h^{-1}$) or in a glucose-limited chemostat at a low growth rate ($0.06\ h^{-1}$), the cells respond by in-

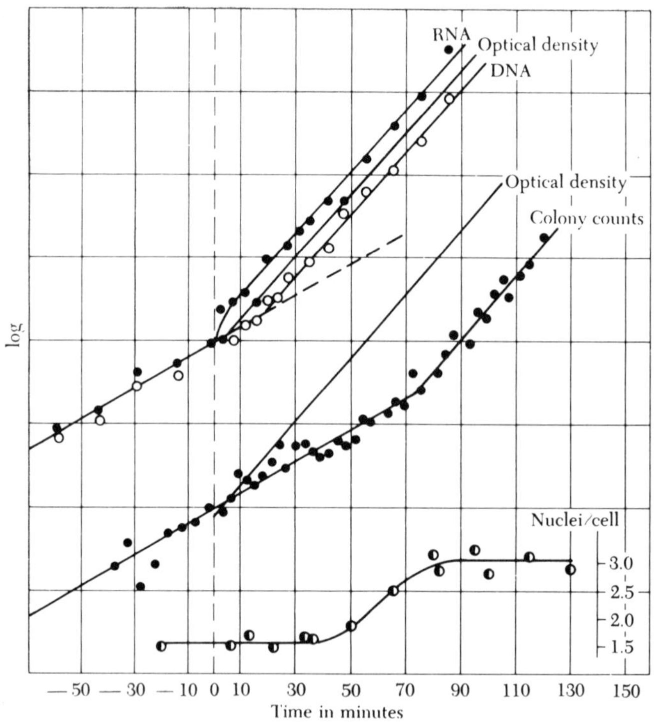

Fig. 4. Nutritional shift-up experiment with *Salmonella typhimurium* at 37 °C from glucose-salt-mineral medium to complete broth at time 0. The distance between horizontal lines corresponds to one doubling. In the lower right, the average number of nuclei per cell obtained from direct counts on stained preparations are given[3]

creasing the rate of RNA synthesis, as measured by guanine incorporation, by as much as 40-fold in less than 3 min. The rate of protein synthesis, as measured by tryptophan uptake, increases by 7-fold within the same time period. Since the protein synthesis rate and the RNA synthesis rate increased simultaneously and immediately this increase was attributed either to an increase in ribosomal efficiency of extant ribosomes or to an increase in the number of ribosomes engaged in protein synthesis. A further discussion of these possibilities will be pursued in Sect. 4.4, entitled: „Ribosomal Efficiency and Extra RNA."

Bremer and Dennis[35] studied the transient growth period of *E. coli* following shifts from one batch culture medium to another, in the same manner as Maaloe and Kjeldgaard[3]. Their results indicated that the stable RNA synthesis rate increased immediately in a step. This observation is consistent with a model of response in which extant RNA polymerase shifts from the mRNA producing genes to rRNA and tRNA genes. After the step increase, the stable RNA synthesis rate continued to increase exponentially. The authors claimed that this increase was attributable to an increase of de novo RNA polymerase.

Maaloe and Kjeldgaard[3] had previously postulated that de novo synthesis of RNA polymerase is not necessary for new RNA synthesis and, in all likelihood, the RNA polymerase is under feedback inhibition control during pre-shift growth, perhaps by nucleotides. A more thorough review of the transient RNA polymerase activity will be presented in the RNA Polymerase Section.

Nierlich[36, 37] has studied the incorporation of radioactive guanine by *E. coli* during shift-ups from a glucose plus salts medium to a medium containing salts plus glucose and Casamino acids. The media also contained radioactively labelled guanine. The author attempted to differentiate between *net* RNA synthesis and *total* RNA synthesis during both balanced and unbalanced growth. Net RNA synthesis was measured as the net rate of incorporation of radioactive guanine into trichloroacetic acid (TCA) insoluble material. In an earlier study, Nierlich[38] determined that the introacellular pools of guanosine triphosphate (GTP) rapidly became labelled in the presence of an addition of radioactive guanine to the medium due to the rapid turnover rate of mRNA. Furthermore, since mRNA is unstable and a feedback inhibition of nucleotide synthesis occurs, the net rate of incorporation reflected only net RNA synthesis which was presumed to be stable RNA (tRNA and rRNA).

To measure the total RNA (stable RNA plus mRNA) synthesis rate, Nierlich[39] developed a technique of assaying the initial rate of incorporation of labelled guanine into TCA insoluble material, correcting for the instantaneous concentration of GTP pool sizes.

As stated before, Nierlich[36, 37] used these assays during both balanced and unbalanced growth. His results indicated that the stable to unstable RNA synthesis rate ratio was approximately 1.0 during balanced growth in a glucose medium and 2.0 during balanced growth on a glucose plus amino acid medium. During the first 10 to 15 min of a shift-up between these two media, the stable forms of RNA are preferentially synthesized with a ratio of the synthesis rates of stable to unstable RNA of approximately 3.0. Nierlich concluded that the stable and unstable RNA's are not necessarily coordinately synthesized and that the synthesis of mRNA is not obligately reflected in an increase in protein synthesis. This observation is in agreement with the previously discussed conclusions derived from synchronous cultures.

The previous studies were primarily concerned with transient response to shift-ups in batch cultures. Two examples of transient studies conducted in continuous cultures will be cited for introductory purposes. Other continuous culture investigation will be discussed later.

Young and Bungay[40)] studied the dynamic response of glucose-limited chemostats using *S. cerevisiae.* Following a two-fold step increase in the glucose concentration, an approximately 30 min lag between RNA fraction maximum and growth rate maximum occurred. A first-order time constant of approximately four hours was calculated for RNA readjustment. If the chemostat was pulsed with substrate, an increase in dry cell weight was not apparent for 30–45 min.

Tempest et al.[41)], in an interesting study which focused upon very low growth rates in glycerol-limited chemostats of *A. aerogenes* found that, after a change in dilution rates from 0.004 h^{-1} to 0.24 h^{-1}, an almost immediate increase in specific oxygen uptake rate and specific carbon dioxide production rate occurred. After 10 h, a 40-fold increase in these two parameters over the initial steady-state rates was observed. Their eventual post-shift steady state rates were 2 times the pre-shift rate. A wash-out in cell mass proceeded for 6–8 h prior to the observation of a rapid increase in growth rate. The total RNA content of the cells, shown in Fig. 5, exhibited an immediate increase, reaching a maximum of 2.5 times the pre-shift rate at 10 h. A build-up of the carbohydrate content of the cell occurred shortly after the pertubation as Harvey[6)] also observed. The DNA concentration remained virtually constant for 6 h.

In summary, the authors generally agree that, following a shift-up, the synthesis rate of RNA increases prior to the subsequent increases in protein and DNA synthesis rates. The RNA polymerase activity, which Maaloe and Kjeldgaard[3)] originally postulated to be under feedback control, has been shown to increase during transients.

In the following sections, a more in-depth analysis of the transient events which occur following perturbations will be pursued with an emphasis upon molecular biological regulations.

4.2 Maaloe's Model

The previous discussion was concerned with an overview of macromolecular syntheses which occur during transitions. In the following sections, a review of recent literature which delves into the finer details of the molecular biological events which are observed during shift-ups will be presented. Before proceeding, credit must once again be paid to Maaloe who generated a model of the control of macromolecular synthesis[29)] through which he attempted to summarize the work of his colleagues and help direct research in the field of temporal organization. The primary assump-

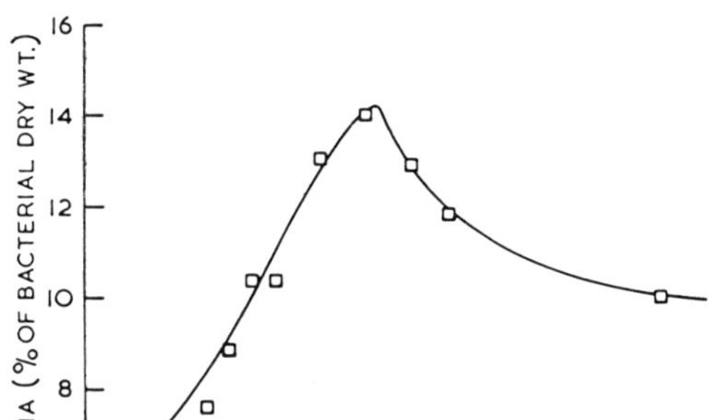

Fig. 5. Effect of changing the dilution rate from 0.004 h^{-1}, in a single step, on the bacterial DNA (○), carbohydrate (•), and RNA (□) contents of glycerol-limited *Aerobacter aerogenes* cultures[41)]

tions of the model result from the following observations from balanced growth: the rRNA concentration is linearly growth associated, whereas the mRNA concentration is independent of growth rate. Since cellular synthesis rates are functions of both the growth rate and the concentration of the synthesized component, the synthesis rates of both rRNA and ribosomal proteins (r-proteins) are functions of the square of the growth rate, whereas the mRNA synthesis rate is a linear function of growth rate. The mRNA per ribosome on the other hand, is independent of growth rate, a fact which seems contradictory to the previous statement but can be reconciled by realizing that mRNA is unstable and its rate of degradation has been found to be independent of growth rate[29)]. Maaloe further assumed that the transcription of a specific operon of the cell is governed both by the polymerase binding affinity of the promoter and the gene dosage.

From these assumptions, Maaloe discussed a class of macromolecular cellular components, such as rRNA, r-proteins, and certain non-repressible enzymes, which are growth-rate regulated. His work is a pioneering effort in legitimizing the concept of growth-rate control even though it lacks rigorous proof and contains some inherent faults, such as the failure to consider seriously the physiological changes which occur during unbalanced growth and their implications on cellular control strategies. It has been included here as an overview and will be returned to in a later section. The data contained in the remaining chapters modify and augment this early model of unbalanced growth. Readers are referred directly to the reference[29] for full details.

4.3 Ribosomal RNA and Ribosomal Proteins

Of the three classes of RNA, rRNA appears to be the macromolecule whose concentration is most sensitive to growth-rate alterations; its concentration is a linear function of growth rate over a wide range of balanced growth rates and its synthesis rate is proportional to the square of the growth rate. Furthermore, it is one of the first fractions of cellular components to be synthesized de novo following shift-ups. For newly synthesized rRNA to become active in protein synthesis, a variety of other events are necessary, including both the synthesis of r-proteins and the maturation of the various subunits of the ribosome into a functioning unit.

Schlief[42] measured the synthesis of r-protein during growth-rate shift-ups. *E. coli* growing in succinate-minimal medium was transferred to glucose minimal medium and the proportion of r-protein to total cell protein (a_r) was found to increase from 0.08 to 0.15 in 2–5 min. These a_r values are the characteristic values of balanced growth in the pre-shift and post-shift media, respectively. This time period coincided precisely with the short lag time associated with rRNA increase.

Harvey[43] also determined that the r-protein and rRNA content in *E. coli* increased simultaneously at a rate which was 3–4 times the pre-shift rate when a glucose-minimal medium was supplemented with amino acids, nucleic acids, and vitamins. Similar results were found by Gullov et al.[44], Carpenter and Sells[45, 46], and Dennis[47].

However, even though authors agree that the differential synthesis of r-protein and rRNA occurs simultaneously, there appear to be discrepancies regarding the synthesis rate of the individual r-proteins themselves. For instance, Gullov et al.[44] and Young and Dennis[48] determined that the ratio of the 30S and 50S ribosomal proteins remained constant following shift-ups while Carpenter and Sells[46] found that, after a 20 min delay, the 50S r-proteins were produced at 3 times the original rate. The discrepancies may be resolved by realizing that Carpenter and Sells were investigating the pool size of the r-proteins as well as the quantity of r-protein incorporated into the ribosomes whereas Gullov et al.[44] and Young and Dennis[48] looked solely at incorporated protein.

Molin et al.[49] conducted an interesting experiment with *E. coli* by stopping and restarting RNA synthesis with rifampicin and observed the kinetic patterns of various r-proteins. They determined that, in glucose minimal medium, 12–15% of the mRNA is r-protein message and that approximately 5 polycistronic units of 7000–12,000 base pair units exist. Four of the units are approximately 10 genes long and one of the units is about 20 genes long. The relatively small sizes of the units, they claim, promote rapid synthesis.

The importance of the shortness of the r-protein genes was also emphasized by Bennet and Maaloe[50] in a study in which fusidic acid (FA), a protein synthesis inhibitor, was added to growing *E. coli* cells. The FA concentrations used in the experiments decreased, but did not stop protein synthesis. The investigators found that even though total protein synthesis slowed down, the ribosomal protein synthesis fraction increased 2-fold in approximately 30 min. They claimed that this increase is not a result of protease activity which decreases bulk protein, but represents de novo synthesis of r-proteins. Bennett and Maaloe attribute this paradoxical effect to the relatively short mRNA coding for r-protein and postulate that a slower rate of mRNA degradation of the short mRNA chains is the result of ribosome blockage.

Sells et al.[51)] measured the accumulation of elongation factor protein. Following the initial increase in 50S and 30S subunit r-protein (6 min), a 30% increase in elongation factor (EF) was observed at 6 min. EF Ts increased considerably faster than EF G and both increased to a lesser extent than r-protein. Sells et al.[51)] concluded that the elongation factors were not polycistronic with each other or with r-proteins. This recent study may, therefore, complicate the emerging picture of the mechanisms involved in regulating protein synthesis during transitions. If elongation factor concentrations limit the initiation rate of translation, a potential protein synthesis bottleneck may occur. More research is necessary in this area relating to the kinetics of protein synthesis during unbalanced growth.

In general, the authors agree that r-protein and rRNA synthesis rates increase simultaneously following shift-ups. This is an important example of differential protein synthesis during unbalanced growth. More research is necessary, however, to fully understand the kinetic limitations of protein synthesis during unbalanced growth.

4.4 Ribosomal Efficiency and "Extra" rRNA

Maaloe and Kjeldgaard[3)], studying *E. coli* growing at rates in excess of 0.4 h^{-1}, found that the ribosomes engaged in protein synthesis were functioning at a rate which was independent of growth rate and postulated a constant ribosomal efficiency (rate of amino acids translated per second per ribosome). Their hypothesis has been modified by other experiments, which will be discussed in which the objectives were to study the influence of growth rate on ribosomal activity.

Nagai et al.[2)] studied the transient responses of *Azotobacter vinelandii* in glucose-limited chemostats following delta pulses of glucose. Their results indicate that the initial growth rate of the culture strongly influences the readjustment time of RNA synthesis and growth rate of the organism. For instance, as shown in Fig. 6, if the pulse perturbation was made to a chemostat with a pre-shift growth rate of 0.1 h^{-1}, the RNA fraction, which showed an initial increase in synthesis after several minutes, achieved its maximum value more than four hours later. The dry cell weight of the culture remained at its original steady state for two hours before increasing. If the initial steady-state growth rate was 0.2 h^{-1}, however, the RNA fraction attained its maximum value in only two hours while the growth rate remained unchanged for just 30 min.

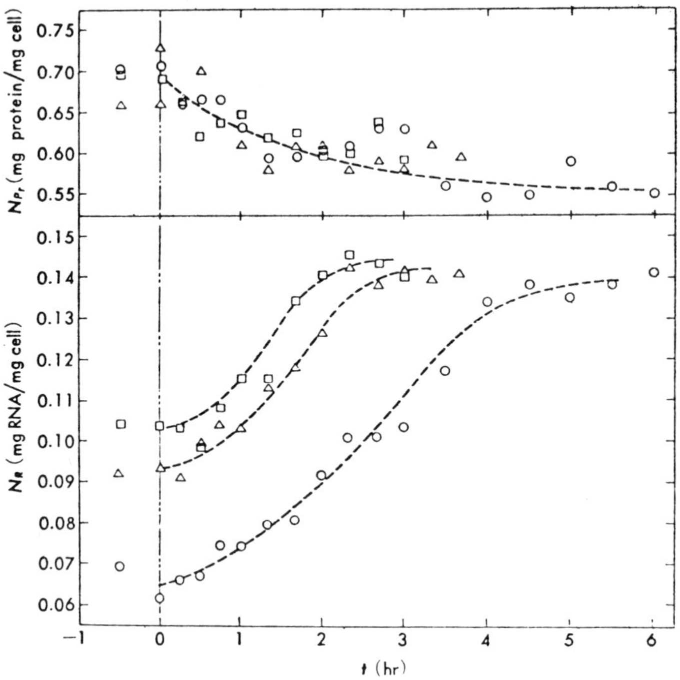

Fig. 6. The synthesis of protein (N_p) and nucleic acid (N_r) after a pulse of glucose to chemostats operating at dilution rates of 0.10 (○), 0.15 (△), and 0.20 (□) h^{-1}. Taken from Nagai et al.[2)]

Mateles et al.[52)] found that, after doubling the dilution rate of an ammonia nitrogen-limited chemostat of *E. coli* from its original steady-state dilution rate of 0.4 h^{-1}, an immediate increase in growth rate of 0.2 h^{-1} occurred. This was followed by an overshoot in the growth rate and a gradual resettling to the new steady state value. However, for small changes in the dilution rate, the growth rate was observed to accommodate the dilution rate change immediately. Mateles et al.[52)] summarized their findings by stating that the protein synthesizing capacity of the cell at pre-transient conditions is not operating at its maximum rate and is therefore inefficient. The data of both Nagai et al.[2)] and Mateles et al.[52)] therefore imply that the pre-shift growth rate may affect the apparent ribosomal efficiency.

More recently, Alton and Koch[53)] studied the transient response of slowly growing *E. coli* (0.05–0.09 h^{-1}) following the removal of phosphate-limitation. The cells were transferred from phosphate-limited chemostats to batch cultures containing excess phosphate, glucose, and amino acids. Their results, shown in Fig. 7, indicated that: (1) a 30 min lag occurred before the new maximum growth rate (1.2 h^{-1}) was achieved, (2) a biphasic response in radioactive uracil up-take rate was consistently observed, with a local minimum rate of 10 min following the perturbation, and (3) the final RNA accumulation rate was achieved in under 100 s following the shift-up. Based on these results, they concluded that the overall ribosomal efficiency was not constant, as

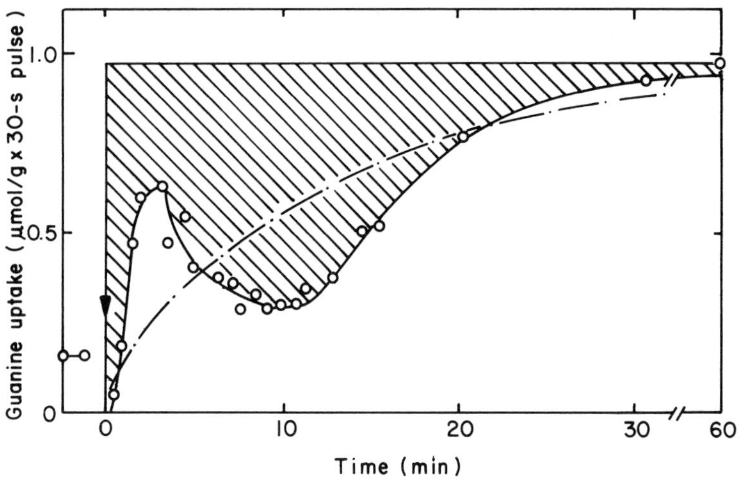

Fig. 7. Rate of RNA accumulation during a shift-up from a phosphate-limited chemostat[53]

Maaloe and Kjeldgaard[3] postulated. They arrived at their conclusions by calculating that 5–6 fold "extra" rRNA was present in these slowly growing cells. They claimed this fact was supported from radioactive tryptophan uptake curves indicating that protein synthesis rates could increase an order of magnitude during the first 10 min following the perturbation. Further, they suggested that the ribosomes actually engaged in protein synthesis functioned at their maximum efficiency and that a pool of inactive, reserve ribosomes or sub-units exists under these slowly growing conditions.

On the other hand, a sulfate-limited chemostat which was rapidly relieved of growth limitation by the addition of sulfate, did not exhibit a sharp RNA and subsequent protein synthesis rate increase[54]. Growth rate increases, as measured by optical density, lagged by 15 min following the removal of glucose limitation and almost 70 min following the removal of sulfate limitation. The authors claim that, in sulfate limitation, the increase in protein synthesis rates is limited by de novo rRNA synthesis rates following a shift-up. Apparently, a reserve pool of ribosomes does not exist during sulfate limitation. Their studies are supported by the previously cited experiments of Dean and Rogers[14] with sulfate-limited chemostats in balanced growth.

Harvey[43] determined the fraction of ribosomes in polysomes as a function of time following a nutritional shift-up. *E. coli* was the microorganism used and culture conditions consisted of batch cultures in which a nutritional shift-up doubled the unrestricted growth rate. He found that the fraction of ribosomes in the polysome fraction increased 50% within the first 5 min following the shift and concluded that a large fraction of reserve ribosomes exist in the cell which can be rapidly engaged in protein synthesis. The author claimed that the rate of protein synthesis is not regulated either by the number of ribosomes or the rate of elongation, but by the rate of initiation of transcription and/or translation.

Forschhammer and Lindahl[55] also suggested that the monosomes and polysomes are in a constant cyclical flux between association and dissociation, but found that the distribution of ribosomal material between polysomes, 70S ribosomes, and free 30S plus 50S particles was independent of growth rate in balanced growth. They disputed Harvey's results[43] with regards to his calculation of peptide chain elongation rates, but made no mention of his ribosome distribution data. It should be mentioned that Forschhammer and Lindahl studied ribosome distribution only in balanced growth and not during transitions. Harvey's conclusion[43] may still be valid despite the apparently conflicting data.

Therefore, in summary, slowly growing cells may have a residual quantity of ribosomes, under some culture conditions, which are not engaged in protein synthesis. Following a transition, this pool of monosomes may rapidly form polysomes which are actively engaged in protein synthesis. Consequently, protein synthesis rates can increase without de novo rRNA synthesis. The prevailing belief among these authors is that the efficiency of ribosomal translation is constant and independent of growth rate; the apparent change in ribosomal efficiency with growth rate results from a change in the distribution of ribosomes between inactive monosomes and active polysomes. Therefore, changes in growth rate alter the number of distribution of ribosomes and not the translation rate of ribosomes engaged in protein synthesis.

4.5 Transfer RNA

The study of tRNA has centered both on the tRNA itself and the synthetase enzymes which form amino acyl-tRNA. For instance, Kitchingham and Fournier[56] determined that, during the amino acid starvation of relaxed controlled *E. coli,* unique, apparently undermodified tRNA species were formed which could compete with isoacceptor tRNA molecules for ribosome loading. These unique species could not be readily interconverted to the regularly observed tRNA forms. They postulated two mechanisms of formation of these unique species: either starvation interrupted post-transcriptional modifications of the tRNA or unique tRNA's are synthesized during periods of environmental stress. The authors do not speculate on the mechanism by which these stressful environmental conditions may produce aberrant tRNA forms. Regardless, their data suggest that an important regulatory mechanism exists which has been previously unstudied.

Parker et al.[57] studied argyl- and valyl-tRNA synthetases during balanced and unbalanced batch growth of *E. coli* and *S. typhimurium.* In balanced growth, the synthetases increased 2.5-fold while growth rate increased 7-fold. During shift-ups from an acetate medium to a glucose plus Casamino acid medium, the synthesis rate of the two enzymes increased immediately and coordinately with an increase in tRNA content. Parker et al. concluded that a growth-linked regulatory control was operating rather than a classical induction or biosynthetic feedback control. During a shift-down to a medium excluding amino acids, the synthetase synthesis rates immediately decreased with a concurrent large increase in ornithine transcarbamylase (OTC) activity. OTC is necessary during amino acid biosynthesis and its increase presumably represents a derepression of its operator.

However, the regulation of the synthetases is not well-understood as is evidenced from the following example. Hirshfield and Zamecnik[58] were able to obtain an *E. coli* mutant resistant to thiosine, an analog of lysine. The mutant possessed decreased lysyl-synthetase activity at low growth rates. However, during shift-ups, the synthesis rate of the enzyme increased 15–20 fold whereas, in the wild type, synthesis rates increased only 1.5–2.0 fold. The eventual concentrations of the enzyme in each strain were comparable at the higher growth rate. The authors suggested that this observation implied that the mutant possessed a repressor type alteration or a change in a degrading enzyme, a hypothesis which would refute the conclusion of Parker et al.[57] regarding a growth-limited regulatory control. It remains a moot point whether the mutant strain of Hirshfield and Zamecnik is a pathological case of synthetase regulation during unbalanced growth.

At the moment, there appears to be scant evidence regarding the regulatory role of tRNA in protein synthesis. Several authors have attempted to draw inferences of regulation from evidence regarding peptide chain elongation rates, the degree of amino acid pool saturation, etc.; but the role of tRNA in regulation is still an area of active research. For instance, many authors assert that tRNA concentrations are independent of growth rate. Maaloe[29] has taken this evidence, along with the observation that the r-protein chain elongation rate is essentially constant for all growth rates, and postulated that tRNA concentrations are constantly maximized and are not control parameters during unbalanced growth. Skjold et al.[59] disagreed, claiming that, since the tRNA to total RNA fraction decreases with increasing growth rate, tRNA actually limits protein synthesis at higher growth rates. Unfortunately, the calculations of Skjold et al. were questionable due to the number and quality of assumptions made. The data of Parker et al.[57] might be used to construct a hypothesis that protein synthesis rates may be controlled by the charging rate of tRNA with amino acids. Previously, Forschhammer and Lindahl[55] based on studies of peptide chain elongation rates, hypothesized that at low growth rates the charged tRNA pool is depleted due to a low availability of ATP under these conditions.

As can be seen, much disaggreement exists among the investigators and the control of tRNA is clearly an area of protein regulation which requires more extensive, fundamental studies before any concrete conclusions can be drawn regarding control mechanisms.

4.6 RNA Chain Elongation Rate and RNA Polymerase

In the discussion of his model Maaloe[29] called for experiments designed to investigate the dynamic activity of RNA polymerase. Pato and von Meyenburg[60] attempted to supply some information regarding the cellular distribution of this key regulating enzyme by following H^3-uracil incorporation into *E. coli* during balanced, unrestricted growth and during shifts from growth on succinate to growth on glucose medium. Throughout these experiments, they measured the ratio of unstable RNA (mRNA) as a function of time. The ratio, designated as "Q", was inferred to be an estimate of the fraction of RNA polymerase allocated to mRNA transcription. In balanced growth, Q decreased linearly with increasing growth rate while the number of polymerase molecules per cell mass increased with growth rate. The RNA chain elongation rate, they claimed, was independent of growth rate. During a shift-up from succinate to glucose medium, Q fell immediately and then continued to decrease more slowly to its eventual balanced growth rate level. The total RNA synthesis rate increased 3-fold during the first 6 min following the transition. Balanced growth was achieved after 30 min. During the transient, the nucleotide pool remained constant. Pato and von Meyenburg[60] concluded that a significant quantity of non-active RNA polymerase exists in the cell during growth on succinate and that the polymerase can be rapidly engaged in RNA synthesis.

Mowbray and Nierlich[61] measured the rate of guanine incorporation into *E. coli* during shifts from a glucose medium to a glucose plus amino acids medium. During balanced growth on glucose, the *E. coli* was found to contain 4800 RNA molecules per unit of cell mass with an elongation rate of 28 nucleotides/s/chain. After 3 min of unbalaned growth, the cells contained 4300 RNA molecules per unit of cell mass with a chain elongation rate of 32 nucleotides/s/chain. At the new steady state, the cells contained 7000 RNA molecules per cell mass with an elongation rate of 32 nucleotides/s/chain. The authors concluded that RNA regulation is mediated by the number of RNA polymerase molecules in the cell, all of which operate at a constant efficiency.

To study the rate of RNA synthesis, Dougan and Glaser[62] measured the incorporation of radioactive guanine into the 16S and 23S ribosomal subunits during balanced growth in succinate and glucose media in batch cultures. They observed that in succinate medium ($\mu = 0.9\ h^{-1}$) 84 nucleotides/s were added. The rate of RNA synthesis is determined by the number of growing RNA chains, which in turn is controlled by the rate of RNA initiation and the RNA chain elongation rate. Since total rate of rRNA synthesis is a function of the square of the growth rate and since Dougan and Glaser[62] determined that the RNA chain elongation rate increased at a rate slower than the square of the growth rate, they concluded that both the initiation rate and the chain elongation rate influence the rate of RNA synthesis during balanced growth.

However, Pato and von Meyenburg[60] and Mowbray and Nierlich[61] claim that the chain elongation rate is constant and therefore only the number of growing chains, controlled by the number of RNA polymerase molecules, can alter the RNA synthesis rate. Therefore, a discrepancy exists among the investigators. The experimental difficulty in calculating RNA chain elongation rates[38, 39, 62] may account for these conflicting conclusions.

In an excellent fundamental study of RNA polymerase synthesis during nutritional shift-ups, Iwakura and Ishihama[63] utilized two strains of *E. coli*, B/R, and K12 W3350. In both strains, polymerase core protein concentrations were linearly growth-associated during balanced growth although the slopes and intercepts of the curves were different. The concentration of the σ-subunit, the initiation factor, was growth-independent and equal for the two strains during balanced growth.

The authors initiated a shift-up by transferring an aliquot of a culture growing on a succinate medium ($\mu = 0.225\ h^{-1}$) to a glucose plus amino acid medium ($\mu = 0.9\ h^{-1}$). The shift-up resulted in large and immediate increases in polymerase core protein concentrations of 1.6–2 fold in both strains. However, the response of *E. coli* K12 was biphasic, with a burst of core polymerase synthesis at 10 min to 2 times the pre-shift rate, a rapid decrease to 1.6 times, and a slow increase to 1.7 times at 40 min, as shown in Fig. 8. Strain B/R exhibited a simple overshoot and a gradual return in polymerase core

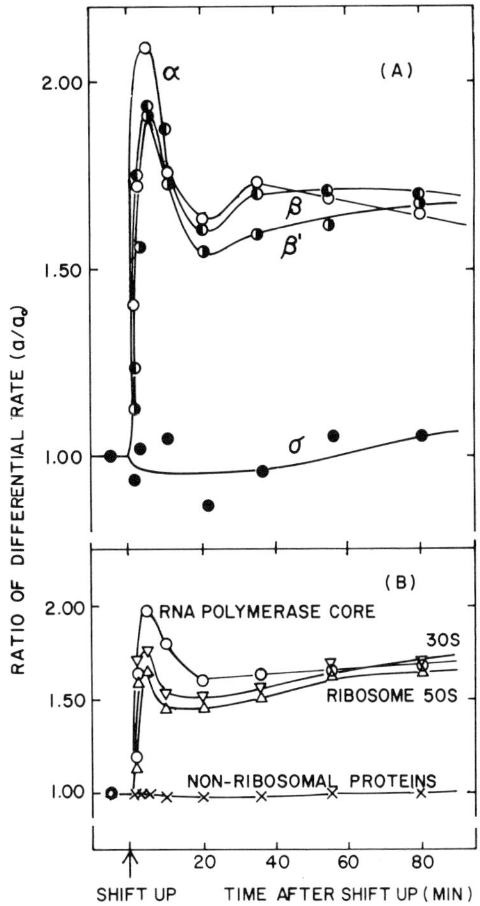

Fig. 8. Change of the differential rate of RNA polymerase and ribosomal during the shift-up of *E. coli* K12[63)]

protein concentration to the typical balanced growth level. The σ subunit concentration in both strains remained constant throughout the shift. Bremer and Dennis[35)] compiled the data of various authors and found that the RNA accumulation pattern for *E. coli* B/R showed a biphasic response with a large initial burst of synthesis within a few minutes followed by a gradual increase in RNA concentration to the post-shift level. Their results appear to confirm those of Iwakura and Ishihama[63)].

Iwakura and Ishihama[63)] measured the protein concentrations and turbidities of cultures during the transients and observed an apparent biphasic response in *E. coli* K12, but a unimodal response in *E. coli* B/R. One must wonder whether these results should be attributed to molecular events or whether they are artifacts resulting from an induction to synchrony[20, 64)].

Iwakura and Ishihama[63)] also studied RNA polymerase activity during nutritional shift-downs which occurred when cells grown on glucose plus amino acids were inoculated into a glucose minimal medium. A differential decrease in both core protein synthesis rate and σ-factor synthesis rate was observed as compared to the synthesis rate of non-ribosomal proteins. The only protein which appeared to be actively and selectively "proteolyzed" was the σ'-subunit. They concluded from their observations that the growth association of RNA polymerase core protein implies that this component is a main control parameter (or limiting factor) of growth rate readjustment and that the σ-factor, which is independent of growth rate, is most likely present in sufficient quantities to satisfy all growth rates. They hypothesized that the selective decrease in σ'-subunit concentration and the concurrent decrease in the rate of transcription may imply that a causality exists between these two parameters. This is an interesting concept which requires further experimental evidence for confirmation.

Boyle and Sells[65] measured RNA polymerase activity, RNA content, and guanosine tetraphosphate (ppGpp) concentration in *E. coli* following a shift from an acetate plus essential amino acids medium to a medium containing glucose plus all the amino acids. They found that the higher RNA synthesis rate was achieved in 2 min even though the β and β' content increased linearly for 15 min, eventually reaching 175% of the original concentration. The ppGpp concentration decreased 50% in 3 min. They concluded that RNA can increase initially without de novo polymerase synthesis, that the 75% increase they observed was in agreement with Iwakura and Ishihama[63] and inferred, with no truly concrete basis, that ppGpp plays a regulating role in RNA polymerase activity. As will be seen in the next section, their conclusions regarding ppGpp had already been refuted by an earlier study.

Hayward et al.[66] constructed a fascinating mutant *E. coli* which was diploid in the rifampicin sensitive region of the genome: one gene was sensitive to rifampicin and one was resistant with the sensitive gene being dominant. The cell was able to grow slowly on rifampicin by manufacturing both sensitive and insensitive RNA polymerases. Following the addition of rifampicin, the cell responded by oversynthesizing polymerase at a rate three times the normal. RNA polymerase eventually constituted 4% of the cell's total protein. Hayward et al.[66] observed a biphasic kinetic response in polymerase activity which they attributed to the interaction of a general control strategy, such as metabolic control[8] and a specific feedback control of RNA polymerase activity mediated, perhaps, through RNA concentration. Iwakura and Ishihama[63] had also observed biphasic kinetics in RNA synthesis.

In general, it appears that the RNA polymerase activity is growth-associated and that RNA polymerase can be synthesized differentially and rapidly following shift-ups. Several authors believe that RNA polymerase reserves may exist in cells during certain culture conditions. Polymerase activity appears to be a key controlling factor of RNA synthesis rates.

4.7 Guanosine Tetraphosphate

The attempt of Boyle and Sells[65] to implicate ppGpp, the "Magic Spot" discussed by Cashel and his co-workers[67, 68] as a controlling factor during shift-ups is logical in light of its role in regulating RNA polymerase activity during amino acid starvation in stringently controlled *E. coli*. In fact, Lazzarini et al.[69] had shown that the ppGpp concentration during balanced, unrestricted growth is inversely proportional to growth rate and RNA concentration. Lazzarini et al.[69] suggested a possible regulatory role for ppGpp based upon this correlation.

However, Friesen et al.[70] proved rather conclusively through a clever experiment that, if ppGpp is a controlling factor during shift-ups, it is certainly not the sole regulator of RNA polymerase activity. They measured ppGpp content during three shift-up experiments: a shift from a Tris-acetate medium to a glucose minimal medium, a shift from a Tris-acetate medium to a glucose plus amino acids and hypoxanthine medium, and a shift from a Tris-acetate medium to Tris-acetate plus 5 amino acids medium. This latter medium supports the same growth rate as the glucose minimal medium. During the shift to the glucose plus amino acid medium, the ppGpp level dropped dramatically in 30–40 s and remained low for 30–40 min. The RNA synthesis rate increased rapidly, as expected. In the shift to the glucose minimal medium, the ppGpp level remained relatively high through the transient and the RNA content increased more slowly than the previous case. On the other hand, during the switch to Tris-acetate plus 5 amino acids, the ppGpp level fell dramatically, but the growth and RNA synthesis rates were comparable to the synthesis rates during the glucose minimal medium shift-up. Therefore, Friesen et al.[70] concluded that ppGpp levels reflect amino acid pool sizes and may not directly control polymerase activity during shift-ups.

4.8 Chromosome Replication

The control of chromosome replication has attracted a large amount of research and, consequently, a vast literature has developed. In this section, a short aside will be made regarding DNA synthesis prior to pursuing a more in-depth discussion of metabolic regulation. As mentioned before, Maaloe and Kjeldgaard[3] reported that a 60–70 min time delay occurs before the cell increases its rate of DNA synthesis following a shift-up. This phenomenon has been called "rate maintenance". Cooper[71] was able to relate this rate maintenance to the well-known Cooper-Helmstetter model of DNA replication[72] which postulates that the replication point traverse time (C) and the time between the end of replication and cell division (D) are constant and independent of growth rate. Two corollaries of the theory are that an initiator compound must be present in a sufficient quantity to initiate replication and that the rate of replication is a function of the number of traversing replicating points. Cooper claimed that if the initiation of DNA replication occurs prematurely following the transition, the cell, which is already in the process of replication, will not express the new synthesis rates until a time of (C+D) elapses. For *E. coli*, (C+D) is approximately 60 min.

Sloan and Urban[73] have very recently extended the above observation to show that rate maintenance will occur in *E. coli* growing in unrestricted growth at doubling times of less than 120 min, but that immediate increases in DNA synthesis rates may occur in cells having pre-shift doubling times exceeding 120 min following a shift to a richer medium. These slowly growing cells will rapidly express DNA synthesis rates equivalent to cells with 60 min doubling times. Sloan and Urban try to explain their results with the above mentioned Cooper-Helmstetter model by hypothesizing that the (C+D) time of very slow growing cells may be accelerated and completed during a small fraction of the growth cycle. This hypothesis, they admit, is tenuous and they state that further study in this area is required.

In summary, until very recently, Maaloe and Kjeldgaard's observations[3)] have remained well-explained. If the recent work of Sloan and Urban[73)] is correct, a complication has arisen. In support of Sloan and Urban's observations, it should be mentioned that they are reminiscent of the observations of investigators who have worked with slow growing chemostats[6, 41, 52)]. Further work in this area, perhaps in synchronous culture, is warranted.

4.9 Enzyme Synthesis

Ultimately, the metabolic regulatory system will manifest itself through a changing enzyme spectrum during unbalanced growth to enable the cell to adjust to the demands of a new environment. In this section, investigations into both fundamental enzyme studies and studies with broader scopes will be discussed to elucidate the manner in which cells respond to fluctuating environments.

Among the first investigators to shed some light on transient enzyme synthesis were Mitsui et al.[74)]. They performed an early and enlightening study by removing *E. coli* cells growing exponentially on glucose, washing and starving them of glucose, and then reinoculating them into an enriched medium containing glucose, Casamino acids, and peptone. Using this technique, they were able to combine a nutritional enrichment with a methodology which produced synchronous cultures. They found, as others previously had, that the RNA increase which occurred immediately was an increase solely in rRNA and tRNA and that protein synthesis was delayed by approximately 25 min. However, when they added a radioactive label at 45 min, the newly formed RNA was mRNA.

For specific studies of individual enzyme synthesis, the lactose (lac) operon has long been a favorite model. Two such studies will be discussed here. Coffman et al.[75)] induced slowly growing chemostat cultures for β-galactosidase production and found that the time of completion of the first polypeptide chain following induction was not a function of growth rate at dilution rates as low as $0.03\ h^{-1}$. In all cases, β-galactosidase production was started within 100 s. They concluded that both the peptide chain elongation rate and the half-life of mRNA are independent of growth rate. Dalbow and Young[76)] also studied the rate of β-galactosidase synthesis at varying growth rates and arrived at a slightly different conclusion: at growth rates exceeding $0.95\ h^{-1}$, the peptide chain elongation rate is constant at 16 amino acids $\cdot\ s^{-1}$. However, at a lower growth rate, the elongation rate decreased (e.g. to 13 amino acids $\cdot\ s^{-1}$ at $\mu = 0.5\ h^{-1}$). Therefore, their study confirms an earlier investigation by Forschhammer and Lindahl[55)] in which the peptide chain elongation rate was found to be independent of growth rate at high growth rates, but dependent on growth rate at lower values. Forschhammer and Lindahl[55)] concluded from their study that, under some conditions, the overall rate of nascent protein synthesis may be a function of both the peptide chain elongation rate and the number of nascent proteins. Apparently, at high growth rates, the protein synthesis rate is governed solely by the initiation rate and, at lower growth rates, both factors are important. Since the constancy of the peptide chain elongation rate was a critical assumption in Maaloe's model[29] the model appears to contain a serious flaw if the above studies are correct in their conclusions.

Dalbow and Bremer[77] also utilized the lactose operon, but maintained the operon in a quasi-constitutive rather than inducible mode by supplying cyclic AMP and the gratuitous inducer, isopropyl-β-D-thiogalactopyranoside (IPTG). By shifting cells in exponential growth from a succinate medium to a glucose plus Casamino acid medium, they observed an immediate decrease in the differential β-galactosidase synthesis rate to 0.4 times the pre-shift rate, even though the ribosomal protein synthesis rate increases 2.1 times. They explained their results by stating that, following a shift-up, stable RNA is synthesized at the expense of mRNA (cf. Pato and von Meyenburg[60]). Furthermore, even though total protein synthesis may be a function of ribosome concentration and may increase following a shift-up, the differential rate of synthesis of a particular enzyme is a function of the fraction of total mRNA coding for that enzyme or in Maaloe's Model. They also calculated that, following the shift-up, the amount of RNA polymerase available for mRNA synthesis decreases due to the rapid increase in rRNA synthesis. They postulate that ribosome synthesis is under an active rather than constitutive control. A logical corollary of their hypothesis is that following shift-ups the rRNA and r-protein genes have much greater polymerase binding affinities than do other genes. This is also a postulate of Maaloe[29] and Rose and Yanofsky[8], whose results are cited below.

The concept of producing a constitutive mutant from an inducible wild type strain was employed by Rose and Yanofsky[8] in an important study of the tryptophan (tryp) operon. The mutant *E. coli* was a repressor-negative strain with a functional tryp operator gene. In balanced growth on a minimal medium with single carbon sources, the tryp-mRNA content increased with growth rate. However, if the carbon sources were supplemented with amino acids, a decrease in the tryp mRNA occurred during increases in growth rate. Tryp mRNA concentration and the activities of the enzymes it codes for are therefore at a maximum in a glucose minimal medium.

During a shift-up from a proline-minimal medium to a glucose-minimal medium, the stable RNA synthesis increased 6-fold in 2 min, but the tryp mRNA content increased at a slower rate for 70 min. In a shift from glucose to glucose plus amino acids, an extremely rapid decrease in tryp mRNA was observed. Rose and Yanofsky[8] attribute this effect to the phenomenon of metabolic regulation, a regulatory mechanism of key importance to non-operator containing (or constituitive) genes. They claim that this control is mediated by the frequency of initiation of transcription through either the availability of RNA polymerase or a binding affinity effect at the promoter site.

From their observations that rRNA synthesis increases rapidly following the introduction of amino acids to the growth medium, an addition which causes multiple repressions of biosynthetic genes, and that tryp mRNA synthesis decreases following the addition, Rose and Yanofsky[8] postulate that the RNA polymerase concentration is a controlling factor of protein synthesis and that genes which compete most effectively for the polymerase are transcribed preferentially.

The discussion in Sect. 4.6 presented evidence which demonstrates that RNA polymerase concentration is indeed a function of growth rate. Sankaran and Pogell[78], in a fascinating study using the intercalating dyes ethidium bromide (EB) and acridine orange (AO), have provided indirect proof that the RNA binding affinity of certain genes may be a function of DNA conformation. The authors studied the inducible enzymes: β-galactosidase, tryptophanase, acid hexose phosphatase, and alkaline phosphatase. The first three enzymes are subject to catabolite repression. For these three enzymes, the presence of inducer plus AO differentially inhibited the synthesis of these enzymes. Cyclic AMP was able to reverse this inhibition. In lac repressor-negative mutants, the inhibition was still observed but in a lac promoter mutant which was no longer sensitive to cyclic AMP, β-galactosidase synthesis was not inhibited by AO. Conversely, the specific activity of alkaline phosphatase, which is not catabolite-sensitive, was actually increased in the presence of AO.

Sankaran and Pogell[78] proposed a model for catabolite-sensitive genes in which:

... *induction of transcription involves a conformational change in the promoter region, which increases its affinity for inter-calating dyes ... The results would be decreased affinity for the formation of the transcription initiation complex and decreased β-galactosidase synthesis. Excess cyclic AMP would partially reverse the inhibition by favoring the formation of the transcription initiation complex and displacement of the intercalating dye molecules.*

This model is shown graphically in Fig. 9 as it pertains to the events occurring at the lac promoter site.

The authors offered two possible explanations for the increase in alkaline phosphatase activity in the presence of AO: either the dye disrupted the normal repression mechanism of this operon or the number of ribosomes per alkaline phosphatase operon increased following the inhibition of catabolite-sensitive genes. From the data of Rose and Yanofsky[8] and others, the phrase "number of ribosomes" in the previous sentence might be replaced by "RNA polymerase activity".

Therefore, even though Sankaran and Pogell[78] concentrated on inducible enzymes, their data suggests that the RNA polymerase binding affinity of a promoter is a function of DNA conformation. Rose and Yanofsky[8] postulated that the transcription of

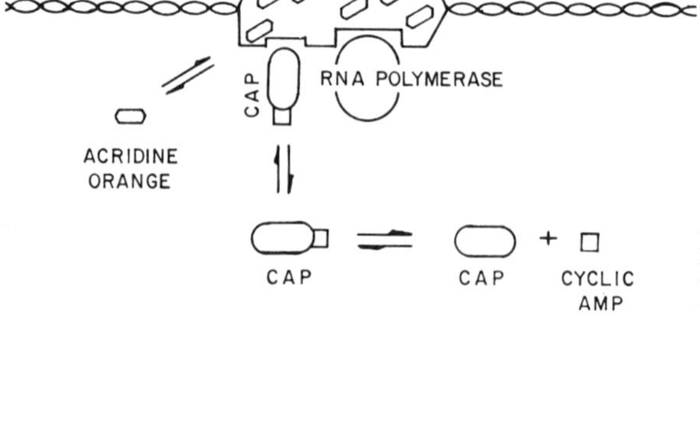

Fig. 9. Model of Sankaran and Pogell[78] where C_{ap} represents the catabolite gene activator protein

non-inducible genes is a function of both RNA polymerase concentration and binding affinity. From various sources, experimental evidence is now available which confirms that both of these parameters are important regulatory mechanisms and that both may change in response to environmental alterations. More data most definitely is needed regarding the dependency of promoter binding affinity on DNA conformation, as this form of regulation may be the most important one from the standpoint of non-inducible genes. As Koch[1)] prophesies:

... I predict that the unwinding and concomitant rewinding of DNA during RNA synthesis will turn out to be a rate-limiting step to the overall process of protein synthesis under in vivo conditions in enteric bacteria.

Jensen and Neidhardt[79)] approached the metabolic regulation of microorganisms in a somewhat different manner and provided a slightly different perspective. They studied the formation of histidase, an inducible catabolic enzyme of histidine degradation, in *Aerobacter aerogenes.* The medium they used was limited in arginine or sulfate and contained excess histidine and salts. When the dilution rates of a chemostat at $D = 0.13\ h^{-1}$ was increased to $D = 0.66\ h^{-1}$, the growth rate rapidly increased to $0.4\ h^{-1}$ and, then after 2 h, increased further to $0.66\ h^{-1}$. Although they did not calculate it, the apparent specific synthesis rate of histidase increased immediately to, but did not overshoot, the post-shift apparent synthesis rate. This histidase activity, which is growth-associated in steady-state chemostats, rapidly increased following the shift and achieved characteristic specific activity by the end of 2 h.

Jensen and Neidhardt[79)] proved rather convincingly that the activity of histidase was inhibited as well as repressed at low growth rates. Following an increase in growth rate, the inhibition was released immediately. The uninhibited histidase activity had the capacity of catabolizing histidine to meet a growth-rate demand of $0.5\ h^{-1}$; this may explain the cell's ability to rapidly achieve higher growth rates with extant enzymes. In fact, they observed that cells in chemostats could adjust almost immediately with respect to growth rate and protein synthesis for small increases in dilution rates (i.e. $D = 0.15\ h^{-1} \rightarrow D = 0.42\ h^{-1}$), even though histidase activity continued to increase during the first 40 min. These results are similar to those observed by other investigators[34, 52)]. All other investigators who had focused either on general macromolecular synthesis or constitutive enzymes have attributed the ability of the cell to immediately increase its growth rate to extent non-functioning RNA which is readily mobilized. Jensen and Neidhardt[79)] have gone a step further and demonstrated that the physiological capacity of the cell, defined as the activity of a key inducible catabolic enzyme, is sufficient to sustain the immediate increase in growth rate.

Jensen and Neidhardt[79)] make some particularly germane comments and predictions in their conclusions. An adaptation to a shift-up requires an efficient mustering of intermediary metabolites, ATP, and reducing power. In the case of a rich medium containing the end products of biosynthesis (e.g. amino acids), intermediary metabolites will not be depleted through biosynthesis. The cell must rid itself of potentially catabolite-repressing compounds through excretory pathways in order to prevent a repression of ATP synthesis which is necessary to permit increases in growth rate. An example they cite is acetate excretion but, of course, this is speculative.

In an interesting experimental series, Harvey[6] utilized two strains of *E. coli*, one which could produce glycogen, a glucose storage compound, and one which could not. He effected shift-ups by transferring cells from glucose-limited steady-state chemostats to growth vessels containing excess glucose and salts and growth supplements, if needed. Following the shift-up, immediate step increases in apparent growth rate could occur to 0.32 h^{-1} with the glycogen-less strain if the pre-shift dilution rate was less than 0.3 h^{-1} However, as others had observed, if the pre-shift dilution rate exceeded 0.3 h^{-1}, long lag times prevailed. With the wild type strain, which could produce glycogen, immediate increases in optical density were observed even for pre-shift dilution rates exceeding 0.3 h^{-1}. The other macromolecular constituents, such as RNA, showed characteristic lag times. He concluded that initial large optical density increases may not be due to growth, but may reflect increases in glycogen storage. It is interesting to speculate in light of the comment by Jensen and Neidhardt[79] that the glycogen synthesizing ability also serves as a rapid means for getting rid of ATP which is rapidly produced following a transient. In this sense, glycogen acts initially as a sink for ATP and secondly as a reserve.

To ascertain the effects of growth supplements on transient responses, the glycogen-less strain, in balanced growth at 0.34 h^{-1}, was transferred to vessels containing various supplements. An immediate increase in total RNA content was observed, followed by a 20 min lag in protein synthesis, in the presence of glucose plus 20 amino acids. If the 20 amino acids were excluded and the supplement contained glucose and either ribosides plus thymidine or glutamate plus aspartate or vitamins, no increase in RNA synthesis or growth rate was observed for over 60 min.

Harvey[6] also studied the synthesis of two growth-associated enzymes, glutamic dehydrogenase and glutamic-oxalacetic transaminase during transients initiated by adding glucose to glucose-limited, minimal medium chemostats (Fig. 10). The pre-shift dilution rate was 0.32 h^{-1} and the glycogen-less strain of *E. coli* was used. The transaminase specific activity rapidly increased to the apparent unrestricted level whereas the glutamic dehydrogenase activity remained more closely coupled to the growth rate during the transient.

Harvey draws the following conclusion. Glutamic dehydrogenase functions in converting ammonia to amino groups which are then transferred into biosynthetic pathways via transamination. After an addition of glucose, the growth rate of the cell apparently becomes limited by the activity of an enzyme in an amino acid biosynthetic pathway. Therefore, additions of amino acids can circumvent this metabolic block and can permit the cell to readjust much more rapidly. The same conclusion was also implied by Koch and Deppe[34] from work with a sulfate-limited chemostat, and Kennel and Magasanik[80] in studies conducted on β-galactosidase induction.

Lilly et al.[81] also were able to obtain increases in enzyme synthesis rate during perturbations from steady states in continuous cultures. The microorganism utilized was *Streptomyces* 17, a thermophile which produced β 1,3-glucanase extracellularly. The steady-state profile of β 1,3-glucanase was inversely proportional to growth rate and the authors claimed that this observation was the result of catabolite repression. If the inducer gentiobiose was added to steady-state cultures, no increase in specific activity was observed even though, in batch cultures, pronounced induction could be seen.

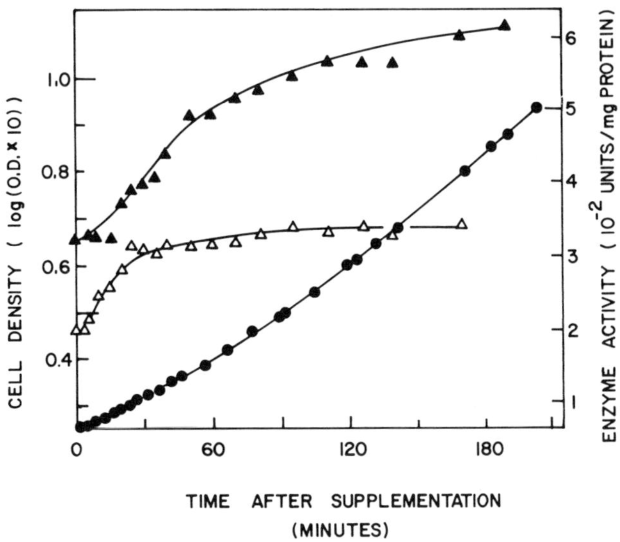

Fig. 10. Increase in the specific activities of glutamic dehydrogenase (▲) and glutamic-oxalacetic transaminase (△, ordinate scale is 0.67 x 10^{-2} units/mg of protein) during the transition from restricted to unrestricted growth. At time zero, 300 ml were removed from a glucose-limited chemostat culture (D = 0.32 · h^{-1}) of *E. coli* and added to a vessel containing 100 ml of basal medium. Glucose (0.05%) was added, and growth was measured by OD readings (•). See Harvey[6)] for details

During a shift-down in dilution rate with the exclusion of gentiobiose from the medium, enzyme synthesis rates were observed to increase as much as an order of magnitude while the specific activity increased 20–30% over steady-state activities. The authors attributed this overshoot to catabolic derepression. However, if gentiobiose was added prior to shift-down, significantly larger increases in enzyme synthesis rates and activities were obtained. The authors suggest that their observations indicate that a complicated interaction occurs between the inducer and cyclic AMP. Their results also suggest that the transient period following shift-down causes a disturbance of the binding of the repressor molecules which allows gentiobiose to act as an inducer; of course, this remains speculative.

Carlsson and Elander[82)] studied the formation of the constitutive extracellular enzyme dextransucrase by *Streptococcus sanguis* during balanced and unbalanced growth in continuous cultures. The enzyme, which is involved in dental plaque formation, is inversely growth-associated in glucose minimal medium and complex medium. Under the former conditions, specific enzyme activity is 5–6 fold less than the latter conditions. However, during a transition, from the simple to the complex medium, a transition which allows for a considerable increase in μ, the specific dextransucrase activity increases to a maximum 20-fold over the preshift specific activity and 2-fold over the post-shift specific activity. Therefore, even though specific activity decreases with increasing μ in balanced growth, a shift from a minimal to a complex medium caused an overshoot in specific activity.

Carlsson and Elander[82] present teleological implications of their data regarding oral ecology and caries production, but do not speculate on the regulatory mechanisms responsible for this microbial response.

Working with an enzyme of central metabolism, acetate kinase and using *E. coli*, Koplove and Cooney[83] examined in detail the production of this enzyme in two types of shift-up experiments. These studies were done in an anaerobic, glucose-limited chemostat such that acetate kinase is actually necessary for the cell to obtain the maximum available ATP for growth. In one case, glucose alone was added and in a second case both glucose and casamino acids were added.

Addition of the glucose alone results in little or no intracellular accumulation of acetate kinase and only a slight increase in its rate of synthesis (Fig. 11 a). Whereas, addition of both glucose and Casamino acids leads to a marked increase in both the specific control of the enzyme on a cell weight basis and the specific rate of synthesis of the enzyme (Fig. 11 b). Studies with chloramphenical addition showed that the measured increase in activity was not due to activation of extant enzyme. These results are interpreted in the context of the metabolic regulation model of Rose and Yanofsky[6] where simultaneous addition of amino acids with glucose, represses the amino acid biosynthetic pathways and allows the available RNA polymerase to be focused on enzymes such as acetate kinase.

Many authors have attempted to model enzyme synthesis during unsteady-state growth with varying degrees of success. An example of an interesting and successful

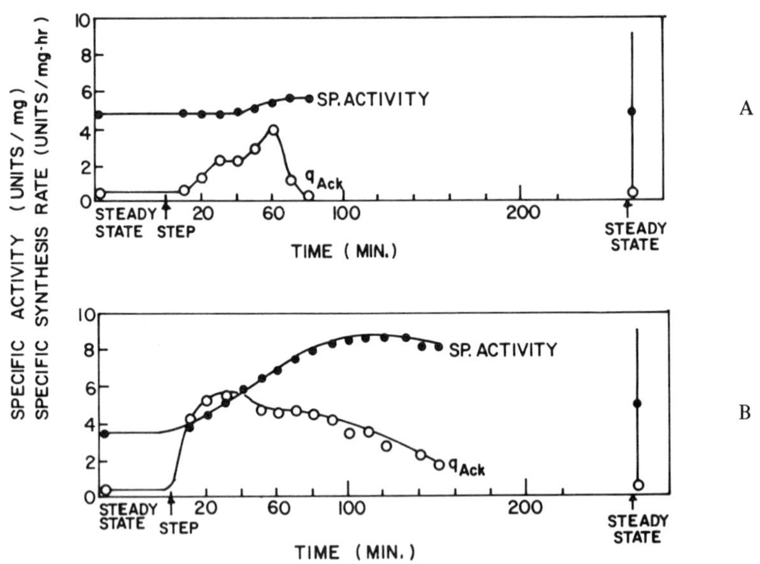

Fig. 11. The specific activity and specific rate of synthesis (q_{ack}) of acetate kinase in *E. coli* following a nutritional shift up by the addition of glucose (A) and glucose plus casamino acids (B) to a glucose-limited chemostat

model is that of Imanaka and his co-workers[84]. A *Monascus sp.* was grown in continuous cultures containing glucose and galactose. High glucose concentration repressed galactose consumption and the production of α-galactosidase, and high galactose concentration inhibited glucose utilization. The manifestation of these effects was a function of dilution rate: at $D < 0.142\ h^{-1}$, both carbon sources were utilized; and at $D > 0.142\ h^{-1}$, only glucose was consumed. A hysteresis effect was observed at this dilution rate: if steady states were established by progressively increasing the dilution rate, one set of data for glucose and galactose concentration, cell mass, and α-galactosidase activity were obtained. If the system was established at $D > 0.142\ h^{-1}$ the dilution rates were lowered rather than raised, an entirely different set of values were obtained.

A very interesting model of this system which accounted for inhibition and induction effects was presented by Imanaka and co-workers[85]. Inhibition effects were incorporated into the familiar Monod model[86] and enzyme synthesis was presumed to be a function solely of mRNA concentration which was a function of repressor and inducer concentrations. The parameters of the model were obtained by trial and error from experimental data. Excellent agreement was obtained for chemostat cultures and for a transition between $D < 0.14^2\ h^{-1}$ to $D > 0.142\ h^{-1}$, the point of hysteresis. Imanaka et al.[87] used this model to predict the optimum reaction system for producing α-galactosidase. Therefore, in some cases which are well-defined with respect to the control of the operon, modelling is a beneficial exercise.

In summary, following a shift-up, the RNA polymerase activity available for the transcription of "general" mRNA decreases as a result of the increased binding of polymerase by the rRNA, the r-protein, and perhaps the tRNA genes. This increase in binding may result either from some form of active binding of RNA polymerase or from an increase in promoter affinity due perhaps to a change in the conformation of the promotor region.

The response of the enzyme synthesis varies from one enzyme to another. The differential synthesis rate of a lac operon enzyme, β-galactosidase, decreases following a nutritional enrichment. Conversely, the differential synthesis rates of other enzymes, such as histidase, glutamic dehydrogenase, glutamic-oxalacetic transaminase and acetate kinase may increase significantly following shift-ups. And, the differential synthesis rates of the tryp operon enzymes in a constitutive mutant may either increase or decrease following a shift-up depending upon the medium formulation.

4.10 Maaloe's Model Revisited - The Potential for Process Improvement

From the various investigations cited thus far, a verification and expansion of Maaloe's model[29] can be attempted. Furthermore, it is now useful to consider translation of their concept into process improvement. Maaloe hypothesized that, following a shift-up, any limitation to RNA synthesis which would occur as a result of RNA polymerase activity would be rapidly relieved by de novo RNA polymerase synthesis. Recent studies have indicated that the polymerase is growth-associated during balanced growth. This indicates that polymerase activity may act as a bottleneck enzyme during a portion of the transient. Also consistent with this hypothesis are the studies showing that the polymerase is indeed preferentially synthesized immediately following a transition.

Maaloe claims that the total RNA synthesis rate is proportional to the square of growth rate. Based on this hypothesis, subsequent investigations have concluded that the rate of total RNA synthesis is a function of both the number of growing chains and the elongation rate of each chain. The latter part of this statement however is still disputed, with several authors claiming that RNA chain elongation rate is constant.

The ribosomal efficiency of ribosomes engaged in protein synthesis is essentially constant and independent of growth rate and any deviations from apparent constancy may be explained in terms of a pool of non-functioning monosomes and subunits. During low growth rates, the cell has the ability to instantly increase its protein synthesis rate without de novo RNA synthesis following a shift-up. This non-functioning rRNA pool may account for this adaptive ability. The available results suggest that no improvement in protein synthesis rate is likely through improvement of ribosome efficiency, but rather is likely to come from increases in the amount of ribosomes made available through de novo synthesis. In Maaloe's model it is presumed that any increase in ribosome activity is a result of de novo rRNA and r-protein synthesis and the failure to account for a pool of inactive ribosomes may be the result of focussing on data obtained from batch cultures, rather than from continuous cultures.

Following a nutritional enrichment in the form of an amino acid addition to a minimal medium, there is rapid and selective synthesis of stable RNA at the expense of mRNA. Examples of mRNA templates whose synthesis is repressed during shift-ups are tryp and lac operon mRNA. However, there exists a class of proteins whose synthesis rate differentially increases following shift-up. These proteins are considered to be constitutive as opposed to inducible. The rate of synthesis of mRNA templates for those proteins may be regulated by the availability of RNA polymerase, ATP, and perhaps charged tRNA groups and the binding affinity of the promoter sites. It is possible that at different times during the shift-up, the limitation may shift among the parameters mentioned. An example of proteins under this form of "metabolic control" is r-proteins.

Certain enzymes which are growth-associated remain correlated to growth rate throughout the transient whereas others are synthesized at rates higher than those dictated by the steady-state growth rate profile. Some authors speculate that certain growth-associated enzymes are, in fact, growth limiting. Once again, it is possible that growth limitation may shift from one enzyme or metabolite to another as metabolite pools well and shrink in response to the increase in biosynthesis rate.

Therefore, when responding to shift-ups which cause multiple transcriptional repressions the cell adapts by selectively synthesizing RNA polymerase. The mechanism by which this selective synthesis is regulated is still speculative. Logical hypotheses may be either that changes in the polymerase binding affinities of the promoters occur or that changes in the polymerase structure itself result which enable it to selectively bind "desirable" operons. The important point, however, is that for at least some enzymes, it is possible to achieve selective and rapid synthesis by imitating unbalanced growth. In order to meet the onslaught of increased protein synthesis, reserve pools of ribosomes are rapidly called to active duty. Interestingly, no investigations were encountered in which the ribosomal content was determined to limit the rate of protein synthesis.

In retrospect, Maaloe's model[29] remains well intact and recent research for the most part have continued to strengthen his speculations. Now, it is time to take advantage of this model and seek out new approaches for the overproduction of not only enzymes but also the products of enzymes. For those enzymes that are positively growth associated and are required for growth, it is possible to achieve preferential synthesis during a shift-up. The initial bottleneck is the availability of RNA polymerase; this suggests that improvement in polymerase availability through either increases in its production (e.g. via genetic manipulation) or decrease in demands for its use (e.g. via environmental production) may lead to further improvements in selective enzyme synthesis. Koplove and Cooney[83] achieved a selective stimulation of acetate kinase by initiating a shift-up by addition of glucose to permit additional growth in a glucose-limited chemostat and of a small quantity of amino acids to represe the amino acid and biosynthetic pathway. The presumably made more polymerase available for preferential synthesis of acetate kinase as well as other essential enzyme synthesis. Thus, the use of unbalanced conditions amy prove to be an important tool for the biochemical engineer. Such unbalanced conditions appear to make better use of the available ribosomes by mobilizing the ribosome components in the pool.

Since it is possible to achieve selective enzyme synthesis, it follows that preferential synthesis of products of the catalized reactions also should occur. A prerequisite for overproduction of metabolites will be the availability of both sufficient metabolic energy and precursors. Two interesting examples of product oversynthesis following shift-ups have been presented. In the first, Welles and Blanch[88] fed glucose intermittantly to anaerobic glucose-limited chemostat of *Saccharomyces cerevisiae* and found that the conversion yield of ethanol from glucose was improved from 0.365 to 0.55 g ethanol/g glucose. Since ethanol production (and the enzymes lending to its formation) is growth associated it is likely that the repeated shift-ups lead to preferential synthesis of those enzymes needed to obtain energy from the production of ethanol.

In studies with *Streptococcus mutans,* Leung et al.[89] added pulses of glucose to glucose-limited chemostats. The cells responded by rapidly converting over 95% of the added glucose to lactate compound with less than 50% at steady state. Some of the rapid oversynthesis of lactate was shown to result from an activation of extant lactate dehydrogenase. However, de novo protein synthesis was required for maximal response. The results are again consistant with their being preferential enzyme synthesis following the shift-up and then further expression of this event by preferential production of specific products.

5 Conclusions

During the past decade, our understanding of the fundamentals of molecular biology has advanced astronomically. This review tries to summarize that segment of molecular biology that relates to the synthesis of enzymes during transients in growth, especially transients following nutritional shift-ups. The overall objective has been to provide the reader with a fundamental understanding of the events that occur during transients so that you may not only visualize transient phenomena, but also obtain a quantitative picture of the limitations and potentials of protein synthesis for differential synthesis of specific and often tightly controlled enzymes. It is now the job of the biochemical engineer to translate these potentials into enhanced enzyme production and to go to the next step in achieving preferential product synthesis.

6 References

1. Koch, A.L.: Microbiol. Physiol. *6*, 147 (1971)
2. Nagai, S. Nishizawa, Y., Endo, I., Aiba, S.: J. Gen. Appl. Microbiol. *14*, 121 (1968)
3. Maaloe, O., Kjeldgaard, N.O.: Control of macromolecular synthesis. New York: W.A. Benjamin, Inc. 1966
4. Herbert, D., Ellsworth, R., Telling, R.C.: J. Gen. Microbiol. *14*, 601 (1956)
5. Ierusalimsky, N.D.: In: Microbial physiology and continuous culture. Powell, E.O., Evans, C.G.T., Strange, R.E., Tempest, D.W. (eds.), p. 23. London: Her Majesty's Stationary Office 1967
6. Harvey, R.J.: J. Bacteriol. *104*, 698 (1970)
7. Goodwin, B.C.: Nature *209*, 479 (1966)
8. Rose, J.H., Yanofsky, C.: J. Mol. Biol. *69*, 103 (1972)
9. Chai, N., Lark, K.G.: J. Bacteriol. *104*, 401 (1970)
10. Rosset, R., Jolson, J., Morier, R.: J. Mol. Biol. *18*, 308 (1966)
11. Dennis, P.P., Bremer, H.: J. Bacteriol. *119*, 270 (1974)
12. Waldron, C., LaCroute, F.: J. Bacteriol. *122*, 855 (1975)
13. Herbert, D.: Symp. So. Gen. Microbiol. *11*, 391 (1961)
14. Dean, A.C.R., Rogers, P.L.: Biochim. Biophys. Acta *148*, 267 (1967)
15. Cooney, C.L., Wang, D.I.C., Mateles, R.I.: Appl. Environ. Microbiol. *31*, 91 (1976)
16. Sykes, J., Young, T.W.: Biochim. Biophys. Acta *169*, 103 (1968)
17. Norris, T.E., Koch, A.L.: J. Mol. Biol. *64*, 633 (1972)
18. Mandelstam, J., McQuillen, K.: Biochemistry of bacterial growth. New York: Halsted Press 1973
19. Harrison, D.E.F., Maitra, P.K.: Biochm. J. *112*, 647 (1969)
20. Beck, C., von Meyenburg, H.K.: J. Bacteriol. *96*, 479 (1968)
21. Sanwal, B.D.: Bacteriol. Rev. *34*, 20 (1970)
22. Clark, P.H., Lilly, M.D.: Symp. Soc. Gen. Microbiol. *19*, 118 (1969)
23. Atkinson, D.E.: Biochem. *7*, 4030 (1968)
24. Dean, A.C.R.: J. Appl. Chem. Biotechnol. *22*, 245 (1972)
25. Matin, A., Grootjians, Hogenhuis, H.: J. Gen. Microbiol. *94*, 323 (1976)
26. Kuempel, P.L., Masters, M., Pardee, A.B.: Biochem. Biophys. Res. Commun. *18*, 858 (1965)
27. Donachie, W.D., Masters, M.: In: The cell cycle. Padilla, G.M., Whitson, G.L., Cameron, I.L. (eds.), p. 37. New York: Academic Press 1969
28. Gorman, J., Tauro, R., Laberge, M., Halvorson, H.: Biochem. Biophys. Res. Comm. *15*, 43 (1964)
29. Maaloe, O.: Dev. Biol. Suppl. *3*, 33 (1969)
30. Kjeldgaard, N.O., Maaloe, O., Schaechter, M.: J. Gen. Microbiol. *19*, 607 (1958)
31. Kjeldgaard, N.O., Kurland, C.G.: J. Mol. Biol. *6*, 341 (1963)
32. Neidhardt, F.C., Fraenkel, D.G.: Gold Spring Harbor Symp. Quant. Biol. *26*, 63 (1961)

33. Koch, A.L.: Nature *205*, 800 (1965)
34. Koch, A.L., Deppe, C.S.: J. Mol. Biol. *55*, 549 (1971)
35. Bremer, H., Dennis, P.O.: J. Theor. Biol. *52,* 365 (1975)
36. Nierlich, D.P.: J. Mol. Biol. *72*, 751 (1972)
37. Nierlich, D.P.: J. Mol. Biol. *72*, 765 (1972)
38. Nierlich, D.P.: Science *158*, 1186 (1967)
39. Nierlich, D.P.: Proc. Nat. Acad. Sci. *60*, 1343 (1968)
40. Young, T.B., Bungay, H.R.: Biotech. Bioeng. *15*, 377 (1973)
41. Tempest, D.W., Herbert, D., Philipps, P.J.: In: Microbial physiology and continuous culture. Powell, E.O., Evans, C.G.T., Strarge, R.E., Tempest, D.W.. (eds.), p. 240. London: Her Majesty's Stationary Office 1967
42. Schlief, R.: J. Mol. Biol. *27*, 41 (1967)
43. Harvey, R.J.: J. Bacteriol. *101*, 574 (1970)
44. Gullov, K., von Meyenburg, K., Molin, S.: Molec. Gen. Genet. *130*, 271 (1974)
45. Carpenter, G., Sells, B.: Biochim. Biophys. Acta *287*, 322 (1972)
46. Carpenter, G., Sells, B.: Eur. J. Biochem. *44*, 123 (1974)
47. Dennis, P.P.: J. Mol. Biol. *89*, 223 (1974)
48. Young, R., Dennis, P.P.: J. Bacteriol. *124*, 1618 (1975)
49. Molin, S., von Meyenburg, K., Gullov, K., Maaloe, O.: Molec. Gen. Genet. *129*, 11 (1974)
50. Bennett, P.M., Maaloe, O.: J. Mol. Biol. *90*, 541 (1974)
51. Sells, B.H., Boyle, S.M., Carpenter, G.: Biochem. Biophys. Res. Commun. *67*, 203 (1975)
52. Mateles, R.I., Ryu, D.Y., Yasuda, T.: Nature *208*, 263 (1965)
53. Alton, T.H., Koch, A.L.: J. Mol. Biol. *86*, 1 (1974)
54. Koch, A.L., Deppe, C.S.: J. Mol. Biol. *55*, 549 (1971)
55. Forschhammer, J., Lindahl, L.: J. Mol. Biol. *55*, 563 (1971)
56. Kitchingham, G.R., Fournier, M.J.: J. Bacteriol. *124*, 1382 (1975)
57. Parker, J., Flashner, M., McKeever, W.G., Neidhardt, F.C.: J. Biol. Chem. *249*, 1044 (1974)
58. Hirshfield, I.N., Zamecnik, P.C.: Biochim. Biophys. Acta *259*, 330 (1972)
59. Skjold, A.C., Juarez, H., Hedgcoth, C.: J. Bacteriol. *115*, 177 (1973)
60. Pato, M.L., von Meyenburg, K.: Cold Spring Harbor Symp. Quant. Biol. *35*, 497 (1970)
61. Mowbray, S.L., Nierlich, D.P.: Biochim. Biophys. Acta *395*, 91 (1975)
62. Dougan, A.H., Glaser, D.A.: J. Mol. Biol. *87*, 775 (1974)
63. Iwakura, Y., Ishihama, A.: Molec. Gen. Genet. *142*, 67 (1975)
64. Boddy, A., Clarke, P.H., Houldsworth, M.A., Lilly, M.D.: J. Gen. Microbiol. *48*, 137 (1967)
65. Boyle, S.M., Sells, B.H.: Arch. Biochem. Biophys. *172*, 215 (1976)
66. Hayward, R.S., Tittawella, I.P.B., Scaife, J.G.: Nature *243*, 6 (1973)
67. Cashel, M.: M. Biol. Chem. *244*, 3133 (1969)
68. Cashel, M., Kalbacker, B.: J. Biol. Chem. *245*, 2309 (1970)
69. Lazzarini, R.A., Cashel, H., Gallant, J.: J. Biol. Chem. *246*, 4381 (1971)
70. Friesen, J.D., Fiil, N.P., von Meyenburg, K.: J. Biol. Chem. *250*, 304 (1975)
71. Cooper, S.: J. Mol. Biol. *43*, 1 (1969)
72. Cooper, S., Helmstetter, C.E.: J. Mol. Biol. *31*, 519 (1968)
73. Sloan, J.B., Urban, J.E.: J. Bacteriol. *128*, 302 (1976)
74. Mitsui, H., Ishihama, A., Osawa, S.: Biochim. Biophys. Acta *76*, 401 (1963)
75. Coffman, R.L., Novis, T.L., Koch, A.L.: J. Mol. Biol. *60*, 1 (1971)
76. Dalbow, D., Young, R.: Biochem. J. *150*, 13 (1975)
77. Dalbow, D., Bremer, H.: Biochem. J. *150*, 1 (1975)
78. Sankaran, L., Pogell, B.M.: Nature *245*, 257 (1973)
79. Jensen, D.E., Neidhardt, F.C.: J. Bacteriol. *98*, 131 (1969)
80. Kennel, D., Magasanik, B.: Biochim. Biophys. Acta *81*, 418 (1964)
81. Lilly, G., Rowley, B.I., Bull, A.T.: J. Appl. Chem. Biotech. *24*, 677 (1974)
82. Carlsson, J., Elander, B.: Caries Res. *7*, 89 (1973)

83. Koplove, H.M., Cooney, C.L.: J. Bacteriol. *134*, 992 (1978)
84. Imanaka, T., Kaida, T., Sato, K., Taguchi, H.: J. Ferm. Technol. *50*, 633 (1972)
85. Imanaka, T., Kaieda, T., Taguchi, H.: J. Ferm. Technol. *51*, 423 (1973)
86. Monod, J.: Ann. Rev. Microbiol. *3*, 371 (1949)
87. Imanaka, T., Kaieda, T., Taguchi, H.: J. Ferm. Technol. *51*, 431 (1973)
88. Welles, J.B., Blanch, H.W.: Biotech. Bioeng. *18*, 129 (1976)
89. Leung, J., Haggstrom, M., Cooney, C.L., Sinskey, A.J.: J. Bacteriol. (submitted)

Stabilized Soluble Enzymes

Rolf D. Schmid
Department of Biotechnology, Henkel KGaA
D-4000 Düsseldorf, West Germany

After an introduction by a chapter on enzyme denaturation by physical, chemical, and biological agents, methods for the stabilization of water-soluble enzymes are discussed in this review. This includes screening for intrinsically stabilized enzymes, addition of stabilizing agents such as substrates, solvents, salts or polymers, and chemical modification (e.g. acylation, reticulation, glycosidation or binding to polymers).

Technical applications of stabilized water-soluble enzymes are indicated, as described in the patent literature; special reference is made to enzymes in detergents.

1 Introduction

While the concept of insolubilized enzyme science and technology is a fascinating part of modern biochemistry – as seen in the preceding chapters of this book – the vast majority of the enzymes on the market are in their native form.

From an economic point of view, benefits-high selectivity, high efficiency, and "natural behavior" – have to be matched with cost which to a significant extent is determined by stability during storage and use.

Thus the stabilization of native enzymes is pertinent to the industrial biochemist interested in such different fields as, e.g., detergent enzymes, food enzymes or enzymes for diagnostic use.

In addition, the potential use of chemically modified water-soluble enzymes has been demonstrated in applications such as, e.g., continuous-flow ultrafiltration reactors containing soluble enzymes or in medicine where enzyme derivates with low antigenicity and long circulation time have been proposed as pharmaceutical agents.

The intrinsic instability of enzymes in aqueous solution and the methods for keeping them stable are described in general terms in most enzymology textbooks (e.g., Bergmeyer, 1970; Dixon and Webb, 1966; Guilbault, 1970). As Euler put it in 1920: "... As dry fat-free powders enzyme preparations have an almost unlimited shelf-life if kept in the dark .... The storability of enzymes in solution, on the other hand, is much less satisfactory various enzymes differing considerably. The storability always decreases markedly with increasing temperature; in pure aqueous solutions rapid inactivation takes place between 50–80 °C which completely destroys the enzymatic activity within one to five hours. Many enzymes are quite well stored in glycerol solution. This solvent is particularly well-suited for the storage of proteolytic enzymes, particularly as these enzymes ... are very well extracted with glycerol from animal secretory organs ..." (Euler, 1920, translation).

In textbooks, however, few, if any, references to original articles on stabilization of enzyme solutions are given. In addition, apart from a short discussion of some additives useful for enzyme stabilization (Wiseman, 1973), no recent evaluation of this field seems to exist except in Japanese (Oshima et al., 1975; Ida, 1976).

This review is an attempt to describe present principles for stabilization of water-soluble enzymes; where appropriate, reference is also made to proteins which do not exhibit enzyme activity. While the citation of original literature is certainly not complete, it is hoped that the pertinent fields of stabilization techniques have been adequately covered.

First, as some base work, the most important factors leading to protein denaturation are enumerated. Then, after a short evaluation of intrinsically stabilized enzymes, literature (except patents) on the stabilization of native water-soluble enzymes by additives is reviewed, followed by a chapter on water-soluble enzymes stabilized by chemical modification. Finally, patent literature in this field is indicated with special references to enzymes in detergents.

The newly emerging field of stabilization of native water-insoluble proteins (e.g., membrane proteins) (Tanford and Reynolds, 1976) is not covered.

2 Enzyme Denaturation

2.1 General Considerations

Though a given amino acid sequence allows for a large number of conformations (Anfinsen and Scheraga, 1975), proteins under physiological conditions assume a distinct "tertiary structure" (the native conformation) of minimum free energy which is a prerequisite for their biological function.

It is now well established by X-ray structure analysis (Fersht, 1976) and many other less direct methods (e.g., Tanford, 1968; Brandts, 1969; Klapper, 1973) that the structure of most water-soluble proteins may grossly be described as a hydrophobic core of non-polar amino acid groups surrounded by a hydrophilic shell of polar solvated amino acid side-chains thus resembling a large micelle.

Contrary to the latter, however, a native protein contains other structure-forming elements which, from an energetic point of view, destabilize the ones mentioned above; thus the polar polypeptide backbone extends through the hydrophobic interior while a significant part of the solvated surface is covered with non-polar amino acid side-chains.

Thus a delicate balance of different types of forces determines whether under given environmental conditions a protein exists in its native or in a denaturated state (Table 1).

Table 1. Important structural features which contribute to the conformation of protein[a]. (After Brandts, 1967)

Native conformation	Denatured conformation	Contribution[a]
Rigid structure with little rotational freedom	Flexible structure	Major
Peptide-peptide hydrogen bonds	Peptide-solvent hydrogen bonds	Major
Non-polar side-chains in protein interior	Non-polar side-chains exposed to solvent	Major
High charge density due to compactness	Low charge density in extended conformation	Minor[b]
Ionizable side-chains with local interactions	Ionizable side-chains fully solvated	Minor[b]
Cystin bridges	Cystein residues	–

[a] Large or small contribution to the difference in free energy between the native and denatured states (Brandts, 1967; Tanford, 1968)
[b] Contribution may become large under specific conditions, e.g., pH-changes

While the different conformational forces indicated in Table 1 may well exceed 100 kcal $mole^{-1}$ on an absolute scale (Brandts, 1969) (Fig. 3), they compensate so strongly that the free energy of denaturation of globular proteins rarely exceeds 15 kcal $mole^{-1}$ (Tanford, 1970; Pace, 1975; Privalov and Khechinashvili, 1974) as indicated in Fig. 1.

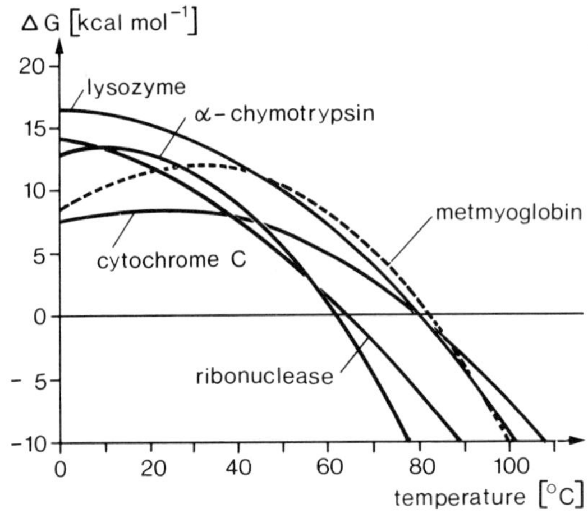

Fig. 1. Free energy of stabilization of 5 globular proteins as a function of temperature as determined by scanning microcalorimetry (Privalov and Khedinashvili, 1974)

Thus native proteins are only marginally stable. Under denaturing environmental conditions, they pass through a variety of conformational states which only rarely have been stable enough for identification as in the case of α-lactalbumin, schematically illustrated in Fig. 2.

As indicated in the sketches of Fig. 2, increasingly denatured protein conformations may be described by their increasing degree of exposure of the polypeptide backbone and non-polar residues to the aqueous environment. An estimation (Tanford, 1968) is given in Table 2.

Table 2

Conformational state[a]	% exposure to solvent	
	Peptide backbone	Non-polar amino-acid side-chains
N	40	40
ID	65	55
RC	75	75
Helical[b]	40	75

[a] Legend for abbreviations see Fig. 2

[b] Catalytically inactive conformational state which is only attained under special environmental conditions (e.g., at high alcohol concentrations)

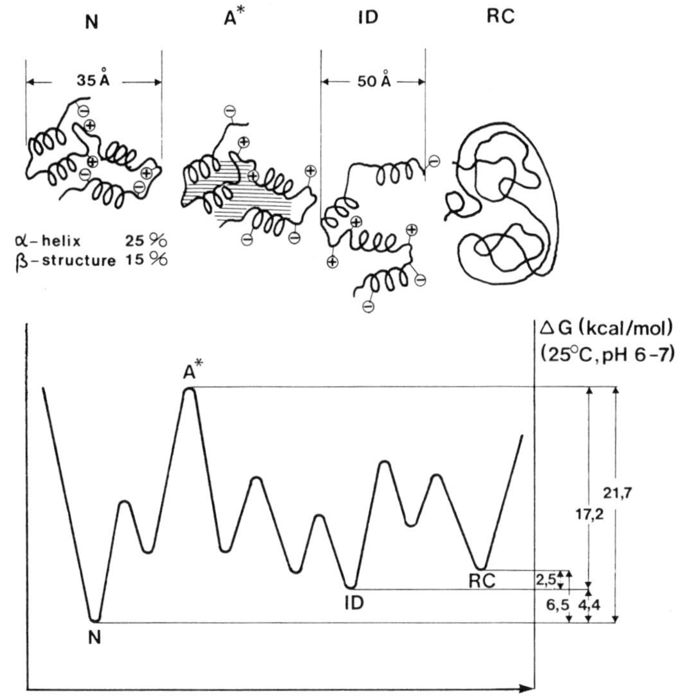

N = native conformation
A* = critically activated state
ID = incompletely disordered conformation
RC = random coil (fully denatured state)

Fig. 2. Schematic illustration of reversible conformational transitions of α-lactalbumin; thermodynamic data and sketches for N, A*, and ID from Kuwajima (Kuwajima, 1977)

In summary, it should be kept in mind that native enzymes exhibit only marginal stability which is easily upset by even subtle environmental changes.

This is a challenge both to the theoretically and practically-minded scientist. For the former, a quote of Tanford (Tanford, 1968) seems appropriate:

"It has not been possible to account for the stability of the native state over all denatured states under native conditions. Were it necessary to make a prediction in the absence of experimental knowledge, one would probably conclude that the native state should not exist." For the latter, a small calculation (Silverstein and Grisolia, 1972) may serve to illustrate the problem:

If only one in 10^4 enzyme molecules [corresponding to $\Delta G^\circ = -5.45$ kcal mole^{-1} (pH 7.0, 25 °C)] exists in the critically activated state A* – see Fig. 2 – exhibiting a half life of one minute before becoming irreversibly denatured, half of the enzyme's activity would be lost in seven days.

While these quotations make obvious that enzyme stabilization in general will be a highly sophisticated endeavor, it is important to realize that due to the subtle balance of opposing forces in a native protein a similar degree of denaturation may be reached by different agents and mechanisms as schematically exemplified in Table 3.

Table 3

Denaturant	Influence on substructures[c] in the native state			
	Polypeptide backbone	Nonpolar residues	Ionic side-chains	H-bonds of side-chains
Organic solvent[a]	Strong destabilizing	Strong stabilizing	Weak destabilizing	Weak effect
Salt[b]	Strong stabilizing	Strong destabilizing	Weak stabilizing	Weak effect

[a] E.g., 100% ethanol
[b] E.g., concentrated $CaCl_2$ solution
[c] As derived from thermodynamic experiments with model compounds (Tanford, 1968)

This is further emphasized in Table 4 where the effects of various physical, chemical, and biological denaturants have been summarized.

It, therefore, seems warranted to briefly depict the action of various denaturants in order to develop a strategy for adequate measures of stabilization.

2.2 Physical Denaturants

Heat

As indicated in Fig. 1 for five globular proteins, denaturation is facilitated at rising temperatures: most enzymes are inactivated at temperatures above the physiological range. While at low protein concentration denaturation is often fully reversible, at high concentrations an irreversible aggregation step ensues (Brandts, 1967) which in some cases may be due to the formation of intermolecular disulfide bridges or other covalent modifications (Tanford, 1968).

In mechanistic terms an estimate of the temperature variance of instability factors such as increased conformational transitions *versus* the stabilizing influence of hydrophobic and peptide backbone hydrogen-bonding may serve to illustrate again the delicate balance between native and denatured states in spite of bonding forces which are large on an absolute scale ($>$ 250 kcal mole^{-1} at 30 °C) (Brandts, 1967) (Fig. 3).

Cold

Many enzymes become less active between temperatures of 0 and 20 °C [e.g., β-fructofuranosidase, pyruvate carboxylase, glutamate decarboxylase (Brandts, 1967, 1969), phosphofructokinase (Kono and Uyeda, 1971), pyruvate kinase (Kuczenski and Suelter,

Table 4. Effects of various denaturants on proteins

Denaturant	Target	Driving force	End product	Ref.
Physical denaturants				
Heat	Hydrogen bonds	Increase of denatured conformations due to increased thermal movement and decreased solvent structure Irreversible covalent modification (e.g., disulfide interchange)	HD Aggregates	Tanford (1968) Brandts (1967) Brandts (1967) von Hippel and Schleich (1969) Edelhoch and Osborn (1976)
Cold	Hydrophobic bonds solvated groups	Altered solvent structure Dehydration	Aggregates Inactive monomers	Brandts (1967) Brandts (1969)
Mechanical forces	Solvated groups void volume	Changes in solvation and void volume Shearing	HD Inactive monomers	Brandts (1969) Tanford (1968) Penniston (1971) Suzuki and Taniguchi (1974)
Radiation	Functional groups (e.g., cySH, peptide bonds)	Decrease of structure-forming interactions after photooxidation or attack by radicals	HD Aggregates	Simic (1978) Grossweiner (1976) Alexander and Lett (1967)
Chemical denaturants				
Acids	Buried uncharged groups (e.g., his, peptide bonds)	Decrease of structure-forming ionic interactions	RC	Tanford (1968)
Alkali	Buried uncharged groups [e.g., tyr, cySH, $(cyS)_2$]	Decrease of structure-forming ionic interactions	RC	Tanford (1968)
Organic H-bond-formers	Hydrogen-bonds	Decrease of structure-forming H-bonds between water and native conformation	RC	Tanford (1968) Tanford (1970) Warren (1970)
Salts	Polar and non-polar groups	"Salting-in"/"salting out"-bias of polar and nonpolar groups in solvent of increased DK	HD	von Hippel and Schleich (1969a, b)
Solvents	Non-polar groups	Solvation of nonpolar groups	Highly ordered Peptide-chains with large helical regions	Tanford (1968) Brandts (1969)

Table 4 (continued)

Denaturant	Target	Driving force	End product	Ref.
Surfactants	Hydrophobic domains (all surfactants) and charged groups (ionic surfactants only)	Formation of partially unfolded substructures including micelle-like regions	ID Large helical regions	Tanford (1968) Schwuger and Bartnik (1978) Tanford and Reynolds (1976)
Oxidants	Functional groups (e.g., cySH, met, try, and others)	Decrease of structure-forming and/or functional interactions	Inactivated enzyme Sometimes disordered structure	Cohen (1968) Westhead (1972)
Heavy metals	Functional groups (e.g., cySH, his, and others)	Masking of groups pertinent to structure or function	Inactivated enzyme	Vallee and Ulmer (1972)
Chelating agents	Cations important for structure or function	Ligand substitution or cation removal	Inactivated enzyme	Bardsley and Chields (1974) Vallee and Wacker (1970)
Biological denaturants				
Proteases	Peptide bonds	Hydrolysis of terminal or other peptide bonds	Oligopeptides, amino acids	Goldberg (1974, 1976) Okunuki (1961)

HD = highly disordered structure; RC = random coil; ID = incompletely disordered structure [cySH = cystein $(cyS)_2$ = cystin, his = histidine, try = tryptophane, met = methionine]

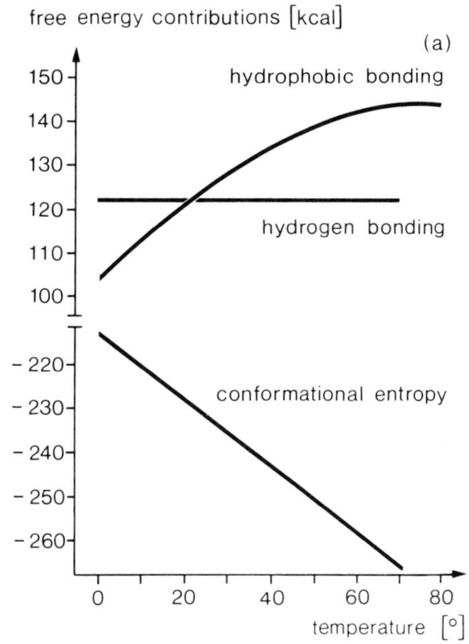

Fig. 3. Estimates of the effect of temperature on the total contributions to the stability of native chymotrypsinogen resulting from hydrogen bonding, hydrophobic bonding, and conformational entropy. (After Brandts, 1967) (a) The increase with temperature in hydrophobic bonding is supposed to be due to the unique features of water as a structured solvent

1970, Hunsley and Suelter, 1969, and others). In some cases a temperature of maximum stability T_{max} well above 4 °C has been established (e.g., 12 °C for chymotrypsinogen, 10–13 °C for chymotrypsin (Brandts, 1967) and examples in Fig. 1] above and below which denaturation is facilitated. Brandts has dicussed the thermodynamic implications of these findings (Brandts, 1969).

For practical purposes, however, it is generally held that enzyme preparations are most stable between 0 and 4 °C (Bergmeyer, 1970, 1977); this may in part be due to the fact that proteolytic enzymes, which are present in most enzyme preparations, exhibit low activity in the cold.

Freezing and thawing may lead to inactivation [e.g., catalase, myosin (Hanafusa, 1970, 1971), chymotrypsinogen (Brandts et al., 1971), aldolase (Südhof, 1960), lactate dehydrogenase and others (Chilson et al., 1965)].

While conformational transitions due to reduced hydrophobic interactions at low temperatures (Brandts, 1969), dissociation of quaternary structures (Kuczenski and Suelter, 1970; Hanafusa, 1971; Kono and Uyeda, 1971; Chilson et al., 1965) and aggregate formation (Cowman and Speck, 1969) have been discussed as likely denaturation mechanisms, structural changes due to dehydration after freezing may play an important role (Darbyshire, 1975).

It may be noted that some enzymes, e.g., lipases, catalases, and peroxidases in frozen food are reported to exhibit high biological activity far below 0 °C (Schormüller, 1967; Kuehnau, 1969).

Mechanical Forces

Enzymes may be denatured by mechanical forces such as high orthostatic (Hüttenrauch, 1976; Kasai and Uchida, 1966) or hydrostatic pressure (Penniston, 1971; Williams and Shen, 1972). Other potentially harmful conditions include ultrasound (Dietrich, 1962; Luca, 1971; El'piner et al., 1960) and shearing (Duerre and Ribi, 1963; Horikoshi et al., 1977; Charm and Wong, 1970). Thermodynamic implications of pressure denaturation have been discussed (Brandts, 1969; Tanford, 1968; Hawley, 1971).

While pressures up to ~1000 bar may stimulate enzyme activity (Williams and Shen, 1972; Penniston, 1971), higher pressures invariably lead to denaturation (Hüttenrauch, 1976; Hawley, 1971). Oligomeric enzymes under high pressure are inactivated through dissociation into monomers (Penniston, 1971).

Denaturation by ultrasound is probably due to surface inactivation (Dietrich, 1962); after ultrasound treatment under reducing conditions, pepsin and trypsin have been reported to retain their activity (El'piner et al., 1960).

The dependence of enzyme activity on mechanical forces was elegantly demonstrated (Berezin et al., 1976) by pressure-induced stretching of trypsin and chymotrypsin attached to polymer fibers.

Radiation

Proteins must be protected from light, since under appropriate conditions photooxidation due to photochemical reactions of susceptible groups (e.g., cystein, tryptophane, histidine) is possible (McLaren and Luse, 1961; Vladimirov et al., 1970; Grossweiner, 1976). Denaturation is further dependent on environmental factors such as pH, temperature, dryness, presence of oxygen and others (Alexander and Lett, 1967).

The effects of radiation on tryptophane residues in proteins have been separately considered (Fontana and Toniolo, 1975).

The susceptibility of functional groups in proteins to ionizing radiation as, e.g., employed in radiation preservation of food has been recently reviewed (Simic, 1978).

2.3 Chemical Denaturants

Acids

In protein chemistry, acid denaturation is a well-established procedure.

While in a few cases high stability has been reported at low pH-values (e.g., lysozyme, ribonuclease, acid proteases), most enzymes in an acidic environment are rapidly transformed into highly disordered inactive conformations (Tanford, 1968). For some proteins (e.g., ferrimyoglobin, hemoglobin, and carbonic anhydrase) it has been found that at acid pH protonation of a buried uncharged histidine side-chain leads to a (denatured) conformation in which this newly charged residue will be exposed to the solvent.

Strong organic acids such as trifluoroacetic acid may act by protonation of peptide groups (Tanford, 1968).

Alkali

Protein denaturation at alkaline pH may be complicated by chemical side reactions, e.g., of cysteine or cystine residues (Noetzold et al., 1977). Exposure of buried tyrosyl groups above pH 10–11 seems to play an important role in conformational transitions. The limited number of studies on this subject have been reviewed by Tanford (Tanford, 1968).

Organic H-Bond-Formers

Compounds such as urea, guanidinium HCl or formamide are potent protein denaturants which probably act by competing more effectively than water with structure forming intra- or interchain hydrogen-bonds pertinent for maintaining the native protein structure (Tanford, 1968; von Hippel and Schleich, 1969; Jencks, 1962). Hydrophobically modified H-bond-formers such as acetamide or alkylureas exhibit still more potent denaturing behavior, thus giving evidence for additional hydrophobic interactions (Warren, 1970, 1976).

Neutral Salts

Many enzymes depend on cofactors such as alkali, alkaline earth, or other ions (e.g., Dixon and Webb, 1966; Vallee and Wacker, 1970). They may be inactivated by low concentrations of "wrong" metal ions competing with the "right" ions for binding sites important for enzyme structure or activity. This point will be discussed in more detail in Sect. 3.2.4.

High concentrations of salts are often employed for protein purification ("salting-out"); precipitation is achieved by decreased protein solubility due to an increase of the ionic strength of the solvent.

In the presence of high salt concentrations, the conformational stability of a protein may be influenced by the binding of salts to charged protein groups (e.g., carboxyl groups) or dipoles (e.g., peptide bonds) ("salting-in") and by reduced solubility of hydrophobic groups in a medium of high ionic strength ("salting-out"). The effect of inorganic salts on proteins is not simply a function of the charge type of the salt, but depends on the actual ions of which the salt is composed as exemplified in Fig. 4 for the reversible thermal denaturation of ribonuclease as a function of various salts added:

KCl and NaCl exert little influence; KSCN and $CaCl_2$ are strong destabilizers, while $(NH_4)_2SO_4$ and K-phosphate show stabilizing effects.

Anions and cations are additively effective.

The denaturing potential of different ions on one protein is often correlated with their influence on solubility ("Hofmeister-series") which is similar for different proteins (von Hippel and Schleich, 1969b).

Whether a particular salt serves as a stabilizer or destabilizer of the native conformation depends on the free energy balance of "salting-out" (exposed non-polar groups) and "salting-in" (simultaneously exposed polar groups) in the native and denatured conformations.

In some cases salts in high concentrations have been shown to interfere with substrate binding (e.g., Ottesen and Svendsen, 1976).

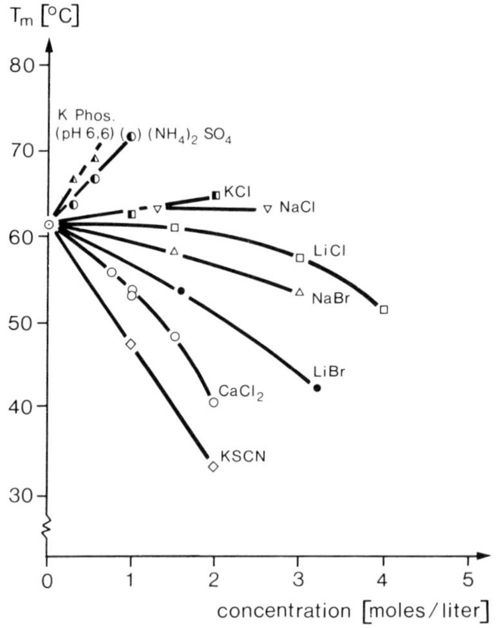

Fig. 4. Midpoint transition temperature (T_m) (temperature at which half of a protein is reversibly denatured) of ribonuclease as a function of concentration of various added salts at pH 7.0. (After von Hippel and Schleich, 1969b)

Organic Solvents

Solvents such as 2-propanol or acetone are widely used as precipitating agents for proteins. At high concentrations most nonaqueous solvents such as monohydric alcohols or dioxane cause an unfolding of native protein structures by solvation of "interior" non-polar residues. This often leads to conformations with large helical regions (Tanford, 1970).

Polyhydric alcohols and some polar aprotic compounds such as DMF or DMSO are far less effective denaturants (Tanford, 1970; Singer, 1962; Jacob and Herschler, 1975); at high concentrations, however, they may also denature proteins (e.g., Klyosov et al., 1975). Several attempts have been made to correlate the denaturing effect of a solvent with its dielectric constant (e.g., Pérez-Villaseñor and Whitaker, 1967; Castañeda-Agullo and Del Castillo, 1959).

The effect of temperature on water structure in the presence of cosolvents and its implications for the mechanism of solvent denaturation have been discussed by Brandts (1967, 1969).

For some proteins such as serum albumin and β-lactoglobulin it has been established that a large number of binding sites for solvent molecules exist (Steinhardt and Reynolds, 1969). Enzymes with large hydrophobic binding sites (e.g., proteases) may bind organic solvent molecules especially well; thus inhibition may occur at very low solvent concentrations due to solvent interference with substrate binding (Svendsen, 1971; Ralston, 1972).

Surfactants

Surfactants are unique in displaying very strong interactions with protein at very low concentrations. Ionic surfactants are more effective than nonionics by virtue of electrostatic *and* hydrophobic binding. While nearly all proteins exhibit massive co-operative binding towards elevated concentrations of ionic detergents, leading to transitions into biologically inactive monomers or into denatured conformations, a few hydrophilic proteins – e.g., serum albumin and α-lactalbumin (which biologically function as carriers for fatty acid anions) – or membrane proteins (Tanford and Reynolds, 1976) – have a small number of binding sites with high affinity for detergents. This is indicated for albumin (Helenius and Simons, 1975) in Table 5.

Table 5. Binding of surfactants to albumin. (After Helenius and Simons, 1975)

	Concentration (mM)			
	Sodium dodecylsulfate	$C_{14}NMe_3$[a]	Triton X-100[b]	Deoxy-cholate
50% saturation of high affinity sites of native albumin	$1 \cdot 10^{-3}$ (10 sites)	$5 \cdot 10^{-2}$ (4 sites)	$5 \cdot 10^{-2}$ (4 sites)	$1.5 \cdot 10^{-2}$ (4 sites)
Critical concentration for massive conformational change	0.3	3	Not observed	
CMC	1	4	0.3	3

[a] Tetradecyltrimethylammonium chloride
[b] p-t-Octylphenol ethoxylate

As indicated in Table 5, nonionic surfactants are less structure-perturbing due to their low monomer concentration and their less directional interactions (no electrostatic binding) (Schwuger and Bartnik, 1978).

Mixed micelle formation between detergent micelles and hydrophobic regions at the protein surface has been discussed (e.g., Tanford, 1968). While at low concentrations of ionic surfactants co-operative binding of monomers is much more important, the interaction between nonionic detergents and a few hydrophilic or – more specifically – membrane proteins follows the latter mechanism more closely. Specific examples include cytochrome b_5, pancreatic colipase, reverse transcriptase (Thompson et al., 1972; Tanford and Reynolds, 1976) and lysozyme (Bernath and Vieth, 1972).

In addition to concentration, the interaction between proteins and surfactants depends strongly on structural properties (chain length, steric requirements, protein structure) and environmental factors (pH, ionic strength, temperature) (Tanford, 1968; Steinhardt, 1975; Schwuger and Bartnik, 1978; Nakaya et al., 1976).

Oxidants

Some amino acids (e.g., cysteine, methionine, tryptophan) are susceptible to photochemical (Westhead, 1972) or chemical (Cohen, 1968) oxidation (Fontana and Toniolo, 1975).

A short reference to photochemical oxidation has been made in Sect. 2b. The chemical oxidant most thoroughly studied is H_2O_2 (Cohen, 1968; Aoshima et al., 1977) but others, e.g., fatty acid peroxides (Wills, 1961; Roubal, 1966) or N-bromosuccinimide are also described in the literature.

An interesting example is the self-catalyzed destruction of lipoxygenase by fatty acid and oxygen, its substrates (Smith and Lands, 1970).

Heavy Metals

Though many enzymes exist which require metal ions for catalytic activity (Vallee and Coleman, 1965), certain heavy metals such as Hg, Cd, and Pb tend to interfere with the biological activity of proteins by binding to ligands such as cysteinyl and histidyl sidechains. For the above-mentioned metal ions the subject has been recently reviewed by Vallee and Ulmer (1972).

Chelating Agents

Metalloenzymes of metal-activated enzymes may be inactivated by chelating agents such as EDTA, 1,10-phenantroline or others due to a) ligand binding inside the enzyme or b) removal of metal to form a coordination complex in solution (Bardsley et al., 1974).

Specific examples include diamino oxidase (Bardsley et al., 1974), glycerylphosphorylcholine diesterase (Baldwin et al., 1969), isocitrate dehydrogenase (Ingebretsen and Sanner, 1976), neutral proteases (Feder et al., 1971), and acetylcholinesterase (Wermuth and Brodbeck, 1973).

2.4 Biological Denaturants

Proteolysis

Due to their polypeptide nature, enzyme proteins are susceptible to proteolytic attack.

In the living cell, this sensitivity is mediated by compartmentation and by a complicated system of reaction sequences as reviewed most recently for microbial enzymes (Switzer, 1977).

From studies on protein turnover *in vivo* it is apparent that the degradation rates of proteins are determined to a large extent by their conformations; additional relationships seem to exist between proteolytic susceptibility and size, charge and structural features such as surface hydrophobic amino acids, helical content, number of disulfide bonds (Goldberg and St. John, 1976; Goldberg and Dice, 1974). Acid proteins are degraded faster than basic or neutral ones (Dice and Goldberg, 1975) and larger proteins are more susceptible to degradation than smaller ones. Thus small basic proteins [e.g., histones, subtilopeptidase (Ottesen and Svendsen, 1970)] are often more stable than others, but exceptions to this general behavior exist.

In general, denaturation often precedes proteolytic degradation. Many references of earlier work on this subject may be found in a review by Okunuki et al. (1961).

3 Enzyme Stabilization

In the preceding section it was stressed that enzymes are only marginally stable and prone to denaturation by a large number of environmental factors due to various opposing forces which cause them to maintain their functional tertiary structure.

In order to enhance soluble enzyme stability, four major routes of work have been pursued in the past:

- screening for enzymes with enhanced intrinsic stability
- addition of stabilizing compounds
- chemical modification
- entrapment in gels, fibers, micelles, and related techniques.

The first three routes will be discussed in the following sections. Entrapment techniques have been extensively reviewed elsewhere (e.g., Mosbach, 1976).

3.1 Intrinsically Stabilized Enzymes

A number of microorganisms are capable of living in hostile environments such as hot springs or saline media of very high ionic strength; marine organisms have been observed at depths which correspond to a pressure of hundreds of bars; and arctic life survives at average temperatures far below 0 °C. In many of these organisms specifically adapted "stabilized" proteins have been identified and characterized (Kleiner, 1978).

A field which has been thoroughly investigated during the last decade is the structure and function of enzymes from thermophilic microorganisms; the proceedings of a recent symposium on the subject have been published (Zuber, 1976) and reviewed (Zuber, 1975). From the available data it seems that no simple structural principle exists which might confer thermal stability to a protein. This conclusion has been corroborated by X-ray structure analysis of the thermophilic protease thermolysin (Matthews, 1976).

It has been found, however, that – when compared to mesophilic counterparts – thermophilic proteins may contain additional metal ions (e.g., thermolysin, α-amylase and aminopeptidase from *Bacillus stearothermophilus*), slightly different amino acid compositions or different structural features (e.g., reduced β-structures) (Zuber, 1976).

From a thermodynamic point of view, enhanced thermal stability might result through i) increased activation energy for denaturation, ii) higher free energy of denaturation (Fig. 5, Curve *1*), iii) a flattened profile of the free energy of denaturation (Fig. 5, Curve *2*), or iv) a shiftened profile of the free energy of denaturation (Fig. 5, Curve *3*).

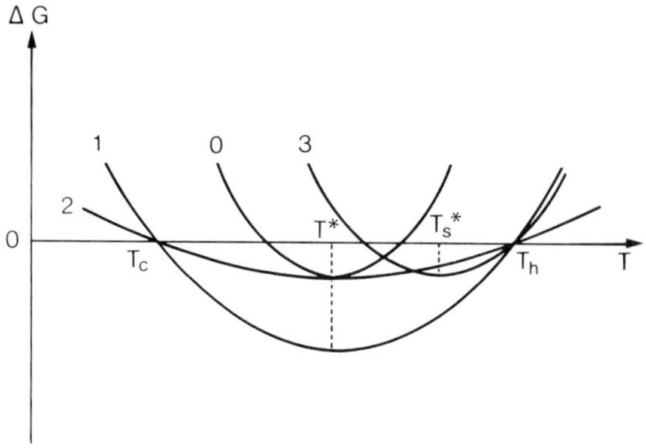

Fig. 5. Hypothetical free energy curves (ΔG *versus* T) of thermostable globular proteins (ΔG is defined as the difference of Gibbs free energy of the native enzyme from that of the denatured one). (After Nojima et al., 1977)
Curve *0*: "mesophile" protein
Curve *1*: protein with higher free energy of denaturation
Curve *2*: protein with larger range of T-stability
Curve *3*: protein with stifted temperature of maximum stability (T)

Barach and Adams have demonstrated that an increased activation energy for denaturation (i) need not be the reason for enhanced thermostability; it was shown that two extracellular proteases – thermolysin (thermostable) and the closely related protease MG60 (thermolabile) – had approximately the same activation parameters for denaturation between 110 and 150 °C (Barach and Adams, 1977).

The other mechanisms of Fig. 5 have been observed: iv) Matthews (Matthews, 1976) has described mutant lysozymes whose free energy of denaturation corresponds to Curve *4*; iii) Nojima (Nojima et al., 1977) concluded from the observation that the free energy of denaturation of thermostable phosphoglycerate kinase was independent of temperature (0–60 °C) that this enzyme behaves according to Curve *2*; ii) many examples exist for thermostable enzymes with higher free energies of denaturation than their thermolabile counterparts (e.g., Hocking and Harris, 1976; Nojima et al., 1977). These differences, however, are below ~10 kcal $mole^{-1}$, thus indicating that only small additional bond energies are required to transform a "mesophilic" (T-optimum ~300 K) into a "thermophilic" (T-optimum ~350 K) protein. Thus subtle changes in the common types of bonding such as additional salt bridges, more hydrogen bridges or increased hydrophobic bonding appear to be perfectly adequate for stabilizing proteins at high temperatures.

It has also been inferred (e.g., Barach and Adams, 1977) that thermostability might reflect structural flexibility in allowing rapid and accurate protein *renaturation* rather than maintenance of the native structure during heating.

This is corroborated by the finding that enhanced thermostability may parallel stability against high ionic strength and other denaturants (Fig. 6) (Zuber, 1976).

Enhanced stability is most probably not due to a single structural change. Thus, thermophilic α-amylase which requires Ca^{2+} for enhanced stability is more stable to urea than its mesophilic counterpart even in the absence of Ca^{2+}, thus indicating stabilizing structural features in addition to metal salt bridges (Yutani, 1976) (Fig. 6). A similar finding has been reported for the heat stability of Ca-free thermolysin as compared with a Ca-free mesophilic neutral protease (Tojima et al., 1976).

3.2 Enzyme Solutions Stabilized by Additives

The observation that enzymes may be stabilized by the addition of various compounds was already made in the 19th century. Among the first stabilizers identified were polyhydric alcohols such as sucrose or glycerol and salts such as $(NH_4)_2SO_4$. In more recent times findings such as the potential of substrates as stabilizers and the effect of polymers have been systematically investigated.

3.2.1 Substrates

It is generally held that the protein conformation at binding sites is less stable than at other areas; this has been sumarized in saying that a "delicate balance of folding very nearly to the point of instability may be essential at the catalytic site" (Lumry, 1959).

Binding of substrates may lead to stabilization or labilization of an enzyme or to no effect at all. In a recent critical review of conformational adaptability in enzymes, Citri tabulated a large number of enzymes the majority of which were stabilized and a few labilized towards thermal or urea denaturation in the presence of specific ligands (Citri, 1973). Many further examples for the influence of substrates on the differential inactivation of proteins by proteolytic attack, solvent denaturation or other methods are included in his work and in an earlier review (Grisolia, 1964).

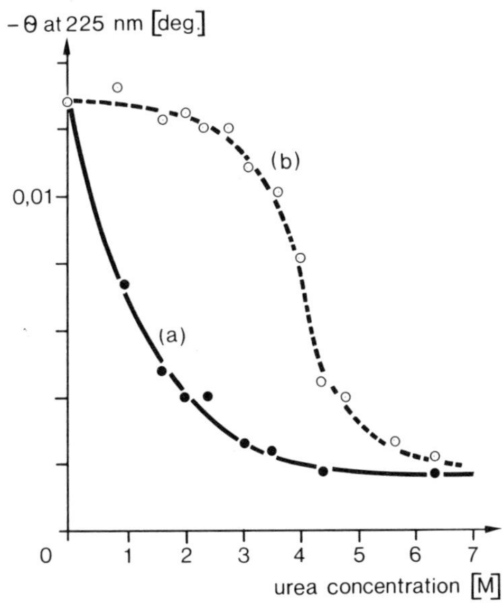

Fig. 6. Comparison of the denaturation of calcium-free mesophile (*a*) and thermophile (*b*) α-amylases. Each calcium-free enzyme was incubated with the respective concentrations of urea at 10 °C for 24 h (Yutami, 1976)

A few specific examples seem appropriate. Lactate dehydrogenase is stabilized towards thermal denaturation by its substrates NAD, NADH, lactate and by effectors such as 3-acetylpyridine-NADH and fructose-diphosphate; it is destabilized by its substrate pyruvate (Citri, 1973).

Pyruvate inhibition of LDH – which possibly parallels the dissociation of the tetrameric native enzyme into subunits – does not take place if the enzyme is in its native environment (muscle subcellular particulate structure) (Ehmann and Hultin, 1973).

Glucoamylase exhibited significantly increased thermal stability in the presence of its substrate analogues glucose, δ-glucono-lactone and lactose; kinetic evidence pointed to substrate stabilization for the two former compounds, while stabilization by lactose was mainly due to solvent effects paralleling the dielectric constant of the solution (Moriyama et al., 1977).

Glucose and fructose stabilized β-glucosidase against denaturation by urea, while maltose did not show any effect (Dean and Rodgers, 1969).

Asparaginase was protected against denaturation by heat, protease and iodine vapor by binding of L-asparagine or L-aspartate (Citri, 1972a, b), and glyceraldehydephosphate dehydrogenase (1.2.1.12) was stabilized towards the action of proteases in the presence of NAD, but labilized by NADH and AMP (Citri, 1973). A general model has been proposed to account for the effects of substrates on protein stability (Fig. 7) (Silverstein and Grisolia, 1972).

If substrate binding leads to a conformation of higher internal energy, denaturation through the critically activated state A* is facilitated; conversely, if by substrate binding a conformation of lower internal energy is obtained, the resulting complex is better protected against denaturation.

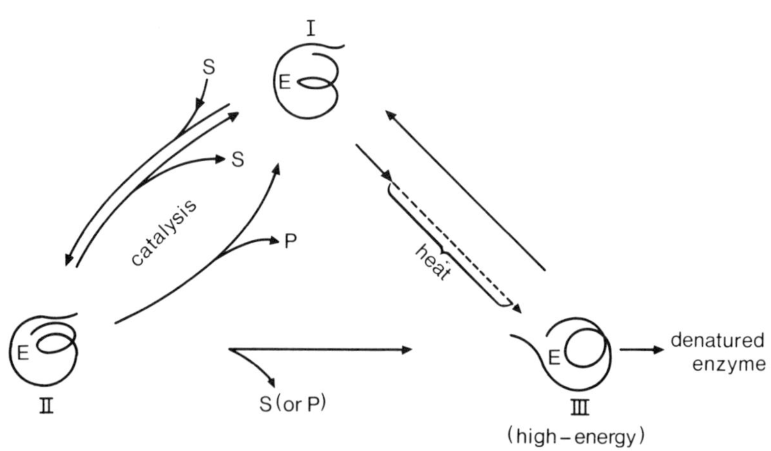

Fig. 7. Model for inactivation of enzymes accelerated by substrate. I represents a low-energy enzyme conformer; II the ES complex; III high-energy enzyme conformer. Heating minimizes energy differences between all enzyme conformers leading to a shift in concentration of I and II toward III (Silverstein and Grisolia, 1972)

Covalently bound substrates (e.g., in acyl enzymes) may also stabilize against denaturation (Bernhard and Rossi, 1968).

3.2.2 Solvents

As discussed in Sect. 2.3, high concentrations of most organic solvents lead to protein denaturation. At solvent concentrations below ~20%, however, stabilizing effects have been observed. Additionally, some solvents such as, e.g., glycerol exhibit a stabilizing effect on proteins even at very high concentrations.

Some of these findings will be discussed in this section; reports referring to activation of enzymes by solvents have been omitted.

Solvents acting as cryoprotectants will be discussed in a separate paragraph.

Polyhydric Alcohols

It has been known since the end of the last century that glycerol or sugar solutions may protect proteins from denaturation. References of this early work may be found in several publications (e.g., Yasumatsu et al., 1965; Gerlsma, 1968). Table 6 gives recent findings which show increased thermal or storage stability for various enzymes in polyhydric alcohol solutions.

The stabilization mechanism of these agents has not yet been firmly established. The storage stability of enzymes in concentrated polyalcohol solutions is certainly influenced by the repression of bacterial growth due to high osmotic pressure (Yasumatsu et al., 1965). A second mechanism – at least for the stabilizing action of sugar solutions on carbohydrate-transforming enzymes – might involve stabilization by substrate or substrate analogues as discussed in Sect. 3.2.1.

Studies on the effect of alcohols on the thermal transition of ribonuclease (Gerlsma, 1968) and chymotrypsinogen A (Gerlsma, 1970) have indicated that the stabilizing effect of polyhydric alcohols is due to a diminished H-bond rupturing capacity as compared to monovalent alcohols. Measurements of the dielectric constant (DK) and thermodynamic considerations suggest a similar mechanism in the case of glucoamylase; solvents which decrease the DK but, due to weak solvation properties towards non-polar residues, show better ligand binding to the native enzyme form (e.g., to the active center) than to the denatured enzyme have been shown to protect this enzyme in solution (Moriyama et al., 1977).

Reduced surface energy of polyalcohol solutions may also be involved in the stabilization mechanism (Timasheff et al., 1976).

Conformational changes in the presence of polyhydric alcohols have been discussed for, e.g., aldehyde dehydrogenase (Bradbury and Jakoby, 1972), phosphorylase b (Damjanovich et al., 1972), pyruvate kinase (Ruwart and Suelter, 1971), aconitate isomerase (Klinman and Rose, 1971), lipoxygenase, alcohol dehydrogenase (Eriksson et al., 1974) and ribonuclease (Bello, 1969). In some cases ligand binding (Dean and Tanford, 1977; Moriyama et al., 1977) or substrate solvation (Haddad et al., 1977) have been inferred. A systematic study (Myers and Jakoby, 1973) on the effect of glycerol, sucrose and ethylene glycol on the kinetic constants of 16 different enzymes indicated in

Table 6. Native enzymes stabilized by polyhydric alcohols

Enzyme	Polyhydric alcohol	Enzyme stability	Ref.
Glucoamylase (3.2.1.3)	10–100% glycerol	T ↑ (10 min, 60 °C)	Moriyama et al. (1977)
Protease	50– 70% glycerol	S ↑ (28 d, 33 °C)	Yasumatsu et al. (1965)
α-Amylase (3.2.1.1)	50– 70% sorbitol	T ↑ (10 min, 80 °C)	
Aldehyde dehydrogenase (1.2.1.3)	30% glycerol	I ↑ (20 min, iodoacetate)	Bradbury and Jakoby (1972)
Ribonuclease (3.1.4.22)	10– 25% glycerol 25% erythritol 10– 40% sorbitol	T ↑ (transition temperature) D ↑ (urea)	Gerlsma (1968)
Chymotrypsinogen A	10– 30% erythritol 10– 30% sorbitol	T ↑ (transition temperature)	Gerlsma (1970)
Chymotrypsin (3.4.21.1)	50– 90% glycerol	P ↑ (24 h, 25 °C)	Sokolova et al. (1971)
Trypsin (3.4.21.4)	30– 50% ethylene glycol	D ↑ (urea) T ↑ (3 h, 35 °C)	
β-Amylase (3.2.1.2)	50% glycerol	S ↑ (48 h, 37 °C)	Banks (1969)

T = heat stability
S = storage stability
P = stability towards proteolysis
D = stability towards denaturants
I = stability towards inhibitors
↑ = improved

most cases substantial changes which were not readily predictable functions of pH or concentration or nature of the polyhydric compound.

It has been pointed out (Bello and Bello, 1976) that impurities such as aldehydes which may be present in glycerol and ethylene glycol may lead to ambiguous results due to side-reactions.

Other Solvents

A number of authors have reported on enhanced enzyme stability after addition of organic solvents (Table 7).

The properties of proteins in organic solvents have been discussed (Singer, 1962; Tanford, 1968; Brandts, 1968). A general mechanism based on the reduced activity coefficient of water at low temperature in the presence of small amounts of solvents ("chlatrate melting") has been proposed (Brandts, 1968).

George (George et al., 1969) observed ~50% residual activity in cold alcoholic solutions of malate dehydrogenase after 24 weeks of storage as compared to 18% in the alcohol-free control. Addition of 40% DMSO led to 100% residual activity after nine moths. Many other solvents led to enhanced storage stability.

At high solvent concentrations, some hydrolytic enzymes catalyze the reverse reaction due to the law of mass action. A few examples are indicated in Table 8.

3.2.3 Cryoprotectants

Some solvents such as glycerol or DMSO are powerful cryoprotectants which help to maintain native protein conformations at temperatures below the freezing point of water. They are being extensively used in protein crystallography (Petsko, 1975) and enzymology (Douzou, 1973, 1975, 1976).

The use of polyhydric alcohols (Ruwart and Suelter, 1971; Eriksson and Svensson, 1974), other solvents (Douzou, 1973; Ruwart and Suelter, 1971), amino acids (Hanafusa, 1970, 1971), proteins and polymers (Chilson et al., 1965; Darbyshire, 1975) as well as other agents (Chilson et al., 1965; Kono and Uyeda, 1971; Cowman and Speck, 1969) as enzyme cryoprotectants has been described in the literature.

Similar principles for effecting stability of proteins in the cold have been observed in nature. Thus glycerol acts as cryoprotectant in hibernating insects (Nordin et al., 1970); sugars, amino acids and similar compounds convey frost resistance to plants (Santarius, 1971); arctic fish have developed specific glycoproteins as "antifreeze" agents (Feeney and Osuga, 1977).

3.2.4 Salts

Many references to the stabilizing effects of salts such as $CaCl_2$ or $(NH_4)_2SO_4$ may be found in the literature of enzyme stabilization. For discussion purposes it seems appropriate to consider separately effects due to low ($\lesssim \cdot 1\ M$) and high ($\gtrsim \cdot 1\ M$) salt concentrations.

Table 7. Native enzymes stabilized by solvents

Enzyme	Solvent	Enzyme stability	Ref.
Metapyrocatechase (1.13.11.2)	10% acetone 10% ethanol	Protection from O_2-inactivation	Nozaki et al. (1963)
Benzylalcohol-dehydrogenase (1.1.1.1)	5–10% acetone 5–10% ethanol	S ↑ (24 h, 5 °C)	Takemori (1967) Katagiri et al. (1967)
	20% acetone 10–20% ethanol	S ↑ (24 h, 5 °C)	Takemori et al. (1967)
Homogentisicase (1.13.11.5)	10–20% acetone	S ↓ (20–26 h, 4 °C)	Takemori et al. (1967)
β-Galactosidase (3.2.1.23)	5–20% methanol ethanol 2-propanol n-propanol	T ↑ (10 min, 47 °C) at and below 10% for ethanol, 2-propanol T ↓ (10 min, 47 °C) for all other concentrations	Shifrin and Hunn (1969)
Lactate dehydrogenase (1.1.1.27)	40% DMSO 10% ethanol 10% methanol	S ↑ (9 m, 4 °C) S ↑ (36 w, 4 °C)	George et al. (1969)
Protease E 30	5% $CHCl_3$	S ↑ (4 °C, 20 °C)	Nachev et al. (1971)
Trypsin (3.4.21.4)	12% ethanol	R ↑ (10 °C)	Pohl (1968)

R = renaturation
T = heat stability
S = storage stability
↑ = improved
↓ = decreased

Table 8. "Reversed hydrolysis" of enzymes in organic solvents

Goal	Enzyme	Solvent	Ref.
Formation of urea from $(NH_4)_2CO_3$	Urease (3.5.1.5)	Acetone, DMF, and others	Butler and Reithel (1977)
Formation of ester from N-Ac-tyrosine and alcohol	Subtilopeptidase B (3.4.21.14)	Ethanol, glycerol	
	α-Chymotrypsin (3.4.21.1)		Ingalls et al. (1975)
Formation of sucrose from glucose and fructose	Invertase (3.2.1.26)	Acetone, pyridine	Kelly et al. (1976)
Formation of glycerides from fatty acids and glycerol	Lipase (3.1.1.3)	Glycerol	Iwasaki et al. (1976)

Low Concentrations

At low salt concentrations, stabilizing effects are due to specific interactions between the salt cations and the metalloenzymes. Metalloenzymes – a term often opposed to metal protein complexes with weak binding forces – are proteins which exhibit strong binding of metals (Scrutton, 1973; Dixon and Webb, 1966; Vallee and Wacker, 1970; Lontie and Vanquickenborne, 1974; Darnall and Birnbaum, 1976).

As indicated in a recent survey of about 80 metalloenzymes (Vallee and Wacker, 1970), metals such as Ca, Zn, Mn, Fe, Mo, and Cu may participate in functions as diverse as oxygen transport [e.g., Cu in hemocyanins (Lontie and Vanquickenborne, 1974)], zymogen activation [e.g., Ca in trypsinogen (Darnall and Birnbaum, 1976)], catalytic function [e.g., Zn in carboxypeptidase (Chlebowski and Coleman, 1976)] or maintenance of tertiary structure [e.g., Ca in α-amylase (Siegel, 1973)].

For metals participating in the catalytic function of enzymes, incorporation into the active site must be considered as well as participation in the formation of the enzyme-substrate complex (Dixon and Webb, 1966; Scrutton, 1973), removal of substrates (e.g., fatty acids as Ca-salts during lipase action), or allosteric effects (Dixon and Webb, 1966). Stabilization by added salts may be due to interaction with either of these mechanisms.

In maintaining the tertiary structures of some enzymes and proteins, Ca plays a most important role (Siegel, 1973). As an example, α-amylase from *Bacillus caldolyticus* is stabilized by firmly bound Ca ions (Heinen and Lauwers, 1976). After removal of Ca with complexing agents, thermal stability is lost, as indicated in Fig. 8.

It may be seen from this example that stability often critically depends on specific ions, since Sr cannot substitute for Ca.

Interestingly, α-amylase from *B. subtilis* contains 4 Ca ions and no disulfide linkage, while α-amylase from human saliva contains only 1 Ca ion, but several disulfide bridges. It is thus tempting to speculate that stabilization by Ca leads to analogous effects as stability due to disulfide bonds (Siegel, 1973).

Thermolysin, a thermostable Zn protease from *Bacillus stearothermophilus*, has been shown by X-ray structure analysis to contain its four Ca ligands bound to asparagine

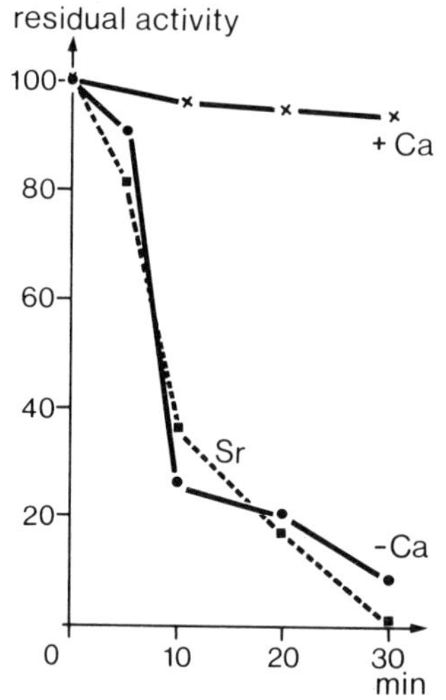

Fig. 8. α-Amylase stability at 70 °C in absence or presence of Ca or Sr (Heinen and Lauwers, 1976)

and glutamine carboxyl groups (Matthews, 1976). It has been inferred that the bridging function of Ca within the polypeptide chain reduces the flexibility of the polypeptide backbone thus enhancing stability; in addition, introduction of Ca might counteract a destabilizing accumulation of negative charges (Fontana et al., 1976). For thermostable enzymes, stabilization by metal bridges is especially noteworthy since metal complexation is entropy driven (replacement of water from metal hydrate) thus being favored by increasing temperature.

While some metalloenzymes often undergo marked and sometimes apparently irreversible alterations in their three-dimensional structure (Vallee and Wacker, 1970) after removal of the metal, others yield stable, metal-free inactive apoenzymes to which activity can be fully restored by readdition of metal ions.

Thus a variety of mechanisms exists by which addition of appropriate metal salts in small concentrations may confer increased stability to metalloenzymes.

High Concentrations

While addition of small concentrations of salts such as $CaCl_2$ may greatly enhance enzyme stability, large concentrations of the same salt may have a significant destabilizing effect as indicated in Fig. 4 for the thermal denaturation of ribonuclease.

According to the "Hoffmeister-series" of salt interaction with proteins (von Hippel and Schleich, 1969) discussed in Sect. 2.3, high concentrations of some other salts as, e.g., phosphates may unspecifically stabilize against denaturation. $(NH_4)_2SO_4$ is a potent stabilizer which is widely used as an additive for enzyme storage.

3.2.5 Polymers

The addition of proteins as stabilizing agents for enzyme solutions is a procedure well established in the art. Recently, synthetic polymers have also been investigated (Table 9).

Table 9. Native enzymes stabilized by added polymers

Enzyme	Polymer	Enzyme stability	Ref.
Glucose oxidase (1.1.3.4)	PVP/PVAc	T ↑ (4 h, 50 °C)	O'Malley et al. (1973)
	PVP	T ↑	
	PVP/PVA	T ↓	
	PVA	T ↑ (4 h, 50 °C)	Hixson (1973)
α-Amylase (3.2.1.1)	Dextran 250	I ↓ (amylase antibody)	Ceska (1971)
	Ovalbumin	Dil ↑	Bernfeld et al. (1965)
β-Amylase (3.2.1.2)	Bovine albumin	T ↑ (1 h, 37 °C)	Takeda and Hizukuri (1972)
	PVA	D ↑	
	Serum albumin	T ↑ (18 h, 35 °C)	Walker and Whelan (1960)
		T ↑ (48 h, 37 °C)	Banks and Greenwood (1969)
Catalase (1.11.1.6)	Polysaccharide phosphate	S ↑ (120 min, 24 °C)	Shataeva et al. (1976)
Trypsin (3.4.21.4)	Dextran	pH →	Laurent (1971)
α-Chymotrypsin (3.4.21.1)	Polymetacrylate gel	T ↑ (180 min, 60 °C)	Martinek et al. (1977)
Lactate dehydrogenase (1.1.2.3)	Dextran	pH →	
	Gelatin	S ↑ (36 w, 4 °C)	George et al. (1969)
	Bovine serum albumin	Dil ↑	Bernfeld et al. (1965)
Malate dehydrogenase (1.1.1.38)	Gelatin + ethanol or glycerol	S ↑ (5 m, 4 °C)	George et al. (1969)
Glucose-6-phosphate-dehydrogenase (1.1.1.49)	Ficoll	$T_{1/2}$ (6 d, 25 °C)	Pollak and Whitesides (1975)
Aldolase (4.1.2.13)	PVP	T ↓ (35 °C)	Jancsik et al. (1975, 1976)
	PVA		
	Serum albumin	Dil ↑	Bernfeld et al. (1965)

For footnotes see page 66

Enhanced activity of α-amylase in the presence of 2% dextran 250 was observed, but inhibition of the enzyme by specific antibodies was greater than with the polymer-free enzyme solution, thus indicating only weak interaction between enzyme and polymer (Ceska, 1971).

In grafting experiments with glucose oxidase, Hixson observed increased thermal stability of the enzyme after addition of 1% PVA (Hixson, 1973). In similar experiments with the same enzyme the stabilizing effect of polymers increased with increasing hydrophobicity (PVAc > PVP-PVAc > PVP > PVP-PVA) and was independent of the molecular weight of the polymer. Investigation of the thermal stability as a function of the enzyme-polymer ratio showed distinct breakpoints below which no effect of the polymer could be seen (Fig. 9).

It was concluded that significant but little specific interactions between enzyme and polymer took place (O'Malley and Ulmer, 1973).

In contrast to this finding, a linear stabilizing influence of yeast polysaccharide concentration on catalase has been described (Shataeva et al., 1976); bacterial polysaccharides have also been used to stabilize leucine aminopeptidase during storage (Zaikina et al., 1970).

From a mechanistic point of view, the stabilizing effect of polymers on enzymes may be interpreted as an exclusion of the enzyme from part of the solvent thus counteracting detrimental effects of the environment. Kinetic evidence for this hypothesis has been elaborated for three enzymes in the presence of PEG or dextran (Laurent, 1971; Ölbrink and Laurent, 1974).

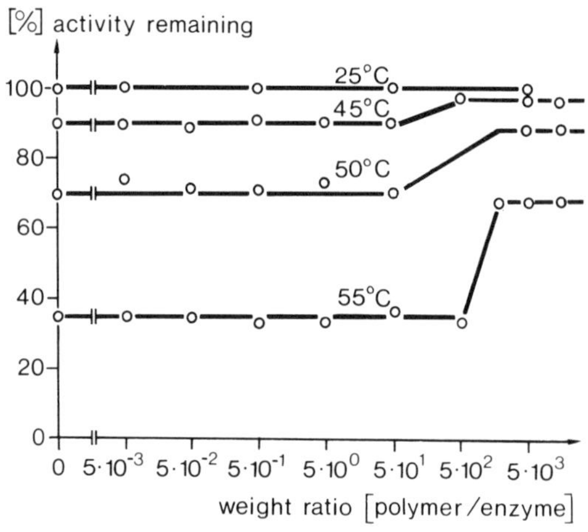

Fig. 9. Effect of 1% acetylated PVA concentration on the enzymatic activity of glucose oxidase at various temperatures with a constant incubation time of 1 h. Enzyme concentration was kept constant at 0.002 mg/ml and the polymer concentration was varied from 0 to 20.0 mg/ml (O'Malley and Ulmer, 1973)

Footnotes to Table 9

PVP = polyvinylpyrrolidone
PVAc = polyvinylacetate
PVA = polyvinylalcohol
PEG = polyethylenglycol
T = heat stability
S = storage stability
pH = pH stability
D = stability towards denaturants
I = stability towards inhibitors
Dil = stability towards dilution
$T_{1/2}$ = half life time (50% residual activity)
↑ = improved
→ = unchanged
↓ = decreased

If interaction is increased by entrapping an enzyme in a charged gel, intermolecular electrostatic interactions and hydrogen bonding due to multipoint noncovalent interaction – above a critical polymer concentration – may result in tremendous stabilization as evidenced for α-chymotrypsin embedded in polymetacrylate gel (Martinek et al., 1977b); in 50% gel (w/w) the increase in thermostability at 60 °C is 10^5 times that of the native enzyme (Fig. 10).

3.2.6 Miscellaneous

Sulfur-groups in proteins are labile to environmental factors such as oxidants, metal ions, alkylating agents, nucleophiles and radiation. The chemistry and biochemistry of cystein- and cystin-groups in proteins has recently been reviewed in a book (Torchinskii, 1974).

The addition of small amounts of thio-compounds such as mercaptoethanol or dithiothreitol is a standard laboratory method to suppress inactivation of sulfur containing enzymes during work-up and purification. Choice of concentration of these compounds is critical, however, since at higher levels SS-bonds may be split.

Thiol compounds such as glutathione, cysteamine, and others have been often found to stabilize enzymes under denaturing conditions such as radiation (Shapiro and Kollmann, 1968; Kopp et al., 1966) heat (Wiseman and Williams, 1971), dilution (Walker and Whelan, 1960), cold (Chilson et al., 1965) or in presence of peroxides (Wills, 1961). In the case of ionizing radiation, the compounds act as efficient radical scavengers (Shapiro and Kollmann, 1968).

Vitamin E and ascorbic acid (Tappel, 1971; Shapiro and Kollmann, 1968) have been added as antioxidant stabilizers to enzyme solutions. Glucoamylase has been protected

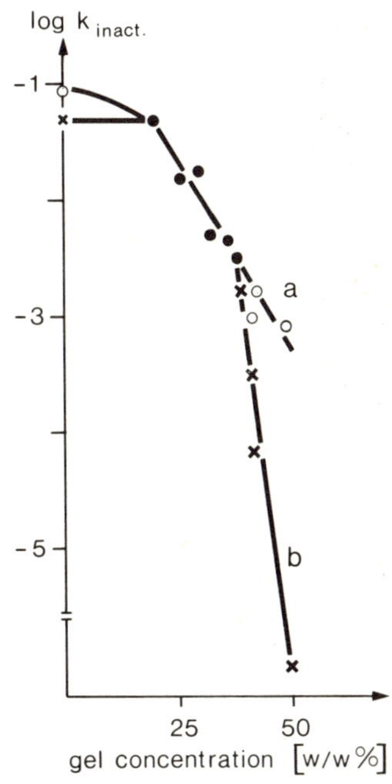

Fig. 10. Dependence of thermoinactivation of chymotrypsin on the concentration of polymethacrylate gel at degrees of conversion (%): *a* 0; *b* 75. The points at 0 and 20% w/w gel correspond to thermoinactivation in solution without a monomer and in the presence of a soluble polymer, respectively. Conditions: 60 °C, pH 8.0. (After Martinek et al., 1977)

from detrimental oxidative effects of H_2O_2 by adding quinine sulfate as H_2O_2 stabilizer (Cho and Bailey, 1977).

Peroxidase was protected from heat and radiation by addition of its coenzyme, heme (Rosén and Nilsson, 1971).

β-Amylase was stabilized against heat denaturation by addition of the nonionic surfactants Tween 20 and Triton X 100 and by bovine albumin (Takeda and Hizukuri, 1972); Triton X 100 stabilized β-amylase, glucoamylase, and α-amylase against dilution inactivation.

A limited number of amino acids were able to protect lysozyme from denaturation by ultrasound; competition with the protein for available surface at the (denaturing) air-water interface has been postulated to account for this finding (Dietrich, 1962).

3.3 Chemically Modified Soluble Enzymes

The lore of enzyme immobilization elaborated during the last decade has led to outstanding success in enzyme stabilization which manifests itself in an increasing number of enzyme reactors used for large-scale technical processes.

Among the various procedures of enzyme immobilization which have been reviewed in this series and elsewhere (e.g., Mosbach, 1976), less attention has been paid to methods which lead to water-soluble enzyme conjugates. However, chemically modified soluble enzymes show a number of potential advantages: soluble modified enzymes often exhibit higher catalytic activity than enzymes insolubilized on the same type of matrix (Svensson, 1976; Axén et al., 1970; Wykes et al., 1971) as indicated in Fig. 11 for chymo-

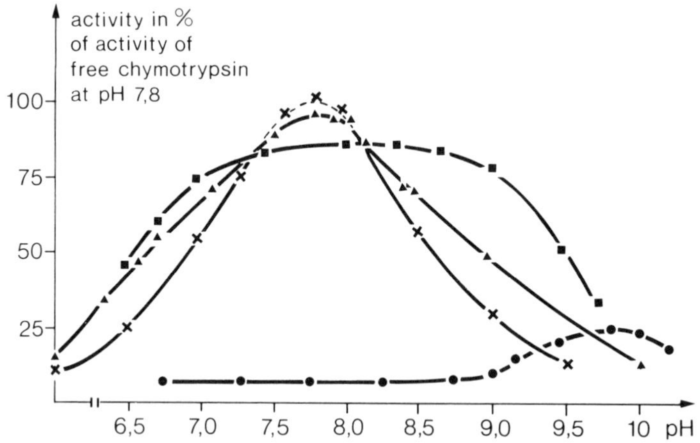

Fig. 11. Curves showing the relation of activity to pH for chymotrypsin, insoluble chymotrypsin-Sephadex and dextranase-solubilized chymotrypsin-Sephadex estimated against N-acetyl-L-tyrosine ethyl ester: (x) free chymotrypsin; (●) insoluble chymotrypsin-Sephadex synthesized from Sephadex G200 activated by cyanogen bromide at pH 10.3; (▲) chymotrypsin-Sephadex digested by dextranase for 6 min; (■) chymotrypsin-Sephadex digested by dextranase for 18 h (Axén et al., 1970)

trypsin bound to insoluble dextran as compared to the same product after solubilization with dextranase.

They may be more stable in liquid formulations, such as, e.g., detergents; they could be used in the continuous degradation of polymeric substrates in an ultrafiltration module which separates the low molecular weight products from the reaction mixture, and they might show favorable properties – e.g., enhanced stability and reduced antigenicity – in the medical field.

Along these lines, various investigations have been published during the last years. In addition, much knowledge pertinent to this field has been gathered from chemical modification of proteins.

In the following discussion, only those reports (with a few exceptions) will be mentioned which refer to the stability of chemically modified water-soluble proteins. At first, monofunctionally substituted proteins will be considered. Secondly, soluble proteins reticulated with bifunctional agents will be discussed. Finally, soluble conjugates of proteins with polysaccarides and synthetic polymers are being considered.

3.3.1 Monofunctionally Substituted Proteins

Much work has been dedicated to the chemical modification of proteins (e.g., Hirs, 1967; Cohen, 1968; Glazer, 1970, 1975). Most of these efforts have been aimed at the elucidation of amino acid sequences and at the identification of residues at the catalytic and binding sites of proteins. An outstanding example is the work on subtilisin, which has recently been reviewed (Svendsen, 1976).

However, only a few authors have included stability tests in their investigations. This brief discussion of published information will be concerned with pertinent data for

- proteins reacted with amino acid derivatives
- acylated proteins
- alkylated proteins
- proteins with miscellaneous substituents.

Enzymes Reacted with Amino Acid Derivatives

The method of preparation and results have been reviewed (Sela and Arnon, 1967).

In the classical procedure protein amino groups attack a N-carboxyamino acid anhydride to form polyamino-acylated proteins with various degrees of substitution and length of side-chain.

Solubility of the products is a function of the polarity of the amino acid substituents. Thus, polyglycyl ribonuclease and polyglycyl chymotrypsin with respectively 58 and 49 additional glycine residues substituted per mole of enzyme remained fully soluble between 0 °C and nearly 100 °C, while the same enzymes substituted with comparable amounts of valine or t-leucine formed insoluble aggregates at or slightly above room temperature and earlier than the native enzymes (Krausz and Becker, 1968). Solubility also depends on pH and ionic strength: polytyrosyl trypsin (with 28 additional tryosine residues) was only slightly soluble between pH 5 and pH 9,5 (Glazer et al., 1962), whereas poly-DL-alanyl trypsin was soluble in the range of pH 3–9 (Epstein et al., 1962);

polyornithyl and polyglutamyl chymotrypsin (up to 820 additional ornithyl and 170 glutamyl residues, respectively) (Goldstein, 1972) and polyalanyl amylase (64 additional alanine residues) also showed good solubility in this pH region (Isemura et al., 1964).

Enzymatic activity was not markedly influenced in some cases (Glazer et al., 1962; Epstein et al., 1962) but more in others (Wellner et al., 1963; Goldstein, 1972). A thorough discussion of the microenvironmental effects of chemical modification on the kinetic properties of polyamino-acylated chymotrypsin has been published (Goldstein, 1972).

In a recent patent up to 20-fold increases of the activity of neutral proteases after substitution with alanyl-, phenylalanyl-, or β-phenylpropionyl-phenylalanyl-residues have been reported (US 3886042, 1975). Amino acid polymerized enzymes were found to exhibit enhanced stability against proteolysis and other denaturants, as indicated in Table 10.

Whereas in the case of trypsin, substitution of lysine residues is a likely explanation for reduced susceptibility to autolysis, this is not the case with substituted chymotrypsin. It is likely that steric hindrance by bulky side-chains plays an important role, since it has been shown with polytyrosyl trypsin that – once denatured – activity is more slowly restored as compared to the native enzyme (Glazer et al., 1962).

It is interesting to note that although no conformational changes could be observed in polyalanylated amylase by means of optical methods, this enzyme derivative was more susceptible to urea denaturation than the native enzyme (Isemura, 1968).

Acylation

Although acylation – which is a standard procedure in the chemical modification of proteins – often leads to high yields of products and high residual activities, only few workers have concerned themselves with studies on the stability of the acylated products. Examples are presented in Table 11.

In most cases, enhanced thermal stability was observed which was dependent on the amount of substitution and the type of acylating agent used.

In a systematic study on the effects of various acylating agents on the thermal stability of α-amylase, it was suggested that stabilization might be more likely due to inductive and mesomeric effects of the substituent than due to steric effects (Hora, 1973). Thermal stability has been reported to increase in the sequence no substituent ($-NH_2$) $\leq$ dimethylcarbamoyl, ethoxyformyl $<$ formyl $\ll$ acetyl $<$ dimethylpropanoyl.

In addition, hydrophobicity of substituents might play a critical role, since it has been shown that very high chain lengths of acyl groups (palmitylation) reduce the thermal stability of acylated α-amylase as indicated in Fig. 12.

The palmitylated amylase has been noncovalently immobilized on a filter membrane and used for the continuous hydrolysis of starch (Urabe and Okada, 1972).

It proved possible to define a "compensation temperature" T_c for the thermal denaturation of acetylated α-amylase; the acetylated enzymes are more stable above while the native protein is more stable below T_c (Fig. 13) (Urabe et al., 1973).

The degree of stability, but not T_c, is influenced by the degree of acetylation. This has been interpreted to indicate that native and acetylated α-amylase both have the

Table 10. Enzymes reacted with amino acid derivatives

Enzyme	Substituent introduced	Reagent	Characterization of enzyme		Stability data	Ref.
			Degree of substitution (%)	Specific activity (compared to native enzyme)		
Trypsin (3.4.21.4)	Polytyrosyl	N-carboxy-DL-alanine anhydride	55 (lys) 20 additional tyr average chain length 2.5	Activity against BAEE and casein unchanged Slightly different activities against synthetic polypeptides	T ↓ (10 min, 100 °C + 70 min, 0 °C) D → (5.2 *M* urea, 35 °C) P ↑ (95 min, 35 °C) I → (soy bean trypsin inhibitor)	Glazer et al. (1962)
	Polyalanyl	N-carboxy-DL-alanine anhydride	75 (lys) 112–130 additional ala average chain length 9.6–11.1	Activity against BAEE and casein slightly reduced	I → (serum inhibitor) P ↑ (36 h, 24 °C) T ↑ (150 h, 38 °C)	Epstein (1962)
Chymotrypsin (3.4.21.1)	Polyalanyl	N-carboxy-DL-alanine anhydride	85 (lys)	Activity against BAEE slightly reduced	T ↑ (43 h, 38 °C)	Epstein et al. (1962)
Thermolysin (3.4.24.4) Neutral Proteases	Acetylalanyl- Acetylphenylalanyl β-Phenylpropionyl-phenylalanyl	Acetyl-aminoacid-N-oxy-succinimide, -N-imidazole		Activity against flurylacryloyl-gly-leu-NH_2 up to 20 fold enhanced	S ↑ (24 h, 4 °C)	US 3886042 (1975)
Ribonuclease	Polyalanyl	N-carboxy-DL-alanine anhydride	94–130 additional ala	Decreased activity against RNA, similar activity against low molecular weight substrates, dependent on pH	D ↓ (6 *M* urea)	Wellner et al. (1963)

Table 10 (continued)

Enzyme	Substituent introduced	Reagent	Characterization of enzyme		Stability data	Ref.
			Degree of substitution (%)	Specific activity (compared to native enzyme)		
α-Amylase (3.2.1.1)	Polyalanyl	N-carboxy-D,L-alanine anhydride	45 (-NH_2) 25–64 additional ala Average chain length 3.2–6.1	Unchanged activity against phenyl-α-maltoside Decreased activity against starch	P → (60 min, 110 °C, subtilopeptidase A) D ↓ (20 h, 2.25–7.5 *M* urea)	Isemura et al. (1964)

T = heat stability
S = storage stability
P = stability towards proteolysis
D = stability towards denaturants

I = stability towards inhibitors
↑ = improved
→ = unchanged
↓ = decreased

lys = lysyl residues
ala = alanyl residues
tyr = tyrosyl residues
-NH_2 = primary amino groups
BAEE = N-benzoyl-L-arginine ethyl ester

Table 11. Monofunctionally substituted enzymes – acylation

Enzyme	Substituent introduced	Reagent	Characterisation of enzyme activity			Stability data	Ref.
			Degree of substitution (%)	Specific activity (% of native enzyme)	Others		
α-Amylase (3.2.1.1)	Citroyl	Acid chloride				T ↑ (100 °C)	Jap. 6810986 (1968)
	Acetyl Butyryl Caprylyl Palmityl	Acid-p-nitrophenyl-ester	12–65 (lys) 12–48 8–65 8–65	80–100	Little changes in K_m V_{max} decreases with increasing substitution Alkaline shift of pH-activity profile	T ↑ (20 min, 75 °C)	Urabe and Okada (1972)

	Acetyl	*p*-Nitrophenyl-acetate	6–42 (lys)	80–100		T ↑ (20–360 min, 70–75 °C)	Urabe et al. (1973)
	Formyl Acetyl 2,2-Dimethyl-propanoyl Ethoxyformyl N′,N′-Dimethyl-carbamoyl	N-Hydroxy-succinimide	14–50 (lys)	91–100		T ↑ (30 min, 60 °C)	Hora (1973)
Trypsin (3.4.21.4)	Citraconyl	Acid anhydride	75 (lys)	High	Alkaline shift of pH-activity profile	P ↑ (24 h, 37 °C; 13 d, 24 °C)	Barker (1976)
	Acetyl Propionyl Butyryl Citraconyl Itaconyl	Acid anhydride	49–85 (lys)	High	Alkaline shift of pH-activity profile Reduced pI Activity reduced against BAEE and casein, highly reduced against hemoglobin	P ↑ (24 h, 25 °C, pH > 6) pH ↑ (pH > 6)	Terminiello et al. (1955)
	Acetyl	Acetic anhydride	85–90 (lys)	High		P ↑ (45 h, 25 °C) T ↑ (45 h, 45 °C)	SriRam et al. (1954)
	Propionyl Citraconyl	Acid anhydride				I ↑ (human serum inhibitor)	Bier et al. (1955)
Subtilo-peptidase B (3.4.21.14)	Succinyl	Acid anhydride	90 (lys)		Activity against BTEE and BLEE unchanged	P ↓ (24 h, 25 °C)	Gounaris and Ottesen (1965)
Aspara-ginase (3.5.1.1)	Succinyl	Acid anhydride	16–22	90	K_m for glutamine unchanged Unchanged pH-activi-ty-profile No oligomerisation	P ↑ (30 min, 37 °C, trypsin)	Holcenberg et al. (1975)
	Succinyl	Acid anhydride	up to 95 (lys)	100		D → (3,5 *M* GuHCl)	Shifrin et al. (1973)

Table 11 (continued)

Enzyme	Substituent introduced	Reagent	Characterisation of enzyme activity			Stability data	Ref.
			Degree of substitution (%)	Specific activity (% of native enzyme)	Other		
Luciferase (2.8.2.10)	Succinyl	Acid anhydride			For dimer with succinylated α- or α+β-monomers	pH ↓ (pH 5–8) T ↓ (20 min, 45 °C)	Meighen et al. (1971)
β-Casein	Acetyl Propionyl Butyryl Isobutyryl Valeryl Hexanoyl Succinyl	Acid anhydride	88–97 (-NH_2)		Homogeneous	S ↑ (pH 6–9, 0.025 *M* $CaCl_2$) (except valeryl)	Hoagland (1968)
Bovine serum albumin	Succinyl Acetyl Nitroguanyl Guanyl Trichloro-acetyl Amidinat					D ↓ (30 h, room temp.) D ↑ (mercaptoethanol in presence of > 3 *M* urea) I ↑ (antiserum)	Habeeb (1966)

T = heat stability
S = storage stability
pH = pH stability
P = stability towards proteolysis
D = stability towards denaturants

I = stability towards inhibitors
↑ = improved
→ = unchanged
↓ = decreased

BAEE = benzoyl-L-arginine ethyl ester
BTEE = benzoyl-L-tyrosine ethyl ester
BLEE = benzoyl- -leucine ethyl ester
lys = lysyl residues
–NH_2 = primary amino groups

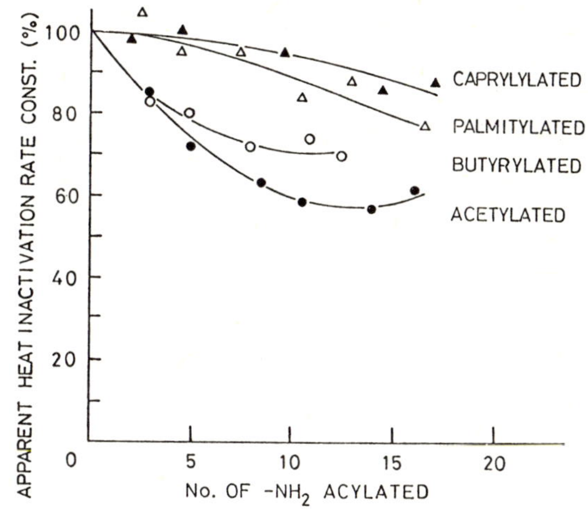

Fig. 12. Effect of acylation on the thermosensitivity of bacterial α-amylase at 75 °C. ● Acetylated α-amylase; ○ butyrylated α-amylase; ▲ caprylylated α-amylase; △ palmitylated α-amylase (Urabe and Okada, 1972)

same transition state for heat inactivation, but not the same ground state, since lower values of ΔH* and ΔS* have been observed for the acetylated enzyme.

Reduced susceptibility to proteolysis and inhibitors has been observed for some acylated enzymes (Barker et al., 1976; Bier et al., 1955; Holcenberg et al., 1975). Blocking of lysine groups by acylation is a likely mechanism for reduced proteolysis by trypsin; the fact that chymotrypsin is similarily stabilized by the same treatment points to additional – probably sterically determined – mechanisms.

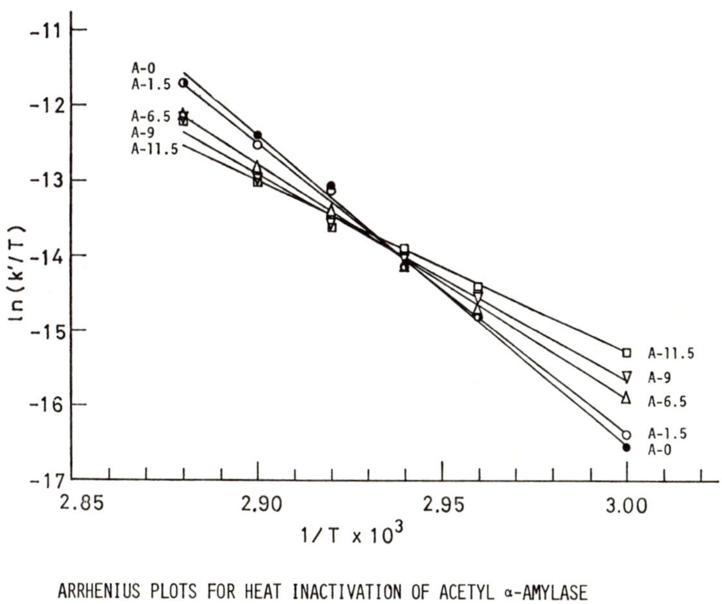

Fig. 13. Arrhenius plots for heat inactivation of native and acetylated α-amylases. ● A-0 native α-amylase; ○ A-1.5 1.5 NH_2-groups acetylated; △ A-6.5 6.5 NH_2-groups acetylated; A-9 9 NH_2-groups acetylated; □ A-11.5 11.5 NH_2-groups acetylated (After Urabe et al., 1973)

Acetylation of trypsin increases its interaction with serum trypsin inhibitor, while acylation with substituents of increasing chain length led to reduced interaction (Sri Ram et al., 1954).

In an investigation aimed at the intravenous application of asparaginase for the treatment of leukemia, the effect of acylation on the half life of this enzyme in serum has been studied by Holcenberg (Holcenberg et al., 1975); this study gives many references on the earlier work done on this subject. Stability under *in vivo* conditions seems to be largely influenced by the type of asparaginase used and by its isoelectric point after modification.

Alkylation

Reductive alkylation is a well investigated, gentle method for the modification of proteins (Means and Feeney, 1968; Feeney, 1977; Galembeck et al., 1977). Only few investigations, however, refer to the stability of the alkylated proteins thus elaborated (Table 12).

In a thorough study of the stability of alkylated glycogen phosphorylase b, it has been observed that thermal stability of the modified enzyme was largely influenced by the degree of substitution and by the type of substituent (Fig. 14).

Low substitution (10% of the total primary amino groups) and moderate hydrophobicity (valeryl residues) led to highest stability with regard to heat and cold inactivation, while hydrophilic alkylation with glyceraldehyde/$NaBH_4$ had little effect on the stability of phosphorylase b.

The high specific activities of the modified products which in the presence of AMP and Mg^{2+} formed the same crystals as the native enzyme point to only slight changes in conformation (Shatsky et al., 1973). This view is supported by the allosteric properties of the butylated enzyme derivative (Wang and Tu, 1969).

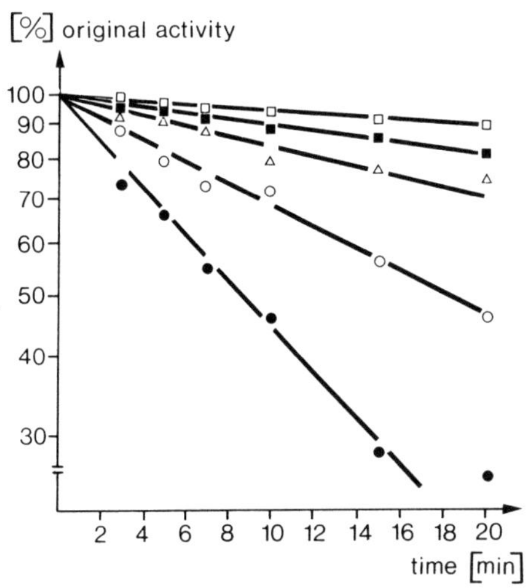

Fig. 14. Kinetics of thermal inactivation of phosphorylase b derivatives of the various aldehydes. All the derivatives were modified at an aldehyde concentration of 0.5% and had been heated at 50 °C for 3 h to remove the less stable derivatives. ● acetaldehyde-modified; ○ propionaldehyde-modified; ■ butyraldehyde-modified; □ valeraldehyde-modified; and △ heptaldehyde-modified phosphorylase b (Shalsky et al., 1973)

Table 12. Monofunctionally substituted enzymes – reductive alkylation

Enzyme	Substituent: Introduced	Substituent: Reagent	Characterization of enzyme activity: Degree of substitution (%)	Characterization of enzyme activity: Specific activity (% of native enzyme)	Characterization of enzyme activity: Other	Stability data	Ref.
Glycogen phos-phorylase (2.4.1.1)	Butyl Propyl	Aldehyde/ $NaBH_4$	10 (lys)	56– 97 62–100	Homogeneous after heat treatment Kinetics for LAMP changed	T ↑ (60 min, 51 °C)	Wang and Tu (1969)
	Ethyl Propyl Butyl Valeryl Heptyl Isobutyl 1,2-Hydroxy-propyl Glyceryl	Aldehyde/ $NaBH_4$	4–14 (lys)	High	Homogeneous after heat treatment	T ↑ (20 min, 50 °C) (10 h, 0 °C)	Shatsky et al. (1973)
Trypsin (3.4.21.4)	Polyethylen-glycolmono-methyl ether	Aldehyde/ $NaBH_4$					US-Pat. 4002531 (1976)

T = heat stability
↑ = improved
lys = lysyl residues

Improved thermal stability has been reported for trypsin alkylated with polyethylene glycolaldehyde/$NaBH_4$ (US-Patent 4002531, 1976).

Proteins with Miscellaneous Substituents

Lactate dehydrogenase (LDH) with a large amount of ϵ-amino groups of lysine modified by imidate exhibited improved thermal and pH-stability (Tuengler and Pfleiderer, 1977); a similar effect has been observed for subtilopeptidases (US-Patent 3770587, 1973) as indicated in Table 13.

The reason for these observation is not yet understood; it is interesting to note, however, that a thermostable LDH differed from thermally unstable LDH elaborated by the same organism cultivated at low temperatures only by a high exchange of lysines by arginine.

Amidination of LDH with a long-chain (hydrophobic) imidate resulted in a product with reduced solubility and thermostability (Kapmeyer and Pfleiderer, 1977).

Tryptic attack on acetamidinated LDH was inhibited, whereas the enzyme derivative was susceptible to proteolysis by chymotrypsin and proteinase K (Tuengler and Pfleiderer, 1977). Guanidinated mercuripapain is much more slowly attacked by trypsin than by the native enzyme (Shields et al., 1959). These findings are in line with the known substrate specificites of the proteases investigated.

3.3.2 Reticulation by Glutaraldehyde and Other Bifunctional Agents

Reticulation with bifunctional reagents is an established procedure in protein chemistry which has proven helpful in such different fields as structure elucidation of proteins (Fasold et al., 1971; Hirs, 1967) and membranes (Peters and Richards, 1977), or enzymology [e.g., for aggregation of multiple enzyme systems (Mattiasson et al., 1974)] and enzyme (Zaborsky, 1974a; Olson and Stanley, 1974), and protein technology (Bjorksten, 1951).

Reticulating agents have been found to be valuable for increasing enzyme stability. By the formation of cross-links they impose strong steric restrictions on attacking proteases. In addition, they may act as a clamp to keep the protein in its native conformation under denaturing conditions.

Reaction conditions must be carefully controlled in order to produce water-soluble intramolecular-linked monomers instead of less soluble oligomeric complexes or insoluble highly reticulated polymers.

Glutaraldehyde

Glutaraldehyde is a widely used reagent in biochemistry (Russell and Hopwood, 1976). While extensive literature and some reviews (Zarborsky, 1974a; Olson and Stanley, 1974; Broun, 1976) exist on the use of glutaraldehyde in enzyme technology, including data for the preparation of water-soluble products, only few of these communications refer to stability data of the reticulated proteins thus produced (Table 14).

Zarborsky's and Broun's reviews give an excellent survey of the methodology and mechanism of the reaction; therefore only a few additional remarks seem warranted.

Table 13. Monofunctionally substituted enzymes – miscellaneous substituents

Enzyme	Substituent introduced	Reagent	Characterization of enzyme activity			Stability data	Ref.
			Degree of substitution (%)	Specific activity (% of native enzyme)	Other		
Subtilopeptidase A (3.4.21.14)	Ethylamine Methyl Ethyl	Ethylenimine Methylhalogenide	90 (lys)		Increased pI	T ↑ (10 min, 50/60 °C in presence of detergent)	US-Pat. 3770587 (1973)
Subtilopeptidase B (3.4.21.14)	Guanidinium Amidino	0-Methylisourea Alkylacetamidate				S ↑ (pH 11)	
Mercuripapain	Guanidinium	0-Methylisourea	88 (lys)	High	Unchanged activity against amides	P ↓ (24 h, 40 °C, leucine aminopeptidase)	Shields et al. (1959)
Lactate dehydrogenase (1.1.1.27)	Acetamidino	Methylacetimidate	73 (lys)	80	K_m for pyruvate unchanged V_{max} for pyruvate slightly reduced	T ↑ (35 min, 55–70 °C) pH ↑ (120 min, pH 10–12) P ↑ (300 min, 25 °C, trypsin) P → (proteinase K)	Tuengler and Pfleiderer (1977)
	ϵ-(N-2,4-dinitrophenyl) aminocaproamidinate	Imidate	39 (lys)	42	Activity depends on extent of modification K_m for NADH higher, for lactate lower, for pyruvate, NAD unchanged V_{max} reduced Reduced solubility of highly substituted products	T ↑ (180 min, 55 °C)	Kapmeyer and Pfleiderer (1977)

Table 13 (continued)

Enzyme	Substituent introduced	Reagent	Characterization of enzyme activity			Stability data	Ref.
			Degree of substitution (%)	Specific activity (% of native enzyme)	Other		
Asparaginase (3.5.1.1)	Acetamidino β-dimethylamino-propionamidinate	Imidate	41–77 (lys)		K_m unchanged pI slightly changed	I ↑ (antiserum) $T_{1/2}$ ↑ (24 h, mice)	Hare and Handschumacher (1973)
Myoglobin	Carboxymethyl carboxamido-methyl	Bromoacetate Iodoacetamide	36–62 (lys)			pH ↓ (pH 5.2–12.8, 23 °C) T → (25 → 85 °C, 1 °C/min) but improved renaturation	Clark et al. (1967)

T = heat stability
S = storage stability
pH = pH stability
P = stability towards proteolysis
I = stability towards inhibitors

$T_{1/2}$ = half life time (50% residual activity)
↑ = improved
→ = unchanged
↓ = decreased
lys = lysyl residues

Table 14. Soluble enzymes reticulated with glutaraldehyde

Enzyme	Coreactant	Characterisation of enzyme product			Stability data	Ref.
		Degree of reticulation (~%)	Specific activty (% of native enzyme	Other		
Papain (3.4.22.2)	–	75 (lys) 40 (his) 10 (arg)	65	Monomer	T ↑ (40 min, 60 °C) (10 min, 80 ° C)	Royer et al. (1977)
Subtilopeptidase (3.4.21.14)	–	30 (lys)	Reduced (casein and BAEE); dependant on degree of reticulation		T ↑ (25 h, 40/50 °C)	Boudrant et al. (1976)
Glucose oxidase (1.1.3.4)	Albumin				T ↑ (30 min, 55 °C)	Atallah and Hultin (1977)
Catalase (1.11.1.6)	Albumin			Similar K_m	T → (75 min, 55 °C)	
Uricase (1.7.3.3)	Albumin		60	Oligomers	T ↑ (6 h, 38.5 °C)	Remy et al. (1976)
Chymotrypsin (3.4.21.1)	Casein-hydrolysate		12		T ↑ (30 min, 50 °C)	DAS 1944048 (1970)
Bacterial protease	Casein-hydrolysate		115		P ↑ (18 d,room temp.)	
	Glutamic acid		30		T ↑ (4 h, 60 °C)	
Glycogen-phosphorylase B (2.4.1.1)	$NaBH_4$	10 (lys)	42	V_{max} 65% Similar K_m for glycogen and glucose-1-P Different K_a for AMP Mixture of oligomers	C ↑ (4 h, 0 °C) D ↑ (30 min, 1–4 *M* urea) T ↑ (90 min, 51 °C)	Wang and Tu (1969)

T = heat stability
C = cold stability
P = stability towards proteolysis
D = stability towards denaturants
↑ = improved
→ = unchanged
BAEE = N-benzoyl-L-arginine ethyl ester
lys = lysyl residues
his = histidyl residues
arg = arginyl residues

It is well established that reactant concentration, pH, ionic strength, temperature, reaction time and lysine content of the protein contribute to the type and yield of products formed. Low concentrations of the reagents generally favor intramolecular cross-linking and formation of water-soluble products, but specific conditions have to be worked out for each enzyme.

Reduction of the primary reaction product with $NaBH_4$ (Wang and Tu, 1969) or co-reticulation with a water-soluble protein – e.g., albumin (Atallah and Hultin, 1977; Remy et al., 1976; DAS 1944048, 1970) – under appropriate conditions enhances water-solubility. This principle has been applied in the production of water-soluble immuno-reagents, molecular weight markers, or multienzyme systems (Avrameas, 1976; Payne, 1973; Atallah and Hultin, 1977).

An elegant, though elaborate method to suppress formation of oligomers has been described (Royer et al., 1977): papain and glutaraldehyde or trypsin and diimidates were reacted after binding of the enzymes to thiolated Sepharose; after completion of the reaction the enzymes were liberated by treatment with dithiothreitol.

Wang's work (Wang and Tu, 1969) on glycogen phosphorylase b demonstrates that even modest (10% of lysine) reticulation of allosteric enzymes may "freeze" a preferred conformational state: the glutaraldehyde-treated enzyme – apart from exhibiting different substrate binding kinetics – is much more resistant to heat, cold, and to high urea concentrations than the native or the butylated protein (Fig. 15).

Glutaraldehyde-reticulated trypsin has been employed for the continuous hydrolysis of casein in an ultrafiltration reactor for 11 days at 30 °C (Boudrant et al., 1976).

Other Reticulating Agents

Next to glutaraldehyde, diimidates are the reagents most often encountered in the recent literature on enzyme reticulation. Though most of this work has been directed to

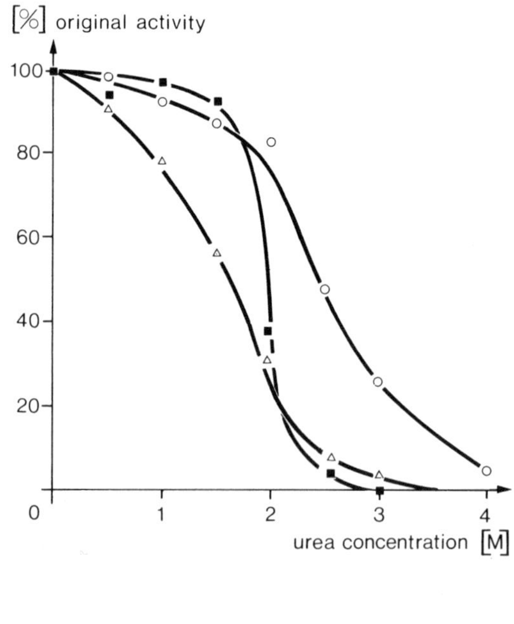

Fig. 15. Urea inactivation of phosphorylase b. Native (△), glutaraldehyde-stabilized (○), and 1% butyraldehyde-stabilized phosphorylase b (■) were incubated at 30 °C (pH 6.9) with various concentrations of urea. After 30 min enzyme activities were measured. Samples incubated in the absence of urea were taken as 100% activity (Wang and Tu, 1969)

problems such as protein or membrane structure (Peters and Richards, 1977), some reports include stability data (Table 15).

Increases in thermal stability have been observed in some cases. Reticulation of two native invertases of different thermal stabilities has been reported (Woodward and Wiseman, 1976). The more thermostable enzyme (from *Candida utilis*) after diimidate reticulation showed enhanced stability to heat treatment, while the less thermostable enzyme (from baker's yeast) showed unchanged stability in the reticulated form. The decrease of stability observed in diimidate reticulated trypsin (Royer et al., 1977) is probably due to a secondary effect: reticulation is effected with carrier-bound protein to suppress intermolecular cross-linking and involves detachment from thiolated Sepharose by means of dithiothreitol which also could attack cystin-bridges of trypsin thus partially destroying the enzyme's activity.

Besides diimidates, disulfonylchlorides (Herzig et al., 1964) and a diisocyanate (Snyder et al., 1974) have led to reticulated enzymes with increased stability under denaturing conditions.

Recently a promising new approach has been independently described by several authors: the enzyme is first modified with a monomer of second functionality which is then reticulated in a second reaction step (Fig. 16) (Reiner et al., 1975, 1977; Martinek et al., 1977a; Torchilin et al., 1978).

While thiolation of amino groups, followed by reticulation through dialkylation, may be used, an alternative route employs acroleyl residues on lysyl groups or carbodiimides on carboxyl groups the second functionality of which is reticulated. The "complementary surface" thus achieved confers very high stability to the soluble enzyme as has been shown for chymotrypsin (Torchilin et al., 1978).

3.3.3 Grafting to Polysaccharides

Many proteins contain carbohydrate residues covalently linked to "surface" amino acids (Pazur and Aronson, 1972). In addition to interesting biological properties, glycoproteins often exhibit increased stability towards proteolysis, heat, storage (Pazur and Aronson, 1972; Pazur et al., 1970; Christensen et al., 1976) or denaturants (Nakamura and Hayashi, 1974), which in many cases seems to be due to the carbohydrate part of the molecule. In addition, most glycoproteins exhibit high water solubility. Thus it seems

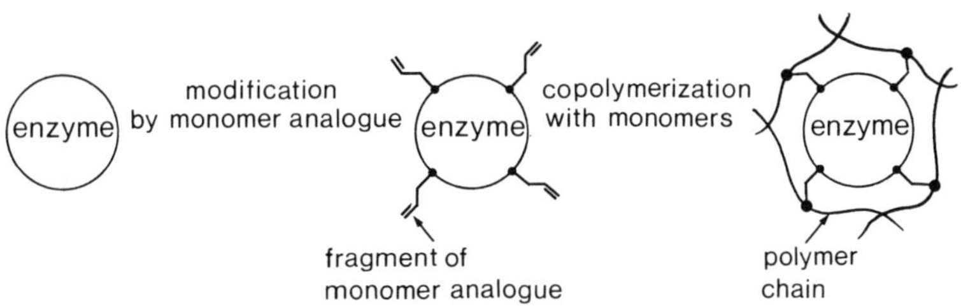

Fig. 16. Schematic representation of copolymerization of the enzyme modified by a monomer analogue with the monomer proper (Martinek et al., 1977)

Table 15. Soluble enzymes reticulated with reagents except glutaraldehyde

Enzyme	Reagent	Characterisation of enzyme product			Stability data	Ref.
		Degree of reticulation (~%)	Specific activity (% of native enzyme)	Other		
Ribonuclease A (2.7.7.16)	1) N-Acetylhomocysteinthiolactone 2) Reticulation			Mixture	T ↑ (2 h, 30–90 °C) D ↑ (7 *M* urea)	Reiner et al. (1977)
	Dimethyladipimidate			Monomer	T ↑ (55 h, 65 °C)	Zaborsky (1974b)
Chymotrypsin (3.4.21.1)	Dimethyladipimidate			Monomer	T ↑ (5 h, 45 °C)	Zaborsky (1974b)
	1) Succinylation 2) 1-Ethyl-3-(3-dimethylaminopropyl)-carbodiimide 3) Reticulation	80 (lys)			T ↑ (30 min, 50 °C)	Torchilin et al. (1978)
Trypsin (3.4.21.4)	Dimethyladipimidate	9 (lys)	69	Monomer	T ↓ (65 °C)	Royer et al. (1977)
	Dimethylsuberimidate	9 (lys)	57	Monomer	T ↓ (65 °C)	
Invertase (3.2.1.26)	Dimethylsuberimidate				T ↑ (10 min, 70 °C) C. utilis enzyme T → (10 min, 70 °C) S. cerevisiae enzyme	Woodward and Wiseman (1976)
Lysozyme (3.2.1.17)	Phenol-2,4-disulfonylchloride		High[a]		D ↑ (1 h, 40 °C, $NaBH_4$)	Herzig et al. (1964)
	α-Naphtol-2,4-disulfonylchloride		High[a]		D ↑ (thioglycollate reduction and reoxidation in presence of urea)	
α-Galactosidase (3.2.1.22)	Hexamethylendiisocyanate				T ↑ (120 min, 50 °C) P ↑ (120 min, 37 °C, trypsin)	Snyder et al. (1974)

[a] Depending on $-NH_2/-SO_2Cl$ ratio

T = heat stability
P = stability towards proteolysis
D = stability towards denaturants

↑ = improved
→ = unchanged
↓ = decreased

lys = lysyl residues

promising to obtain stabilized water-soluble enzymes through covalent attachment to carbohydrates.

As indicated in Table 16, most investigators have been using dextran activated by s-triazine, CNBr or periodate. Synthetic procedures are mild and yields – in terms of specific activity recovered – are generally very high.

Various synthetic procedures have recently been reviewed (Porath and Axén, 1976; Lilly and Dunnill, 1976); a very detailed study on the CNBr-activation method of various dextranes has been published (Marshall and Humphreys, 1977a). In a study on Sephadex G-200-immobilized chymotrypsin solubilized by dextranase (Axén et al., 1970) solubility characteristics of such products have been critically evaluated. In a few cases charged carriers as, e.g., CMC, DEAE-dextran or alginic acid, have been investigated (Table 16). In the last-mentioned case recovery and reuse of the conjugated enzyme is facilitated by pH-controlled solubility characteristics.

Dextran conjugated enzymes may be purified by affinity chromatography on concanavalin-A-Sepharose (Marshall, 1977).

As indicated in Table 16, enzymes attached to polysaccharides may exhibit changed kinetic properties and altered pH-activity-profiles due to steric constraints and microenvironmental effects different from the native protein. This is clearly demonstrated in a comparison of α-amylase bound to CMC (anionic), dextran (neutral), and DEAE-dextran (cationic) (Wykes et al., 1971). In addition, the chemical nature of the support can strongly affect enzyme properties as has been shown by comparing the different kinetic properties of subtilopeptidase bound to DEAE-dextran or DEAE-Sephadex (Svensson, 1976).

In nearly all reports published so far enzymes attached to polysaccharides exhibited improved stability. Continuous hydrolysis of casein in an ultrafiltration reactor by means of dextran-conjugated trypsin was performed for two weeks at 20 °C, thus indicating reduced autolytic degradation (O'Neill et al., 1971); 42% residual activity of polyaldehyde-dextran trypsin after 34 days at 30 °C has been observed, while native trypsin had lost all its activity after three days (Foster, 1975); 99% residual activity of a lysozyme-dextran conjugate after 30 min at 100 °C has been reported, whereas native lysozyme had lost 80% of its activity under these conditions (Christensen, 1976).

In natural glycoproteins each carbohydrate chain is attached to the polypeptide backbone through a single linkage (Pazur and Aronson, 1972). Synthetic polysaccharide-conjugated proteins probably contain additional intramolecular cross-links which convey additional rigidity to the native conformation. For dextran-bound trypsin this has been elegantly demonstrated by the increased susceptibility of this compound to urea denaturation before and after dextranase treatment (Fig. 17).

Polysaccharide-conjugated proteins are less prone to proteolytic attack. Since most glycosidation procedures lead to reduced availability of lysine residues in the conjugated products, proteolysis by proteases with lysine specificity (e.g., trypsin) is strongly reduced. In addition, strong reticulation with a carrier may mask sites important for the binding of proteolytic enzymes.

Christensen pointed to a number of technical applications for polysaccharide-conjugated proteins (Christensen, 1976). These include solubilization of proteins of low

Table 16. Soluble enzymes grafted to polysaccharides

Enzyme	Polysaccharide	Activation	% yield	Characterisation of enzyme product		Stability data	Ref.
				Specific activity (% of native enzyme)	Other		
Chymotrypsin (3.4.21.1)	CMC	Azide		High (ATEE)		T ↑	Mitz and Summaria (1961)
	Sephadex G 200 (after coupling solubilized with dextranase)	CNBr	3.5–9.0[a]	High (ATEE, casein)		pH ↑ (>pH 9)	Axén et al. (1970)
	Dextran DEAE-cellulose	2-Amino-4,6-dichloro-s-triazine	4.6–5.2[a]	High (ATEE, casein) Low (ATEE, casein)		P ↑ (75 h, 40 °C)	O'Neill et al. (1971)
	Dextran T 10		20[a]	55 (casein)		P ↑ (4 d, 37 °C)	Vegarud and Christensen (1975)
	Dextran 500	CNBr	0.5–1[a]		K_{cat} and K_m unchanged (ATEE)	T ↑ (55–60 °C)	Lasch et al. (1976)
Trypsin (3.4.21.4)	Dextran T 70	IO_4^-	3[a]	High (BAPNA) Low (casein)	Changes in pH-activity profile K_m nearly unchanged	pH ↑ (>pH 8) P ↑ (34 d, 30 °C)	Foster (1975)
	Dextran	CNBr	11[a]	7 (casein) 53 (TAME)	Changes in pH-activity profile K_m unchanged	P ↑ (120 min, 37 °C) D ↑ (120 min, 37 °C, 8 *M* urea) T ↑ (120 min, 60 °C) D ↑ (60 min, 37 °C, 1% SDS)	Marshall and Rabinowitz (1975, 1976)

						I	↑ (5 inhibitors, 7 min)	
Microbial protease	Dextran/albu-min-hydrolysate or glycine	CNBr	98–120[b]			T	↑ (20 min, 60 °C)	DOS 1944048 (1970)
α-Amylase (3.2.1.1)	Dextran-2000 CMC	2-Amino-4,6-dichlo-ro-s-triazine	61–65.5[c] 6.4–19.2[c]	60–68 10–60	Changes in pH-activity profile	T	↑ (4 h, 50/70 °C, pH 7)	Wykes et al. (1971)
						T	↓ (4 h, 50 °C, pH 9) (except CMC-derivative)	
	DEAE-dextran 2000		90[c]	33		T	↑ (70 h, 70 °C, pH 7) (CMC derivative)	
	Dextran	CNBr	50[b]			T	↑ (60 min, 65 °C)	Marshall and Rabinowitz (1975)
β-Amylase (3.2.1.2)	Dextran	CNBr	50[b]			T	↑ (60 min, 60 °C)	Marshall and Rabinowitz (1975)
Amyloglu-cosidase (3.2.1.3)	CMC in presence of caseinhydro-lysate	Azide	100[b]			T	↑ (15 min, 70 °C)	DOS 1944048 (1970)
β-Glucosidase (3.2.1.21)	Dextran T 10 FITC-dextran-3					T	↑ (40 min, 60 °C)	Vegarud and Christensen (1975) Christensen et al. (1976)
Lysozyme (3.2.1.17)	Dextran T 10		20–83[a]	16–40		T	↑ (30 min, 100 °C)	Vegarud and Christensen (1975)
						D	↑ (15 min, 100 °C, 0,01% SDS)	
						P	↑ (5 d, 37 °C, chymotrypsin)	
	Alginic acid	CNBr			Soluble at pH > 4 Insoluble at pH < 3	S	↑ (7 cycles of use)	Charles et al. (1974)

Table 16 (continued)

Enzyme	Polysaccharide	Activation	% yield	Characterisation of enzyme product		Stability data	Ref.
				Specific activity (% of native enzyme)	Other		
Ribonuclease	Dextran		2.7[a]			P ↑ (3 d, 50 °C, trypsin)	Christensen et al. (1976)
Glucose oxidase/ catalase	CMC in presence of albuminhydro-lysate	Azide	72[b]			T ↑ (10 min, 80 °C)	DOS 1944048 (1970)
Catalase (1.11.1.6)	Dextran	CNBr		100		T ↑ (60 min, 52 °C) P ↑ (90 min, 37 °C, trypsin) $T_{1/2}$ ↑ (mice)	Marshall and Rabinowitz (1976) Marshall and Humphreys (1977a)
Penicillinase (3.5.2.6)	Dextran	CNBr	96–98[b]			T ↑ (8–15 d, 45 °C)	DOS 2312824 (1976)
	Dextran	CNBr	73[b]			T ↑ (4 h, 37 °C)	US 3887432 (1975)
	Sucrose-epi-chlorohydrine-copolymer		60–86[b]			S ↑ (7 cycles of use)	
Casein Heparin	Dextrin					P ↑ (1 h, 37 °C, trypsin)	Vegarud and Christensen (1975)
Serum albumin	Dextrin					P ↑ (1 h, 50 °C, trypsin)	
Streptokinase (2.7.1.72)	Dextran 70	CNBr	90			Increased therapeutic effectiveness	DOS 2423831 (1974)

For footnotes see p. 89

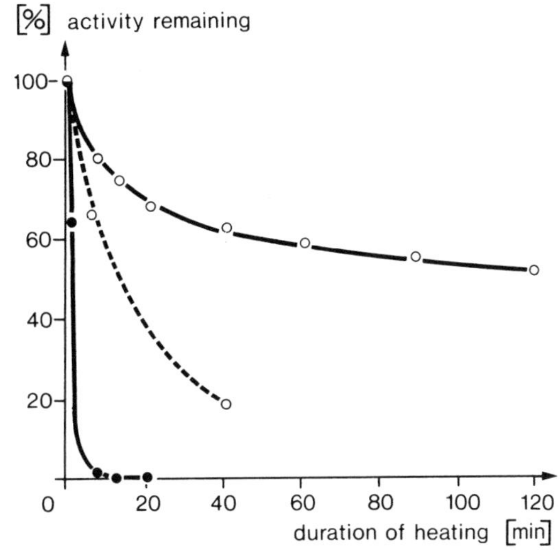

Fig. 17. Inactivation of trypsin (•) and dextran-trypsin conjugate (○) at 37 °C and pH 8.1 in the presence of 5 mM 2-mercaptoethanol. The broken line shows the rate of inactivation of dextran-trypsin conjugate after treatment with dextranase (Marshall and Rabinowitz, 1976)

water-solubility and masking of antigenic determinants. The latter has been observed with dextran-bound streptokinase after intravenous application (DOS 2423831, 1974).

Penicillinase conjugated to dextran and other carbohydrate polymers has been industrially investigated for the continuous hydrolysis of penicillins to 6-amino penicillanic acid (6-APA) in an ultrafiltration reactor (DOS 2312824, 1976; US 3887432, 1975).

3.3.4 Grafting to Synthetic Polymers

Relatively few studies have been reported in which enzymes have been grafted to water-soluble synthetic polymers. A very simple procedure involves the addition of an enzyme to polymeric acid anhydrides such as ethylene maleic acid anhydride copolymer, but other activated polymers have been also employed (Zaborsky, 1974a; Mosbach, 1976). Trypsin has been conjugated to acrolein acrylic acid copolymer. The product exhibited reversible solubility properties as a function of pH thus permitting recovery and reuse (Van Leemputten and Horisberger, 1976). In some cases polymer-bound enzymes exbibited kinetic properties different from the native enzymes which may reflect steric and diffusional restrictions as well as changes in microenvironmental charge distribu-

Footnotes to Table 16

[a] % of original enzyme (protein)
[b] % of original activity
[c] % uptake of added protein

T = heat stability
S = storage stability
pH = pH stability
P = stability towards proteolysis
D = stability towards denaturants
I = stability towards inhibitors

$T_{1/2}$ = half life time (50% residual activity)
↑ = improved
↓ = decreased
ATEE = N-acetyl-L-tyrosine ethyl ester
TAME = N-tosyl-L-arginine methyl ester
BAPNA = N-benzoyl-D,L-arginine p-nitranilide

tion. Thorough discussions of this subject have been published (Goldstein, 1973, 1976; Svensson, 1974; Ollis and Datta, 1976).

Stability aspects are given in Table 17.

In an attempt to synthesize enzyme derivatives with low toxicity and low metabolism after intravenous application, increased stability of α-D-N-acetyl-hexosaminidase to proteolytic attack and increased half life in rabbits after conjugation to activated polyvinylpyrrolidone has been observed; however, stability towards heat or acid inactivation was similar or even slightly decreased as compared to the native enzyme. Contrary to this observation, antibody conjugates of the same enzyme showed increased heat stability (Geiger et al., 1977). Antigenicity and catabolism of injected catalase was decreased after grafting to polyethylene glycol (Abuchowski et al., 1977).

Polymer-grafted enzymes have been patented for use in detergents, penicillin hydrolysis and other process.

4 Technical Applications

Since the stability of enzymes during storage and use is an important prerequisite for any technical application, great efforts have been made in this respect. This is clearly indicated in the patent literature which is summarized in this section. Reflecting the author's interests, stabilization of enzymes in detergents is documented in greatest detail. Other areas such as patent literature on medical, diagnostic or food enzymes are covered only to limited extent which, however, permits a first evaluation of the present state of the art in these fields.

In addition to patents claiming enhanced enzyme stability by the addition of compounds or by covalent modification, some recent patent literature on intrinsically stabilized enzymes is also included (Table 18).

4.1 Enzymes in Detergents

In a patent application issued in 1913, Röhm claimed the use of pancreatic enzymes in laundry detergents (DRP 283923, 1915). In spite of further incentives in this direction – e.g., the proposal to use urease in detergents in order to remove urea stains (Helv. 131571, 1929) – the high price of enzyme preparations and their low stability under washing conditions prevented the large-scale technical application of this finding – except during world war II, when interest in detergent enzymes as soap substitutes was high due to Germany's strive for economic independance.

It was not until the advent of microbial enzyme production by submerged cultivation in the late fifties that enzymes gradually became standard ingredients in detergents. A competent review on this development has been published (Langguth, 1971); some newer developments are covered in two recent publications (Berg et al., 1976; Schreiber et al., 1974).

Table 17. Soluble enzymes grafted to polymers (except carbohydrates)

Enzyme	Polymer	Activation	Yield (% protein in product)	Characterization of enzyme product: Specific activity (% of native enzyme)	Characterization of enzyme product: Other	Stability data	Ref.
Chymotrypsin (3.4.21.1)	EMA				Changes in pH-activity profile	P ↑ (7 d, room temp.)	DOS 1948177 (1970)
	EMA		54		K_m increased K_{cat} increased	pH ↑ (pH>9) T ↑ (15 min, 50–75 °C)	Goldstein (1972)
	MAA/acrylic acid copolymer			60 (casein)	K_m increased Little change in K_{cat}	T ↓ (>45 °C)	Lasch (1976)
Trypsin (3.4.21.4)	PEG				No antibody formation	$T_{1/2}$ ↑ (90 d) P ↑ I ↑ (soybean trypsin inhibitor)	F. F. Davis et al. (1977)
	EMA				Changes in pH-activity profile		DOS 1948177 (1970)
	PVP	Reaction sequence including N-hydroxy-succinimide formation	4	High (BAPA)	Little change in K_m	P ↑ (24 h, 35 °C)	von Specht et al. (1973)
	Acrolein-acrylic acid copolymer			80 (BAEE) reduced (whey)	Insoluble below pH 4	T ↑ (30 min, 65 °C)	Van Leemputten and Horisberger (1976)
Subtilopep-tidase (3.4.21.14)	EMA		30–60			pH ↑ (clupein sulfate) (10 min, 20–65 °C)	Svensson et al. (1974)

Table 17 (continued)

Enzyme	Polymer	Activation	Yield (% protein in product)	Characterization of enzyme product: Specific activity (% of native enzyme)	Characterization of enzyme product: Other	Stability data	Ref.
Microbial-protease	EMA in presence of hydrolysed albumin			98		T ↑ (60 min, 60 °C) D ↑ (15 min, 60 °C, perborate detergent)	DOS 1944048 (1970)
Glucose oxidase (1.1.3.4)	PA/PVA	Carbodiimide	29–45	45–55		T ↓ (4 h, 50 °C)	Hixson (1973)
β-D-N-Acetyl-hexosaminidase A	PVP	Reaction sequence including N-hydroxysuccinimide formation	11	High	K_m unchanged	pH ↑ (pH 2–3) T → (1 h, 0–60 °C) P ↑ (30 min, pronase, cathepsin) $T_{1/2}$ ↑ (rabbits)	Geiger et al. (1977)
Penicillinase (3.5.2.6)	MAA/VME					S ↑ (7 cycles)	US 3887432 (1975)
Asparaginase (3.5.1.1)	PEG	4,6-Dichloro-s-triazine-derivative		High		S ↑ (30 d, 37 °C) $T_{1/2}$ ↑	Vieth and Gilbert (1976)
Catalase (1.11.1.6)	PEG 1900 PEG 5000	4,6-Dichloro-s-triazine-derivative		93–95	No antibody binding	T → (5 min, 0–65 °C) P ↑ (100–150 min, room temp.) $T_{1/2}$ ↑ (50–72 h, mice)	Abuchowski et al. (1977)
Carbonic anhydrase (4.2.1.1)	N-(sym-Trinitroaryl) polyacrylamide-acrylhydryzide-copolymers			>100	K_m decreased	T ↑ (6 h, 60 °C) S ↑ (6 w, 0–2 °C)	Epton et al. (1977)

Bovine serum Albumin	PEG 1900 PEG 5000	4,6-Dichloro-s-triazine-derivative	$T_{1/2}$ ↑ (16–20 d, rabbits)	Abuchowski et al. (1977)

MAA = maleic acid anhydride
EMA = ethylene-maleic acid anhydride-copolymer
PEG = methoxypolyethyleneglycol
PVP = polyvinylpyrrolidone
PA = polyethylene
PVA = polyvinylalcohol
VME = vinyl methyl ether

T = heat stability
S = storage stability
pH = pH stability
P = stability towards proteolysis
D = stability towards denaturants
I = stability towards inhibitors
$T_{1/2}$ = half life time (50% residual activity)
↑ = improved
→ = unchanged
↓ = decreased

Table 18. Enzymes selected for increased intrinsic stability

Company document year	Enzyme	Organism	Remarks
Purdue Research Foundation US 2695863 (1954)	α-Amylase	*Bacillus stearothermophilus*	Thermostable enzyme in food
CPC International Inc., US 3826714 (1974)	Glucoseisomerase	Thermophilic Bacilli	Thermostable enzyme in food
Societa Italiana Resine SIR SpA, US 3813319 (1974)	Protease	Thermophilic Eumycetes group	Thermostable enzyme for detergents
Novo Terapeutisk Lab. A/S, US 3827938 (1974)	Protease	*Bacillus firmus* strain NRS 783	Alkalistable enzyme for detergents
Kyowa Hakko Kogyo Co., Ltd., US 3838009 (1974)	Protease	*Bacillus subtilis* strains	Surfactant – resistant alkaline enzyme for detergents

Table 18 (continued)

Company document year	Enzyme	Organism	Remarks
Hayashibara Comp., US 3827940 (1974)	α-1,6-Glucosidases	Strains of bacteria belonging to the genus Agrobacterium, Azotobacter, Bacillus, Erwinia, Lactobacillus, Peciococcus, Leuconostoc, Micrococcus, Mycobacterium, Nocardia, Sarcina, Serratia, Staphylococcus and Streptococcus	Heat-resistant enzymes for conversion of starch
Hayashibara Comp., US 3804715 (1974)	α-1,6-Glucosidases	Strains produced from *Pseudomonas amyloderamosa* ATCC 21262, *Escherichia intermedia* ATCC 21073, *Agrobacterium tumefaciens* IFO 3058, *Azotobacter indicus* IFO 3744, *Bacilius cereus,* IFO 3001, *Erwinia aroideae* IFO 3057, *Leuconostoc mesenteroides* IFO 3426, *Mycobacterium phlei* IFO 3158, *Micrococcus lysodeikticus* IFO 3333, *Pedicoccus acidilactici* IFO 3884, *Sarcina lutea* IFO 3232, *Serratia indica* IFO 3759, *Staphylococcus aureus* IFO 3061 or *Streptococcus faecalis* IFO 3128	Heat-resistant enzymes for maltose formation
Monsanto Co., US 3846239 (1974)	α-Galactosidase	*Bacillus stearothermophilus*	Thermostable enzyme as feed supplement
Glaxo Laboratories Ltd., US 3880742 (1975)	β-Glucanase	*Penicillium emersonii*	Thermostable enzyme for poultry feeding and brewing
Ajinomoto Co., Inc., US 3871963 (1975)	Protease	*Bacillus sp* (FERM P-1396)	Thermostable enzyme for detergents
K. J. S. Villadsen and K. P. Vestberg US 3960665 (1976)	Protease	*Bacillus firmus var. Arosia*	Alkalistable enzymes for detergents with enhanced stability towards perborate and polyphosphate

Various groups of enzymes have been proposed for the use in laundry detergents; they range from oxidoreductases [tyrosinase, peroxidase, and other oxygenases (Helv. 127227, 1928), glucose oxidase (DOS 1918729, 1969), lipoxygenase (DOS 1944904, 1971), transferases, isomerases, and lyases (Brit. 1151748, 1969; Belg. 697480, 1967), to hydrolases, urease (Helv. 129861, 1929), diastase (US 1882279, 1932), proteases, esterases, carbohydrases, nucleases (Belg. 697480, 1967)] and a large number of proteases and amylases. Only the latter are presently incorporated in laundry (proteases) and dish-washing (amylases) detergents. In addition, during the last few years the use of lipases for special detergents (e.g., presoaks) has been proposed in the patent literature.

By acting on substrates such as blood smears, egg yolk, cocoa, grass spots, and many other proteinaceous stains, proteases significantly improve the washing performance of laundry detergents (Duesing, 1967; Liss and Langguth, 1969; Berg et al., 1976) as has been convincingly demonstrated by comparing the washing performance of standard detergent solution including protease before and after addition of a protease inhibitor (Schreiber, 1972).

As schematically indicated in Table 19, wash liquors provide a very hostile environment for proteins; many patents and other literature on the stabilization of enzymes in detergents thus exist.

Table 19

Conditions for washing process	"Stability" (intrinsic properties) of ... type proteases			
	Serine	Metallo	Thiol	Carboxyl
High temperature (60 °C)	+	++	0	+
Alkaline pH (9–11)	+	–	0	–
Complexing agents present (e.g., pentasodium triphosphate)	+	–	+	+
Oxidants present (e.g., H_2O_2, HOCl)	+	+	–	+
Surfactants present (anionic, cationic, nonionic)	0	0	0	0

++ Some enzymes exhibit high stability
+ Some enzymes are relatively stable
0 Enzymes are generally not stable
– Enzymes are unstable

The deleterious effect of wash liquor ingredients such as H_2O_2, complexing agents (builders) and surfactants has been thoroughly investigated (Enkelund, 1973; Stauffer and Treptow, 1973; Stauffer and Etson, 1969; Jaag, 1969; Wieg, 1969; Langguth and Mecey, 1969; Horikoshi, 1971; Delecourt, 1965; Duesing, 1967; Schreiber, 1972) and recently been reviewed (Berg et al., 1976). Today, the proteolytic enzymes used in detergent powders exhibit excellent stability during storage and use (Schreiber et al.,

1974; Berg et al., 1976). However, storage stability of enzymes in liquid detergents is a considerable problem. For convenience, these topics will be considered separately.

Enzyme Stability During the Washing Process

As indicated in Table 19, enzymes must fulfill stringent requirements in order to exhibit activity during the washing process. Among the types of proteolytic enzymes so far identified, serine proteases are best adapted to these harsh conditions. A variety of microbial serine proteases have been proposed (which often differ biochemically and immunologically); technical grade subtilopeptidases (Ottesen et al., 1970; Ottesen and Svendsen, 1970) play a most important role due to their high intrinsic stability.

Apart from attempts to improve the intrinsic enzyme stability by microbial selection (for examples see Table 18), efforts have been made to stabilize detergent proteases by additives or chemical modification (Table 20); since detergent enzymes are low-priced products, the latter approach is of little economic relevance.

Table 20. Stabilization of enzymes during washing process

Company Document Year	Enzyme(s)	Stabilizer
Monsanto Co., DOS 1948177 (1970)	Protease Lipase Amylase	EMA-copolymers (covalent attachment)
Colgate-Palmolive, DOS 2046351 (1971)	Protease	Lactose
Koninklijke Nederlandsche Gist-En Spiritusfabriek N.V., Belg. 763681 (1971)	Protease	Polymers soluble under alkaline conditions
Hitachi, Ltd., Jap. 7220235 (1972)	Amylase	Hydroxylammonium chloride Thiourea Thioacetamide (chlorine scavenger)
Pfizer Inc., US 3770587 (1973)	Protease	Substituted amino groups
Procter & Gamble, US 3893954 (1975)	Protease	Hydroxylamine (chlorine scavenger)

Calcium ions act as powerful stabilizers (Vaeck et al., 1971; Schreiber, 1972; Stauffer and Treptow, 1973) even in the presence of chelating agents such as pentasodium triphosphate.

Enzyme Stability During Storage of Detergent Powders

The early use of enzyme powders in detergents led to dust problems in the production process. Also reduced stability due to autolysis and detrimental effects to other detergent ingredients in the presence of moisture were encountered. These problems have been fully overcome by granulation techniques and enzyme "prilling" which consists of spray-drying of a molten mixture of enzymes in appropriate fatty alcohol ethoxylates and other ingredients. The techniques are described in the literature (Koch, 1969, 1973) and have recently been briefly reviewed (den Ouden, 1977; Berg, 1976).

In a few cases additives have been described to improve storage stability in detergent powders (Table 21); tests which included direct measurements of residual enzyme activity are covered separately (Table 22).

Enzyme Stability During Storage of Liquid Detergents

A recent development in the field of laundry detergents is the use of concentrated liquid detergent formulations allowing automatic dosage from a storage tank coupled to the washing machine. Liquid detergent formulations presently employed contain high concentrations of surfactants which – from the standpoint of enzymology – pose a very difficult problem for stabilization techniques.

Table 21. Stabilization of enzyme detergent powder during storage/wash tests

Company Document Year	Enzyme(s)	Stabilizer
F. Leidholdt, DDR 14296 (1959)	Protease Lipase Urease	Proteins Na-Sulfate Na-Pyrophosphate
Colgate-Palmolive, DOS 1814532 (1969)	Protease	Chlorophenol *p*-Chlorometaxylenol Chloromethoxypropylmercuric Acetate (microbicides)
Colgate-Palmolive, DOS 1914755 (1969)	Protease	Hydroxylammonium salts Hydrazonium salts
Procter & Gamble, DOS 1913869 (1970)	Protease Urease	Lactose
Salkin, Fran. 2029842 (1970)	Protease	PVP/CMC Coating with waxy materials
Procter & Gamble, US 3600318 (1971)	Protease	Na-Chloride
Fuji Photo Film Co., DOS 2042650 (1971)	Protease Lipase	Ca-chloride Microencapsulation
Dr. U. Wolf, DOS 1944904 (1971)	Amylase Tannase Protease	Alkali metal salts Alkaline earth metal Salts Heavy metal salts in trace amounts
Knappsack AG, DOS 1958105 (1971)	Protease	Methylcellulose CMC

Table 22. Stabilization of detergent powder during storage/enzyme tests

Company Document Year	Enzyme(s)	Stabilizer	Most critical test conditions	% residual activity of best example
Kyowa Fermentation Ind. Co., Ltd., Jap. 6908064 (1969)	Protease	Urea	50 d, 37 °C	98
Procter & Gamble, DOS 1811000 (1969)	Protease	Protein	123 d, 32,2 °C, 80% rel. humidity	95
Procter & Gamble, DOS 1810996 (1969)	Protease	Protein	90 d, 43 °C	79
Farbwerke Hoechst AG, DOS 1802465 (1970)	Protease	Polyglycol ethers	4 w, 20 °C 76% rel. humidity	96
Ajinomoto Co., Inc., DOS 2064480 (1971)	Protease Amylase Lipase	N-Acylamino Acids	14 d, 37 °C	95
Novo Terapeutisk Lab. A/S, DOS 2137042 (1972)	Protease α-Amylase Hemicellulase	Proteins Poly(vinyl-pyrrolidone)	8 w, 30 °C	92
Colgate-Palmolive, DOS 2212141 (1972)	Protease	Nonionic surfactants Na-Tripoly-phosphate	4 w, 43 °C	83
Henkel & Cie GmbH, DOS 2032768 (1972)	Protease	Polyglycols Polyglycol ethers	8 w, 30 °C	67
Procter & Gamble, US 4011169 (1977)	Peroxidase Protease Lipase α-Amylase	Aminated celluloses	6 w, 38 °C	98

Extensive work has been dedicated to this goal including enzyme tests (Table 23) and others (Table 24). High residual activities of proteolytic enzymes after several months of storage in liquid formulations have been mentioned (Christensen et al., 1978; Jensen, 1977). Up to now, however, no enzyme-containing liquid detergents have appeared on the market.

Food Industry

In the food industry, a large variety of hydrolytic enzymes and some oxidoreductases are presently being used. The monograph by Reed on this subject is highly recommended (Reed, 1976).

Food processing involves routine steps deleterious to enzyme activity such as, e.g., heating and freezing. In addition, enzyme inactivation by inhibitors in food (e.g., soy

Table 23. Stabilization of enzymes in liquid detergents/enzyme tests

Company Document Year	Enzyme(s)	Stabilizer	Most critical test conditions	% residual activity of best example
Procter & Gamble, DOS 1808834 (1969)	Proteases	Monoalcohols Diethyleneglycolmonobutyl ether	22 w, 38 °C	100
Koninklijke Nederlandsche Gist-En Spiritusfabriek N.V., DOS 2015504 (1970)	Protease	Saccharose Glucose Lactose Ca-Salt	30 d, 37 °C	97
Procter & Gamble, DOS 1964088 (1970)	α-Amylase	Ca-Salt Glycols	8 w, 38 °C	100
Procter & Gamble, DOS 2036340 (1971)	Amylase	Ca-Salts Mg-Salts K-thiosulfate	10 w, 38 °C	100
Citrique Belge N.V., Belg. 758105 (1971)	Protease	Salts of esters of Sulfonated Polycarboxylic acids	63 d, 20 °C	91
Procter & Gamble, DOS 2057754 (1971)	Protease	Ca-Salt Mg-Salt	4 w, 35 °C	95
Procter & Gamble, US 3557002 (1971)	Amylase Protease	Monoalcohols	22 w, 26 °C	97
Henkel & Cie GmbH, DOS 2058826 (1972)	Protease	Di/polyamines Glycerol-α-monomethyl ether	12 w, 37 °C	98
Henkel & Cie GmbH, DOS 2038107 (1972)	Protease	Glycerol-α-monomethyl ether Glycols Alkylolamines	24 w, 37 °C	76
Chemische Werke Hüls AG, DOS 2044778 (1972)	Protease	Polydiols	28 w, 20 °C	98
Henkel & Cie GmbH, DOS 2038103 (1972)	α-Amylase	Lactose Mannitol Sorbitol Dulcite Arabinose Maltose	2 m, 20 °C	85
Albright & Wilson Ltd., Fran. 2169752 (1973)	Protease	Xylenesulfonate (hydrotrop)	180 d, 20 °C	100
Pabst Brewing Comp., US 3717550 (1973)	Protease Amylase	Polyhydric alcohols	36 d, 20 °C	85
Procter & Gamble, DOS 2709476 (1977)	Protease	Polyacids Ca-Salts	2 w, 35 °C	66

Table 24. Stabilization of enzymes in liquid detergents/wash tests

Company Document Year	Enzyme(s)	Stabilizer
Smillie, US 3472783 (1969)	Protease Amylase	Glycerol Nonionic surfactants
Albright & Wilson, DOS 1930636 (1970)	Protease Lipase	Propylene glycol
Kortman & Schulte N.V., DOS 1942849 (1970)	Protease	Na-Phosphates
Röhm GmbH, DOS 1957010 (1971)	Protease	Polyglycols
Kronwitter & Co., DOS 2060485 (1972)	α-Amylase Protease	Glycocoll
Akzo N.V., DOS 2150142 (1972)	Protease Amylase Lipase	Dimethyl ether Methylvinyl ether
Witco Chemical Corp., US 3697451 (1972)	Protease Amylase Lipase	Diethanolamine salts Triethanolamine salts
Barrett, US 3746649 (1973)	Amylase Protease	Polyglycols
A. E. Staley, US 3761420 (1973)	Protease	Glycerol
CPC International Inc., US 3860536 (1975)	Protease Amylase	Propylene glycol
Miles Lab., Inc., US 4021377 (1977)	Protease Amylase	Diglycols Alkylolamines Amines Citric acid

trypsin inhibitor) must be taken into account. Thus enzyme stabilization during storage and use is an important goal for workers in this field. A small number of recent patent applications on this subject are included in Table 25.

Textile Industry

Starch still plays an important role in the sizing of textiles in order to obtain a smooth, closed yarn surface (Oettel, 1966). For removal of starch size – an important prerequisite for further textile treatment such as dyeing – amylases are the preferred agents. A few patents relating to the stabilization of amylase desizing baths are indicated in Table 26.

Table 25. Food/feed

Company Document Year	Enzyme(s)	Stabilizer	Remarks
Fermco Laboratories, Inc., US 3006815 (1956)	Glucose oxidase catalase	Coating with a thin layer of mineral oil and an insulating layer of water-soluble methylcellulose	For food industry
Rohm and Haas Comp., US 2979440 (1961)	Diastase	Glycerol	For food industry; ~84% residual activity (20,000 p.s.i.)
US Secretary of Agriculture, US 2982697 (1961)	Tomatopectase	Na-Sorbate K-Sorbate Sorbic acid	For food industry
P. P. Muset, US 3133001 (1964)	Hepatocatalase	Coating with sugar	For food industry
J. Fukumoto and H. Negoro, US 3272717 (1966)	Amylase	Cationic surfactants	To improve the thermal stability of amylase for starch-liquefying; 94% residual activity (120 m, 70 °C)
Baxter Laboratories, Inc., US 3296094 (1967)	Papain	Glycerol Protein	For food industry
D. Morita, Jap. 7121785 (1971)	Protease Amylase Lipase Cellulase	$CaCO_3$ $MgCO_3$	As feed additive
Procter & Gamble, US 3819528 (1974)	α-Amylase	Glycols	For food industry; 37% residual activity (8 w, 37,8 °C)
Hayashibara Comp., Brit. 1346221 (1974)	α-1,6-Glucosidase	Ca-Salts	For degradation of starch; 97% residual activity (1 h, 50 °C)
Dainippon Pharmaceutical Co., Ltd., US 3927200 (1975)	Lyase	CMC Glycerol Monoethanolamine	For food industry; 66% residual activity (4 w, 50 °C)

Table 26. Textile

Company Document Year	Enzyme(s)	Stabilizer	Remarks
Wallerstein Co., US 2567747 (1951)	Amylase	Dithiobiuret	For desizing textiles
Schweizerische Ferment AG, Helv. 289582 (1953)	Amylase	Ca-Salt	For desizing textiles
Diamalt AG, DP 941190 (1956)	Amylase	Ca-Salt	For desizing textiles
Standard Brands Inc., US 3524798 (1970)	α-Amylase	Ca-Salt	For desizing textiles; 93.6% residual activity (28 d, 37.8%)

Enzymes in Cosmetic Applications

Many potential applications of enzymes exist in cosmetics and have been indicated in a recent short review (Kermici, 1977). Another field of application is dental hygiene, where numerous proposals for the use of dextranase and other enzymes have been made.

Surfactants are incorporated in many cosmetic formulations. Thus stabilization of enzymes is important. Some patent applications relating to this field are given in Table 27.

Enzymes in Analytical and Diagnostic Applications

Several handbooks on this subject have been published (e.g., Bergmeyer, 1970). Of the numerous examples for stabilization of enzymes during storage, a few are listed in Table 28.

Enzymes in the Medical Field

Many medical applications of enzymes ranging from extra- or intracorporal supplementation of damaged metabolism (Gregoriades, 1976; Chang, 1971) to the production of pharmaceutically active compounds, have been described in the literature.

Some pertinent examples of stabilization procedures are given in Table 29.

4.2 Miscellaneous

In Table 30 a number of patent applications are listed which claim the stabilization of a variety of enzymes; very often, no specific field of application is mentioned.

Table 27. Cosmetics

Company Document Year	Enzyme(s)	Stabilizer	Remarks
Procter & Gamble, DOS 1808834 (1969)	Protease α-Amylase	Cyclic ethers	
Dainippon Pharmaceutical Co., Ltd., DAS 2040440 (1971)	α-Amylase	Glycerol Sorbitol	For dental hygiene
Pacific Biochemical, Inc., Brit. 1232627 (1971)	Protease Amylase	Ca-Salt Polyhydric alcohols Sugars	For dental hygiene
P. August DOS 2064940 (1972)	Lipase	Polyvinylpyrrolidone Leucylglycylglycine	
Sunstar Dentrifice Co., Ltd., Jap. 7349984 (1973)	Dextranase	Amino acids Polyhydric alcohols	For dental hygiene
Dainippon Pharmaceutical Co., Ltd., DOS 2319989 (1973)	Anticariogenic enzymes	Na-Carboxymethyl-cellulose Glycerol Ethanolamine	For dental hygiene; full residual activity (4 d, 80 °C)
Henkel & Cie GmbH DOS 2141764 (1973)	Proteindisulfide reductase Lactate dehydrogenase Glucose-6-phosphate dehydrogenase	Mannitol Saccharose	For hair dressing
Colgate-Palmolive US 3991177 (1976)	Dextranase	Mn-Sulfate Mn-Sulfate	For dental hygiene
Laboratoires Manceau S.A. DOS 2714718 (1977)	Papain	EDTA	For dental hygiene

5 Conclusions

Enzymes are promising catalysts for a variety of technical, analytical, and medical applications, due to their selectivity and their efficiency under mild reaction conditions. In addition, the large number of specifically effective enzymes to choose from as well as the progress made in their technical production and purification permit a bright outlook for enzyme technology; the 1972 enzyme nomenclature (IUPAC/IUB, 1973, 1976) included about 2000 entries, and the actual number is much higher, according to different substrate specificities of enzymes belonging to the same entry; hundreds of enzymes are already commercially available.

Table 28. Analytical

Company Document Year	Enzyme(s)	Stabilizer	Remarks
Gray Industries Inc., DOS 2017571 (1970)	Lactic acid dehydrogenase Desoxyribonuclease Amylase Glutamic acid-oxal = acetic acid transaminase	Irradiation	For clinical diagnosis
Calbiochem, US 3539450 (1970)	Lactate dehydrogenase Malate dehydrogenase Triosephosphate isomerase Glyceraldehydephosphate dehydrogenase Hexokinase Glucose-6-phosphate dehydrogenase	Mannitol Tris(hydroxy-methyl)-amino-methane EDTA Protein Gum arabic	For clinical diagnosis
Boehringer Mannheim, DOS 1930059 (1970)	Glutamate-pyruvate transaminase Lactate dehydrogenase	Poly(vinylpyr-rolidone)	For test reagents
Boehringer Mannheim, DAS 1642582 (1971)	Citrate lyase	Protein Ca-Salt Saccharose	For enzymatic analysis; 76% residual activity (12 w, 4 °C)
Xerox Corp., US 3860484 (1975)	Glucose oxidase	Polyethylene oxide Poly(vinylpyr-rolidone)	For clinical diagnosis
Boehringer Mannheim, DOS 2433454 (1976)	Glycerol kinase	Protein	For enzymatic analysis; 65% resicual activity (18 h, 30 °C)
Miles Laboratories Inc., US 3962037 (1976)	Glycerol kinase Pyruvate kinase	Dextran Dithiothreitol Dithioerythritol	For analytical purposes
J. E. Modrovich, DOS 2711754 (1977)	Lactate dehydrogenase	Dextran Poly(vinylpyr-rolidone) Protein	For clinical diagnostics

Unfortunately most enzymes are rather unstable molecules which – as discussed in Sect. 2 – readily loose their activity during even subtle changes of environmental conditions. Thus, the technical application of enzymes is only feasible if adequate measures of stabilization are developed. Today, extensive information is available on enzyme stabilization. The most prominent method among the recent achievements in this field is the development of immobilization techniques which are predominantly aimed at enzyme insolubilization, thus permitting their use in enzyme reactors.

Table 29. Medical

Company Document Year	Enzyme(s)	Stabilizer	Remarks
Behringwerke AG, Belg. 641705 (1964)	Streptokinase	Proteins	
Baxter Laboratories, Inc., DOS 2059165 (1971)	Streptokinase	Attachment to car- boxymethyldextran Carboxyethyldextran CMC	
Teikoku Hormone Manufacturing Co., Ltd., Jap. 75121485 (1975)	Protease	Polyhydric alcohols Fatty acids Fatty acid esters Hydrocarbons	Protection of the enzyme from inac- tivation by pressure; full residual activity (1515 bar)
Baxter Laboratories, Inc., DOS 2423831 (1974)	Streptokinase L-Asparaginase	Attachment to dextran	Water-soluble carbo- hydrate-enzyme- complexes
Bayer AG, DOS 2312824 (1974)	Penicillin acylase	Covalent attachment of the enzyme to water-soluble dextran	For production of 6-aminopenicillanic acid from penicil- lins
Takeda Chemical Industries, Ltd., Fran. 7439490 (1974)	Protease	Polyglycol	For manufacturing tablets; 97.8% re- sidual activity (2525 bar)
Beecham Group, Ltd., US 3887432 (1975)	Penicillin acylase	Attachment of the enzyme to water- soluble copolymeres of maleic anhydride with vinyl methyl ether, ethylene, styrene or vinyl acetate	For production of 6-aminopenicil- lanic acid from penicillins
Pierce Chemical Comp., US 4002531 (1977)	Asparaginase Phenylalanine-NH_3 Lyase Uricase	Chemical modifica- tion with the aldehy- de derivative of a monoalkylpolyethy- lenglycol	

Table 30. Miscellaneous

Company Document Year	Enzyme(s)	Stabilizer	Remarks
Takeda Chemical Ind., Fran. 1503267 (1967)	Amylase Protease	Sorbitol	
Momotani Juntenkan Co., Ltd., Jap. 7030192 (1970)	Proteases	Glycerol Glycol Sorbitol Ca-Salts	
Protein Biosynthesis Research Institute, SU 340693 (1972)	Enzymes of Bacillus me- sentericus	Ca-Chloride	For production of enzymes
Nagase and Co. Ltd., Jap. 4735192 (1972)	Protease	Sorbitol Na-Borate	80% residual activity (90 d, 37 °C)
Kyowa Fermentation Ind., Jap. 8026976 (1973)	L-Arginase	Amino acids Metal salts	
Toyo Spinning Co., Ltd., Jap. 7306556 (1973)	Uricase	Triton X-100 Tween 80	
Leningrad Chemical- Pharmaceutical Inst. SU 369137 (1973)	Butyrylcho- line esterase	Sugar	
Toyobo Co., Ltd., Jap. 4848681 (1973)	Uricase Catalase	Non-ionic surfactants	
Chugai Pharmaceutical Co., Ltd. Jap. 4843019 (1973)	Proteins	Polyglycols	
J. A. Bjorkesten, Jap. 9030585 (1974)	Not specified	Ca-Salts	
Henkel & Cie GmbH, DOS 2234412 (1974)	Protease	Protein	Process for purifica- tion of protease; full residual activity (20 h, 37 °C)
Tanabe Pharmaceutical KK, Jap. 50036685 (1975)	Catalase	Na-Chloride	
S. Blumberg, B. Holmquist and B. L. Vallee, US 3886042 (1975)	Protease	Acylation of protease with acetyl-L-phenyl- alanyl-N-oxysuccinimi- de, acetyl-(0-benzyl)ty- rosyl-N-oxysuccinimide, acetyl-L-tryptophanyl- N-oxysuccinimide, β- phenylpropionyl-ζ- phenylalanyl-N-imida- zole and other acylat- ing agents	Useful as additives to laundry detergent compositions; as ca- talysts in the chemi- cal modification of steroids; in convert- ing carbon dioxide, nitrogen, and hydro- carbons into amino acids and/or pro- teins; in making cheese; and in de- grading waste ma- terials

Table 30 (continued)

Company Document Year	Enzyme(s)	Stabilizer	Remarks
Du Pont, US 3963578 (1976)	Phosphoenol pyruvate Carboxylase	EDTA Glycerol	
Amano Pharmaceutical Co., Ltd., Japan. Kokai 7626284 (1976)	Cellulase Amylase Protease Lipase	Amino acids Peptides	

The present review is focused on the stabilization of water-soluble enzymes.

In this area, too, much valuable information has been obtained during the last decades in such diverse fields as

- production of intrinsically stabilized enzymes by selection techniques,
- stabilization of enzyme solutions by additives, and
- preparation of chemically modified soluble enzymes.

Improved intrinsic stability of an enzyme, as discussed in Sect. 3.1, is a feature which may hardly be achieved on alternative routes; for example, the suitability of subtilopeptidases for the specific requirements of laundry detergents has been discussed (Sect. 4). The relative ease with which microorganisms may be "programmed" for the formation of enzymes with desired properties, offers a promising perspective for this line of research.

The second - and oldest - technique for the stabilization of enzyme solutions is the addition of selected compounds such as polyhydric alcohols, low or high concentrations of some salt or the incorporation of substrates as discussed in Sect. 3.2. While some agents such as glycerol or ammonium sulfate seem to exert beneficial influences on the stability of most enzymes, other procedures such as the addition of salts, polymers or substrates may lead to enhanced or reduced stability and have to be carefully examined in each specific case. In addition, the presence of stabilizing compounds may lead to important changes in the kinetic properties of the enzyme under study.

The third technique - synthesis of chemically modified water-soluble proteins - has only been recently introduced for the purpose of enzyme stabilization. Many procedures such as acylation, reductive alkylation and conjugation with activated polysaccharides or synthetic polymers have led in high yields to enzymes with high specific activities. However, due to differences in topology, intrinsic stability, and functional group distribution, specific conditions must be worked out for each separate protein. Reaction conditions must be controlled carefully, in most cases in order to prevent side reactions leading to insoluble polymers. This is especially critical in the case of bifunctional agents and activated polymers. Most modification procedures lead to a mixture of various enzyme derivatives, but methods exist for the isolation of homogeneous species (e.g., Wang and Tu, 1969).

Specific activities are usually high for proteins carrying small substituents. An increased degree of reticulation often leads to decreased catalytic efficiency; this is especially valid for hydrolytic enzymes with large substrate binding domains (Svendsen, 1976). Reticulation in the presence of substrate could be an attractive proposition for such enzymes (e.g., Brown and Racois, 1976; Royer and Uy, 1973).

Stabilization of proteins by chemical modification may be due to variations in charge distribution, in the masking of groups susceptible to denaturants, to conveying rigidity to native ternary and quaternary structures or to a combination of these and other factors.

Monofunctional substituents of various chain lengths have been observed to impart different degrees of stabilisation; "excess" hydrophobicity invariably led to decreased stability (Urabe and Okada, 1972; Sri Ram et al., 1954; Glazer et al., 1962; Shatsky et al., 1973; Kapmeyer and Pfleiderer, 1977; Terminiello et al., 1955).

Reticulating agents forming intramolecular linkages imparted high rigidity to ternary and quaternary structures increasing their stability to denaturants.

Covalently linked polymers, especially polysaccharides, were reported to significantly enhance stability to various denaturing agents, probably imparting conformational rigidity.

Various applications of enzymes stabilized by chemical modification have been proposed. Enzyme reactors in the form of ultrafiltration modules have been described for a variety of hydrolytic processes (e.g., Boudrant et al., 1974; Wykes et al., 1971; O'Neill et al., 1971; DOS 2312824, 1975; DOS 1948273, 1970). Increased stability has been reported for polymer-grafted proteases in detergent solutions (DOS 1948177, 1970). Increased half life (Holcenberg et al., 1975) and reduced antigenicity (e.g., Abuchowski et al., 1977) of chemically modified soluble enzymes such as asparaginase, catalase, streptokinase, trypsin, and others have been found. Thus a large number of stabilization procedures exist which have already proven useful in technical applications as indicated in Sect. 4.

Due to the distribution and reactivity of their functional groups and their intrinsic stability, each stabilization of an enzyme has to be investigated as an individual problem. However, the biochemical engineer about to tackle his specific stabilization problem today may choose from a large variety of possibilities available.

Acknowledgement

The author is much indebted to Mr. K. Siekmann for his dedicated and thorough assistance in selecting literature, patents, and in the editorial work of the manuscript.

6 References

Abuchowski, A., McCoy, J.R., Palczuk, N.C., Van Es, T., Davis, F.F.: J. Biol. Chem. *252*, 3582 (1977)

Alexander, P., Lett, J.T.: Effects of ionizing radiations on biological macromolecules. In: Comprehensive biochemistry. Florkin, M., Stotz, E.H. (eds.), Vol. 27, pp. 267–356. Amsterdam, London, New York: Elsevier Publishing Comp. 1967

Anfinsen, C.B., Scheraga, H.A.: Adv. Protein Chem. *29*, 205 (1975)
Aoshima, H., Kajiwara, T., Hatanaka, A., Nakatani, H., Hiromi, K.: Chem. Abstr. *88*, 2245 z (1978)
Atallah, M.T., Hultin, H.O.: J. Food Sci. *42*, 7 (1977)
Avrameas, S.: Meth. Enzymol. *44*, 709 (1976)
Axén, R., Myrin, P.-Å., Janson, J.-C.: Biopolymers *9*, 401 (1970)
Baldwin, J.J., Lanes, P., Cornatzer, W.E.: Arch. Biochem. Biophys. *133*, 224 (1969)
Banks, W., Greenwood, C.T.: Stärke *21*, 177 (1969)
Barach, J.T., Adams, D.M.: Biochim. Biophys. Acta *485*, 417 (1977)
Bardsley, W.G., Childs, R.E.: Biochem. J. *137*, 55 (1974)
Bardsley, W.G., Childs, R.E., Crabbe, M.J.C.: Biochem. J. *137*, 61 (1974)
Barker, S.A., Gray, C.J., Lomath, A.W.: J. Appl. Chem. Biotechnol. *26*, 576 (1976)
Bello, J.: Biochemistry *8*, 4542 (1969)
Bello, J., Bello, H.R.: Arch. Biochem. Biophys. *172*, 608 (1976)
Berezin, I.V., Klibanov, A.M., Samokhin, G.P., Martinek, K.: Meth. Enzymol. *44*, 558 (1976)
Berg, M., Boeck, A., Schmid, R.D., Verbeek, H.: Enzyme als Waschmittelkomponente. In: Waschmittelchemie. Henkel & Cie GmbH (ed.), pp. 155–178. Heidelberg: Dr. Alfred Hüthig Verlag 1976
Bergmeyer, H.U.: Grundlagen der enzymatischen Analyse. Weinheim, New York: Verlag Chemie 1977
Bergmeyer, H.U.: Methoden der enzymatischen Analyse. Weinheim, New York: Verlag Chemie 1970
Bernath, F.R., Vieth, W.R.: Biotechnol. Bioeng. *14*, 737 (1972)
Bernfeld, P., Berkeley, B.J., Bieber, R.E.: Arch. Biochem. Biophys. *111*, 31 (1965)
Bernhard, S.A., Rossi, G.L.: On the substrate induced stabilization of native enzyme protein conformation. In: Structural chemistry and molecular biology. Rich, A., Davidson, N. (eds.), pp. 98–114. San Francisco, London: W.H. Freeman and Comp. 1968
Bier, M., Sri Ram, J., Nord, F.F.: Nature *176*, 789 (1955)
Bjorksten, J.: Adv. Protein Chem. *6*, 343 (1951)
Boudrant, J., Cheftel, C.: Biotechnol. Bioeng. *18*, 1735 (1976)
Boudrant, J., Cuq, J.L., Cheftel, C.: Biotechnol. Bioeng. *18*, 1719 (1976)
Bradbury, S.L., Jakoby, W.B.: Proc. Nat. Acad. Sci. USA *69*, 2373 (1972)
Brandts, J.F.: Heat effects on proteins and enzymes. In: Thermobiology. Rose, A.H. (ed.), pp. 25–72. London, New York: Academic Press 1967
Brandts, J.F., Fu, J., Nordin, J.H.: Chem. Abstr. *74*, 83180k (1971)
Brandts, J.F.: Conformational transitions of proteins in water and in aqueous mixtures. In: Structure and stability of biological macromolecules. Timasheff, S.N., Fasman, G.D. (eds.), pp. 213–290. New York: Marcel Dekker 1969
Broun, G.B.: Meth. Enzymol. *44*, 263 (1976)
Brown, E., Racois, A.: Tetrahed. Lett. *37*, 3317 (1976)
Butler, L.G., Reithel, F.J.: Arch. Biochem. Biophys. *178*, 43 (1977)
Castãneda-Agulló, M., Del Castillo, L.M.: J. Gen. Physiol. *42*, 617 (1959)
Ceska, M.: Experientia *27*, 767 (1971)
Chang, T.M.S.: Nature *229*, 117 (1971)
Charles, M., Coughlin, R.W., Hasselberger, F.X.: Biotechnol. Bioeng. *16*, 1553 (1974)
Charm, S.E., Wong, B.L.: Biotechnol. Bioeng. *12*, 1103 (1970)
Chilson, O.P., Costello, L.A., Kaplan, N.O.: Fed. Proc. Suppl. (Wash.) *24*, 55 (1965)
Chlebowski, J.F., Coleman, J.E.: Zinc and its role in enzymes. In: Metal ions in biological systems. Sigel, H. (ed.), Vol. 6, pp. 2–140. New York: Marcel Dekker, Inc. 1976
Cho, Y.K., Bailey, J.E.: Biotechnol. Bioeng. *19*, 157 (1977)
Christensen, P.N., Holm, P., Sønder, B.: J. Am. Oil Chem. Soc. *55*, 109 (1978)
Christensen, T.B., Vegarud, G., Birkeland, A.J.: Process Biochem. *11* (6), 25 (1976)
Citri, N., Kitron, N., Zyk, N.: Biochemistry *11*, 2110 (1972)
Citri, N., Zyk, N.: Biochemistry *11*, 2103 (1972)
Citri, N.: Adv. Enzymol. *37*, 397 (1973)

Clark, J.F., Gurd, R.N.: J. Biol. Chem. *242*, 3257 (1967)
Cohen, L.A.: Ann. Rev. Biochem. *37*, 683 (1968)
Coleman, J.E., Vallee, B.L.: Compr. Biochem. *12*, 231 (1965)
Cowman, R.A., Speck, M.L.: Cryobiology *5*, 291 (1969)
Damjanovich, S., Bot, J., Somogyi, B., Sümegi, J.: Biochim. Biophys. Acta *284*, 345 (1972)
Darbyshire, B.: Cryobiology *12*, 276 (1975)
Darnall, D.W., Birnbaum, E.R.: The metal ion acceleration of the activation of trypsinogen. In: Metal ions in biological systems. Sigel, H. (ed.), Vol. 6, pp. 252–290. New York: Marcel Dekker, Inc. 1976
Davis, F.F., Abuchowski, A., Van Es, T., Palczuk, N.C., Chen, R., Savoca, K., Wilder, K.: Abstracts of 4th enzyme engineering conference Bad Neuenahr (F.R.G.) 1977
Dean, A.C.R., Rodgers, P.J.: Nature *221*, 969 (1969)
Dean, W.L., Tanford, C.: J. Biol. Chem. *252*, 3551 (1977)
Delecourt, R.: Melliand Textilber. *46*, 721 (1965)
den Ouden, T.: Tenside Deterg. *14*, 209 (1977)
Dice, J.F., Goldberg, A.L.: Proc. Nat. Acad. Sci. USA *72*, 3893 (1975)
Dietrich, F.M.: Nature *195*, 146 (1962)
Dixon, M., Webb, E.C.: Enzymes. London: Longmans, Green & Co., Ltd. 1966
Douzou, P.: Mol. Cell. Biochem. *1*, 15 (1973)
Douzou, P.: Proc. Tenth FEBS Meeting, 99 (1975)
Douzou, P.: Trends Biochem. Sci. *1*, 25 (1976)
Duerre, J.A., Ribi, E.: Appl. Microbiol. *11*, 467 (1963)
Duesing, K.P.: Fette, Seifen, Anstrichm. *69*, 738 (1967)
Edelhoch, H., Osborne, J.C.: Adv. Protein Chem. *30*, 183 (1976)
Ehmann, J.D., Hultin, H.O.: Arch. Biochem. Biophys. *154*, 471 (1973)
El'piner, I.E., Deborin, G.A., Zorina, O.M.: Chem. Abstr. *54*, 13201 c (1960)
Enkelund, J.: Interaction between proteolytic enzymes and detergent components. In: Chemie, physikalische Chemie und Anwendungstechnik der grenzflächenaktiven Stoffe, Bd. III/C, S. 251–265. München: Hanser Verlag 1973 (Kongreßbericht)
Epstein, C.J., Anfinsen, C.B., Sela, M.: J. Biol. Chem. *237*, 3458 (1962)
Epton, R., Marr, G., Morgan, G.J.: Biochem. Soc. Trans. *5* (1), 1 (1977)
Eriksson, C.E., Svensson, S.G.: Lebensm. Wiss. Technol. *7*, 38 (1974)
Euler, H.: Chemie der Enzyme, 2. Aufl. München, Wiesbaden: Verlag J.F. Bergmann 1920
Fasold, H., Klappenberger, J., Mayer, C., Ranold, H.: Angew. Chem. *83*, 875 (1971)
Feder, J., Garrett, L.R., Kochavi, D.: Biochim. Biophys. Acta *235* (1971)
Feeney, R.E., Osuga, D.T.: Trends Biochem. Sci. *2*, 269 (1977)
Feeney, R.E.: Chemical modification of food proteins. In: Food proteins. Feeney, R.E., Whitacker, J.R. (eds.). Adv. Chem. Ser. 160, Washington: ACS 1977
Feeney, R.E., Whitaker, J.R. (eds.): Adv. Chem. Ser. 160, Washington: ACS 1977
Feldmann, F., Butler, L.G.: Biochim. Biophys. Acta *268*, 690 (1972)
Fersht, A.R.: Proteins 7, 352 (1976)
Fontana, A., Boccu, E., Veronese, F.M.: Effect of EDTA on the conformational stability of thermolysin. In: Enzymes and proteins from thermophilic microorganismus. Zuber, H. (ed.), pp. 55–59. Basel, Stuttgart: Birkhäuser Verlag 1976
Fontana, A., Toniolo, C.: Forschr. Chem. Org. Naturstoffe *33*, 311 (1975)
Foster, R.L.: Experientia *31*, 772 (1975)
Galembeck, F., Ryan, D.S., Whitacker, J.R., Feeney, R.E.: J. Agric. Food Chem. *25*, 238 (1977)
Geiger, B., von Specht, B.-U., Arnon, R.: Eur. J. Biochem. *73*, 141 (1977)
George, H., McMahan, J., Bowler, K., Elliott, M.: Biochim. Biophys. Acta *191*, 466 (1969)
Gerlsma, S.Y.: J. Biol. Chem. *243*, 957 (1968)
Gerlsma, S.Y.: Eur. J. Biochem. *14*, 150 (1970)
Glazer, A.N.: Ann. Rev. Biochem. *39*, 101 (1970)

Glazer, A.N., Bar-Eli, A., Katchalski, E.: J. Biol. Chem. *237*, 1832 (1962)
Glazer, A.N., DeLange, R.J., Sigman, D.S.: Chemical modification of proteins. Amsterdam, Oxford: North-Holland Publishing Co. 1975
Goldberg, A.L., Dice, J.F.: Ann. Rev. Biochem. *43*, 835 (1974)
Goldberg, A.L., St. John, A.C.: Ann. Rev. Biochem. *45*, 747 (1976)
Goldstein, L.: Meth. Enzymol. *44*, 397 (1976)
Goldstein, L.: Biochem. *11*, 4072 (1972)
Gordon, J.A., Jencks, W.P.: Biochemistry *2*, 47 (1963)
Gounaris, A., Ottesen, M.: C.R. Trav. Lab. Carlsberg *35*, 37 (1965)
Gregoriades, G.: Meth. Enzymol. *44*, 218 (1976)
Grisolia, S.: Physiol. Rev. *44*, 657 (1964)
Grossweiner, L.I.: Chem. Abstr. *84*, 131833f (1976)
Guilbault, G.G.: Enzymatic methods of analysis. Oxford: Pergamon Press 1970
Habeeb, A.F.S.A.: Biochim. Biophys. Acta *115*, 440 (1966)
Haddad, L.C., Thayer, W.S., Jenkins, W.T.: Arch. Biochem. Biophys. *181*, 66 (1977)
Hanafusa, N.: Chem. Abstr. *73*, 73342c (1970)
Hanafusa, N.: Chem. Abstr. *74*, 19759d (1971)
Hare, L.E., Handschumacher, R.E.: Mol. Pharmacol. *9*, 534 (1973)
Hawley, S.A.: Biochemistry *10*, 2436 (1971)
Heinen, W., Lauwers, A.M.: Amylase activity and stability at high and low temperature depending on calcium and other divalent cations. In: Enzymes and proteins from thermophilic microorganisms. Zuber, H. (ed.), pp. 77–89. Basel, Stuttgart: Birkhäuser Verlag 1976
Helenius, A., Simons, K.: Biochim. Biophys. Acta *415*, 29 (1975)
Herzig, D.J., Rees, A.W., Day, R.A.: Biopolymers *2*, 349 (1964)
Hippel, P.H., von, Schleich, T.: The effects of neutral salts on the structure and conformational stability of macromolecules in solution. In: Structure and stability of macromolecules. Timasheff, S.G., Fasman, G.D. (eds.), pp. 417–574. New York: Marcel Dekker 1969a
Hippel, P.H., von, Schleich, T.: Acc. Chem. Res. *2*, 252 (1969b)
Hirs, C.H. (ed.): Methods Enzymol. *11* (1967)
Hixson, H.F.: Biotechnol. Bioeng. *15*, 1011 (1973)
Hoagland, P.D.: Biochemistry *7*, 2542 (1968)
Hocking, J.D., Harris, J.I.: Glyceraldehyde-3-phosphate dehydrogenase from an extreme thermophile, Thermus aquaticus. In: Enzymes and proteins from thermophilic microorganisms. Zuber, H. (ed.), pp. 121–133. Basel, Stuttgart: Birkhäuser Verlag 1976
Holcenberg, J.S., Schmer, G., Teller, D.C.: J. Biol. Chem. *250*, 4165 (1975)
Hora, J.: Biochim. Biophys. Acta *310*, 264 (1973)
Horikoshi, K.: Chem. Econ. Eng. Rev. *3* (2), 46 (1971)
Horikoshi, K., Takeguchi, N., Morii, M., Sano, A.: Chem. Abstr. *87*, 1154y (1977)
Hüttenrauch, R., Keiner, J.: Pharmazie *31*, 575 (1976)
Hunsley, J.R., Suelter, C.H.: J. Biol. Chem. *244*, 4815 (1969)
Ida, S.: Chem. Abstr. *85*, 1583j (1976)
Ingalls, R.G., Squires, R.G., Butler, L.G.: Biotechnol. Bioeng. *17*, 1627 (1975)
Ingebretsen, O.C., Sanner, T.: Arch. Biochem. Biophys. *176*, 442 (1976)
Isemura, T., Fukushi, T., Imanishi, A.: J. Biochem. *56*, 408 (1964)
IUPAC/IUB: Recommendations (1972) of the International Union of Pure and Applied Chemistry and the International Union of Biochemistry. Amsterdam: Elsevier 1973, Supplement 1: Corrections and Additions (1975), Biochim. Biophys. Acta *429*, 1 (1976)
Iwasaki: Japan. Kokai 7658901 (1976)
Jaag, H.R.: Fette, Seifen, Anstrichm. *71*, 961 (1969)
Jacob, S.W., Herschler, R.: Biological actions of dimethyl sulfoxide. Annals New York Acad. Science *243* (1975)
Jancsik, V., Keleti, T., Biczók, G. Nagy, M., Szabó, Z., Wolfram, E.: J. Mol. Catal. *1*, 137 (1975/76)

Jensen, G.: 8th Spanish Symp. on Surfactants, Barcelona 1977
Kapmeyer, W., Pfleiderer, G.: Biochim. Biophys. Acta *481*, 328 (1977)
Kasai, T., Uchida, T.: Chem. Abstr. *65*, 17299h (1966)
Katagiri, M., Takemori, S., Nakazawa, K., Suzuki, H., Akagi, K.: Biochim. Biophys. Acta *139*, 173 (1967)
Kelly, S.J., Butler, L.G., Squires, R.G.: Enzyme Technol. Dig. *5*, 107 (1976)
Kermici, M.: Enzymes et cosmetique. In: Enzyme technology, accomplishments, and prospects (Conférence inaugurale de congrès ayant trait aux "Enzymes en cosmétologie, parfumerie et aromes alimentaires") – Grasse – 1977
Klapper, M.H.: The apolar bond – a reevaluation. In: Progress in bioorganic chemistry. Kaiser, E.T., Kezdy, F.J. (eds.), Vol. 2, pp. 55–132. New York, London, Sydney, Toronto: Wiley and Sons 1973
Kleiner, D.: Nachr. Chem. Techn. Lab. *26*, 195 (1978)
Klinman, J.P., Rose, J.A.: Biochemistry *10*, 2253 (1971)
Klyosow, A.A., Van Viet, N., Berezin, I.V.: Eur. J. Biochem. *5*, 3 (1975)
Koch, O.: Seifen, Öle, Fette, Wachse *95*, 663 (1969)
Koch, O.: Fette, Seifen, Anstrichm. *75*, 331 (1973)
Kono, N., Uyeda, K.: Biochem. Biophys. Res. Commun. *42*, 1095 (1971)
Kopp, P.M., Read, J.F., Carlesby, A.: Nature *211*, 959 (1966)
Krausz, L.M., Becker, R.R.: J. Biol. Chem. *243*, 4606 (1968)
Kuczenski, R.T., Suelter, C.H.: Biochemistry *9*, 939 (1970)
Kuehnau, J.: Chem. Abstr. *71*, 2171w (1969)
Kuwajima, K.: J. Mol. Biol. *124*, 241 (1977)
Langguth, R.P.: Enzyme detergents. In: Kirk-Othmer encyclopedia of chemical technology, Suppl. Vol., pp. 294–309. New York: Wiley and Sons 1971
Langguth, R.P., Mecey, L.W.: Soap Chem. Special. Aug. 62 (1969)
Lasch, J., Bessmertnaya, L., Kozlov, L.V., Antonov, V.K.: Eur. J. Biochem. *63*, 591 (1976)
Laurent, T.C.: Eur. J. Biochem. *21*, 498 (1971)
Lilly, M.D., Dunnill, P.: Meth. Enzymol. *44*, 717 (1976)
Liss, R.L., Langguth, R.P.: J. Am. Oil Chem. Soc. *46*, 507 (1969)
Lontie, R., Vanquickenborne, L.: The role of copper in hemocyanins. In: Metal ions in biological systems. Sigel, H. (ed.), Vol. 3, pp. 183–200. New York: Marcel Dekker, Inc. 1974
Luca, R., Zamfirescu-Gheorghiu, M.: Chem. Abstr. *75*, 331g (1971)
Lumry, R.: Some aspects of the thermodynamics and mechanism of enzyme catalysis. In: The enzymes. Boyer, P.D., Lardy, H., Myrbäck, K. (eds.), Vol. 1, pp. 157–231. New York, London: Academic Press 1959
Marshall, J.J.: Abstracts 4th enzyme engineering conference Bad Neuenahr (F.R.G.) 1977
Marshall, J.J., Humphreys, J.D.: Biotechnol. Bioeng. *19*, 1739 (1977a)
Marshall, J.J., Humphreys, J.D.: J. Chromatogr. *137*, 468 (1977b)
Marshall, J.J., Rabinowitz, M.L.: Arch. Biochem. Biophys. *167*, 777 (1975)
Marshall, J.J., Rabinowitz, M.L.: J. Biol. Chem. *251*, 1081 (1976)
Martinek, K., Klibanov, A.M., Goldmacher, V.S., Berezin, I.V.: Biochim. Biophys. Acta *485*, 1 (1977a)
Martinek, K., Klibanov, A.M., Goldmacher, V.S., Tchernysheva, A.V., Mozhaev, V.V., Berezin, I.V., Glotov, B.O.: Biochim. Biophys. Acta *485*, 13 (1977b)
Matthews, B.W.: The structure and stability of thermolysin. In: Enzymes and proteins from thermophilic microorganisms. Zuber, H. (ed.), pp. 31–39. Basel, Stuttgart: Birkhäuser Verlag 1976
Mattiasson, B., Johansson, A.-C., Mosbach, K.: Eur. J. Biochem. *46*, 341 (1974)
McLaren, A.D., Luse, R.A.: Science *134*, 836 (1961)
McLaren, A.D.: Adv. Enzymol. *9*, 75 (1949)
Means, G.E., Feeney, R.E.: Biochemistry *7*, 2192 (1968)
Meighen, G.A., Ziegler Nicoli, M., Hastings, S.: Biochemistry *10*, 4069 (1971)
Mitz, M.A., Summaria, L.J.: Nature *189*, 576 (1961)

Moriyama, S., Matsuno, R., Kamikubo, T.: Agric. Biol. Chem. *41*, 1985 (1977)
Mosbach, K.: Meth. Enzymol. *44*, 453 (1976)
Myers, J.S., Jakoby, W.B.: Biochem. Biophys. Res. Commun. *51*, 631 (1973)
Nachev, L., Dobreva, E.: Chem. Abstr. *75*, 52727k (1971)
Nakamura, S., Hayashi, S.: FEBS Lett. *41*, 327 (1974)
Nakaya, K., Ushinvata, A., Nakamura, Y.: Biochim. Biophys. Acta *439*, 116 (1976)
Noetzold, H., Schlegel, B., Breitfeld, D., Freimuth, U.: Nahrung *21*, 697 (1977)
Nojima, H., Ikai, A., Oshima, T., Noda, H.: J. Mol. Biol. *116*, 429 (1977)
Nordin, J.H., Duffield, R., Freedman, N., Gelb, W., Brandts, J.F.: Cryobiology *6*, 373 (1970)
Nozaki, M., Kagamiyama, H., Hayaishi, O.: Biochem. Zschr. *338*, 582 (1963)
Öbrink, B., Laurent, T.C.: Eur. J. Biochem. *41*, 83 (1974)
Oettel, H.: Textilhilfsmittel, Entschlichtungsmittel. In: Ullmanns Encyklopädie der technischen Chemie. Foerst, W. (ed.), Bd. 17, S. 215–216. München, Berlin, Wien: Urban and Schwarzenberg 1966
Okunuki, K.: Adv. Enzymol. *23*, 29 (1961)
Ollis, D.F., Datta, R.: Meth. Enzymol. *44*, 444 (1976)
Olson, A.C., Stanley, W.L.: The use of tannic acid and phenolformaldehyde resins with glutaraldehyde to immobilize enzymes. In: Immobilized enzymes in food and microbial processes. Olson, A.C., Cooney, C.L. (eds.), pp. 51–62. New York, London: Plenum Press 1974
O'Malley, J.J., Ulmer, R.W.: Biotechnol. Bioeng. *15*, 917 (1973)
O'Neill, S.P., Wykes, J.R., Dunnill, P., Lilly, M.D.: Biotechnol. Bioeng. *13*, 319 (1971)
Organon; DOS 1944048 (1970)
Oshima, T., Fujita, S., Imahori, K.: Chem. Abstr. *83*, 159650x (1975)
Ottesen, M., Johansen, J.T., Svendsen, I.: Subtilisin: stability properties and secondary binding sites. In: Structure-function relationships of proteolytic enzymes. Desnuelle, P., Neurath, H., Ottesen, M. (eds.), pp. 175–186. Copenhagen: Munksgaard 1970
Ottesen, M., Svendsen, I.: Meth. Enzymol. *19*, 199 (1970)
Ottesen, M., Svendsen, I.: C.R. Trav. Lab. Carlsberg *38*, 369 (1976)
den Ouden, T.: Tenside Deterg. *14*, 209 (1977)
Pace, N.C.: Stability of globular proteins. In: CRC critical reviews in biochemistry, Vol. 1, pp. 1–43. Cleveland: Chemical Rubber Co. 1975
Payne, J.W.: Biochem. J. *135*, 867 (1973)
Pazur, J.H., Aronson, J.R., N.N.: Adv. Carbohydr. Chem. Biochem. *27*, 301 (1972)
Pazur, J.H., Knull, H.R., Simpson, D.L.: Biochem. Biophys. Res. Commun. *40*, 110 (1970)
Penniston, J.T.: Arch. Biochem. Biophys. *142*, 322 (1971)
Pérez-Villaseñor, J., Whitaker, J.R.: Arch. Biochem. Biophys. *121*, 541 (1967)
Peters, K., Richards, F.M.: Ann. Rev. Biochem. *46*, 523 (1977)
Petsko, G.A.: J. Mol. Biol. *96*, 381 (1975)
Pfizer Inc.: US 3.770.587 (1973)
Pohl, F.M.: Eur. J. Biochem. *7*, 146 (1968)
Pollak, A., Whitesides, G.M.: J. Am. Chem. Soc. *98*, 298 (1976)
Porath, J., Axén, R.: Meth. Enzymol. *44*, 19 (1976)
Privalov, P.L., Khechinashvili, N.N.: J. Mol. Biol. *86*, 665 (1974)
Ralston, G.B.: C.R. Trav. Lab. Carlsberg *39*, 25 (1972)
Reed, G.: Enzymes in food processing. New York: Academic Press 1975
Reiner, R., Siebeneick, H.-U., Christensen, I., Lukas, H.: J. Mol. Catal. *1*, 3 (1975/76)
Reiner, R., Siebeneick, H.-U., Christensen, I., Doring, H.: J. Mol. Catal. *2*, 119 (1977)
Remy, M.-H., Wetzer, J., Thomas, D.: Potentialities of soluble polymers bearing stabilized enzymes. In: Proc. Internat. Symp. Analysis Control Immobilized Enzyme Systems, pp. 284–290. Amsterdam: North-Holland 1976
Rosén, C.-G., Nilsson, R.: Biochim. Biophys. Acta *236*, 1 (1971)
Roubal, W.T., Tappel, A.L.: Arch. Biochem. Biophys. *113*, 5 (1966)
Royer, G.P., Ikeda, S.-I., Aso, K.: FEBS Lett. *80*, 1 (1977)

Royer, G.P., Uy, R.: J. Biol. Chem. *248*, 2627 (1973)
Russell, A.D., Hopwood, D.: Prog. Med. Chem. *13*, 271 (1976)
Ruwart, M.J., Suelter, C.H.: J. Biol. Chem. *246*, 5990 (1971)
Santarius, K.A.: Ber. Deut. Bot. Ges. *84* (7/8), 425 (1971)
Schormüller, J.: Med. Welt *32*, 1843 (1967)
Schreiber, W.: J. Am. Oil Chem. Soc. *49*, 182 (1972)
Schreiber, W., Schindler, J., Gloxhuber, C.: Chem. Ztg. *98*, 539 (1974)
Schwuger, M.J., Bartnik, F.G.: Interaction of anionic surfactants with proteins, enzymes, and membranes. In: Biology. Gloxhuber, C. (ed.), Vol. 7/4. New York: Marcel Decker (to be published)
Scrutton, M.C.: Metal enzymes. In: Inorganic biochemistry. Eichhorn, G.L. (ed.), Vol. 1, pp. 381–437. Amsterdam, London, New York: Elsevier Scientific Publishing Comp. 1973
Sela, M., Arnon, R.: Meth. Enzymol. *11*, 580 (1967)
Shapiro, B., Kollmann, G.: Radiat. Damage Sulphydryl Compd., Proc. Panel 1968, pp. 23–43 (Pub. 1969)
Shataeva, L.K., Zaikina, N.A., Samsonov, G.V.: Appl. Biochem. Microbiol. *12*, 338 (1976)
Shatsky, M.A., Ho, H.C., Wang, J.H.-C.: Biochim. Biophys. Acta *303*, 298 (1973)
Shields, G.S., Hill, R.L., Smith, E.L.: J. Biol. Chem. *234*, 1747 (1959)
Shifrin, S., Hum, G.: Arch. Biochem. Biophys. *130*, 530 (1969)
Shifrin, S., Solis, B.G., Chaiken, I.M.: J. Biol. Chem. *248*, 3464 (1973)
Siegel, F.L.: Calcium-binding proteins. In: Structure and bonding. Dunitz, J.D., Hemmerich, P., Ibers, J.A., Jørgensen, E.C.K., Neilands, J.B., Reinen, D., Williams, R.J.P. (eds.), Vol. 17, pp. 221–268. Berlin, Heidelberg, New York: Springer 1973
Silverstein, R., Grisolia, S.: Physiol. Chem. Phys. *4*, 37 (1972)
Simic, M.G.: J. Agric. Food Chem. *26*, 6 (1978)
Singer, S.J.: Adv. Protein Chem. *17*, 1 (1962)
Smith, W.L., Lands, W.E.M.: Biochem. Biophys. Res. Commun. *41*, 846 (1970)
Snyder, P.D., Jr., Wold, F., Bernlohr, R.W., Dullum, C., Desnick, R.J., Krivit, W., Condie, R.M.: Biochim. Biophys. Acta *350*, 432 (1974)
Sokolova, E.V., Mosolov, V.V., Afanas'ev, P.V.: Biochem. Microbiol. *6*, 683 (1970)
Specht, B.-U., von, Seinfeld, H., Brendel, W.: Hoppe-Seyler's Z. Physiol. Chem. *354*, 1659 (1973)
Sri Ram, J., Terminiello, L., Bier, M., Nord, F.F.: Arch. Biochem. Biophys. *52*, 464 (1954)
Stauffer, C.E., Etson, D.: J. Biol. Chem. *244*, 5333 (1969)
Stauffer, C.E., Treptow, R.S.: Biochim. Biophys. Acta *295*, 457 (1973)
Steinhardt, J.: The nature of specific and non-specific interactions of detergents with proteins: complexing and unfolding. In: Protein-ligand interactions. Sund, H., Blauer, G. (eds.), pp. 412–426. Berlin, New York: de Gruyter 1975
Steinhardt, J., Reynolds, J.A.: Multiple equilibria in proteins. New York, London: Academic Press 1969
Stellwagen, E., Barnes, L.D.: Analysis of the thermostability of enolases. In: Enzymes and proteins from thermophilic microorganisms. Zuber, H. (ed.), pp. 223–227. Basel, Stuttgart: Birkhäuser Verlag 1976
Südhof, H., Wötzel, E.: Klin. Wochenschr. *38*, 1165 (1960)
Suzuki, K., Taniguchi, Y.: Chem. Abstr. *81*, 131925a (1974)
Svendsen, I.: C.R. Trav. Lab. Carlsberg *38*, 385 (1971)
Svendsen, I.: C.R. Trav. Lab. Carlsberg *41*, 238 (1976)
Svensson, B.: C.R. Trav. Lab. Carlsberg *39*, 469 (1974)
Svensson, B.: Biochim. Biophys. Acta *429*, 954 (1976)
Switzer, R.L.: Ann. Rev. Microbiol. *31*, 135 (1977)
Takeda, Y., Hizukuri, S.: Biochim. Biophys. Acta *268*, 175 (1972)
Takemori, S., Furuya, E., Suzuki, H., Katagiri, M.: Nature *215*, 417 (1967)
Tanford, C.: Adv. Protein Chem. *23*, 121 (1968)
Tanford, C.: Adv. Protein Chem. *24*, 1 (1970)

Tanford, C., Reynolds, J.A.: Biochim. Biophys. Acta *457*, 133 (1976)
Tappel, A.L.: Chem. Abstr. *75*, 106800z (1971)
Terminiello, L., Sri Ram, J., Bier, M., Nord, F.F.: Arch. Biochem. Biophys. *57*, 252 (1955)
Thompson, F.M., Lipertini, L.J., Joss, U.R., Calvin, M.: Science *178*, 505 (1972)
Timasheff, S.N., Lee, J.C., Pittz, E.P., Tweedy, N.: J. Colloid Interface Sci. *55*, 658 (1976)
Tojima, M., Urabe, I., Yutani, K., Okada, H.: Eur. J. Biochem. *64*, 243 (1976)
Torchilin, V.P., Maksimenko, A.V., Smirnov, V.N., Berezin, I.V., Klibanov, A.M., Martinek, K.: Biochim. Biophys. Acta *522*, 277 (1978)
Torchinskii, Yu., M., Dixon, H.B.F.: Sulfhydryl and disulfide groups of proteins. New York, London: Consultants Bureau 1974
Tuengler, P., Pfleiderer, G.: Biochim. Biophys. Acta *484*, 1 (1977)
Urabe, I., Nanjo, H., Okada, H.: Biochim. Biophys. Acta *302*, 73 (1973)
Urabe, I., Okada, H.: Proc. IV IFS: Ferment. technol. today, 367 (1972)
Vaeck, S.V., Maes, E.C., de Pauw, L.: Tenside *8*, 188 (1971)
Vallee, B.L., Coleman, J.E.: Compr. Biochem. *12*, 231 (1965)
Vallee, B.L., Ulmer, D.D.: Ann. Rev. Biochem. *41*, 92 (1972)
Vallee, B.L., Wacker, W.E.C.: Metalloproteins. In: The proteins. Neurath, H. (ed.), Vol. 5. New York, London: Academic Press 1970
Van Leemputten, E., Horisberger, M.: Biotechnol. Bioeng. *18*, 587 (1976)
Vegarud, G., Christensen, T.B.: Biotechnol. Bioeng. *17*, 1391 (1975)
Vieth, W.R., Gilbert, S.G.: Enzyme modification, electrophoresis, and applications. Washington DC: National Science Foundation 1976
Vladimirov, Yu.A., Roshchupkin, D.I., Fesenko, E.E.: Chem. Abstr. *73*, 52198x (1970)
Walker, G.J., Whelan, W.J.: Biochem. J. *76*, 264 (1960)
Wang, J.H.-C., Tu, J.-I.: Biochemistry *8*, 4403 (1969)
Warren, J.R., Gordon, J.A.: Biochim. Biophys. Acta *229*, 216 (1970)
Warren, J.R., Gordon, J.A.: Biochim. Biophys. Acta *420*, 397 (1976)
Weaver, L.H., Kester, W.R., Ten Eyck, L.F., Metthews, B.W.: The structure and stability of thermolysin. In: Enzymes and proteins from thermophilic microorganisms. Zuber, H. (ed.), pp. 31–39. Basel, Stuttgart: Birkhäuser Verlag 1976
Wellner, D., Silman, H.J., Sela, M.: J. Biol. Chem. *238*, 1324 (1963)
Wermuth, B., Brodbeck, U.: Eur. J. Biochem. *35*, 499 (1973)
Westhead, E.W.: Meth. Enzymol. *25* B, 401 (1972)
Wieg, A.J.: Process Biochem. *4* (2), 30 (1969)
Williams, R.K., Shen, C.: Arch. Biochem. Biophys. *152*, 606 (1972)
Wills, E.D.: Biochem. Pharmacol. *7*, 7 (1961)
Wiseman, A.: Process Biochem. *8* (8), 14 (1973)
Wiseman, A., Williams, N.J.: Biochim. Biophys. Acta *250*, 1 (1971)
Woodward, J., Wiseman, A.: J. Appl. Chem. Biotechnol. *26*, 580 (1976)
Wykes, J.R., Dunnill, P., Lilly, M.D.: Biochim. Biophys. Acta *250*, 522 (1971)
Yasumatsu, K., Ohno, M., Matsumura, C., Shimazono, H.: Agr. Biol. Chem. *29*, 665 (1965)
Yoshimura, T., Imanishi, A., Isemura, T.: J. Biochem. *63*, 730 (1968)
Yutani, K.: Role of calcium in the thermostability of a α-amylase produced from Baccillus stearothermophilus. In: Enzymes and proteins from thermophilic microorganisms. Zuber, H. (ed.), pp. 91–103. Basel, Stuttgart: Birkhäuser Verlag 1976
Zaborsky, O.R.: Immobilized enzymes. Cleveland: Chemical Rubber Co. 1974a
Zaborsky, O.R.: Stabilization and immobilization of enzymes with imidoesters. In: Enzyme engineering. Pye, E.K., Wingard, L.B., Jr. (eds.), Vol. 2, pp. 115–122. New York, London: Plenum Press 1974b
Zaikina, N.A., Elinov, N.P., Shataeva, L.K., Domorad, A.A.: Chem. Abstr. *73*, 94888c (1970)
Zuber, H.: Nachr. Chem. Techn. *23*, 411 (1975)
Zuber, H. (ed.): Enzymes and proteins from thermophilic microorganisms. Basel, Stuttgart: Birkhäuser Verlag 1976

Note Added in Proof

Since submission of the manuscript, a number of pertinent publications and patents have appeared including two reviews in Russian (Martinek and Torchilin, 1978; Ugarova, 1978).

Enzyme Denaturation

A comprehensive monograph on physicochemical aspects of protein denaturation has been published (Lapanje, 1978). The Lomonossov group (Martinek et al., 1978a) reports almost quantitative reactivation of an "irreversibly" denatured (immobilized) trypsin and discusses the implications of this observation. Reviews on cold lability (Bock and Frieden, 1978) and radiation sensitivity of enzymes (Saito, 1976) have been published.

Several enzymes important in food processing were not denatured by γ-irradiation while their microbial contaminants were reduced significantly (Delincee, 1978a, b).

Pressure inactivation of several enzymes has been investigated (Drickamer, 1977). In the case of lactate dehydrogenase, SH-group oxidation is the rate-determining step (Schmid et al., 1978). Inactivation of this enzyme and of urease has been studied under steady shearing flow (Tirrell and Middleman, 1978). A physical model for shear inactivation of proteins has been proposed (Charm and Wong, 1978).

With regard to chemical denaturants, the addition of alkali to proteins has been shown to result in chemical modification of structure-forming amino acids (Nashef et al., 1977). As established by X-ray structure analysis, guanidinium HCl binds only to the surface of protein (α-chymotrypsin) while urea additionally binds to the hydrophobic interior (Hibbard and Tulinsky, 1978). Lipophilin, a hydrophobic myelin protein, largely resisted denaturation by 6M guanidinium thiocyanate and 8M urea between pH 1.5 and 11, even if heated and reduced; this is thought to result from an impervious hydrophobic core of this protein (Cockle and Epand, 1978).

The stability of five proteases towards SDS and urea has been studied (Hilz and Fanick, 1978). The interaction of proteins with cationic detergents was reviewed in Japanese (Hiramatsu and Aoki, 1978).

Enzyme Stabilization
Intrinsically Stabilized Enzymes

In a recent monograph on microbial life in extreme environments many aspects of native protein stability at high and low temperatures, high pressures, extreme pH-values, and high salinity are discussed (Kushner, 1978).

The implications of intrinsic protein stability for enzyme activity on Jupiter and similar environments has been considered (Siegel and Speitel, 1977). More than a dozen enzymes isolated from the extreme thermophile, *T. thermophilus,* were found to be resistant to high temperatures (Oshima, 1978).

Additives

Covalent substrate complexes of three enzymes were more stable to pH-induced inactivation than the free enzymes (Volini and Wang, 1978). Low concentrations of alcohols and ketones stabilized proteins against surface denaturation, while high concentrations and some other solvents had a destabilizing effect (Asakura et al., 1978). Phosphorylase B was stabilized in its catalytic conformation by the addition of low amounts of alcohols (Dreyfus et al., 1978).

No simple correlation between the DK of various organic solvents and k_{cat} or $K_{M(app)}$ of several hydrolytic enzymes in such solvent systems could be established, but DK-dependant interactions of these solvents with hydrophobic substrate binding domains were observed (Maurel, 1978).

The role of Ca as a stabilizing agent for extracellular proteases has been reviewed (Roche and Voordouw, 1977). LiCl – a protein denaturant – stabilized ribonuclease A against isothermal denaturation by urea through formation of a Li-urea complex (Ahmad and Bigelow, 1978). The stabilization of enzymes by polymers has been extended to polyacrylamide gels with a pore size allowing for tight enzyme chlathrate formation after drying; stability of the enzyme polymer exceeded soluble enzyme stability by 13 orders of magnitude (Martinek et al., 1978b).

Chemically Modified Soluble Enzymes

The thermodynamics of protein cross-links have been investigated for a soluble lysozyme ester derivative (Johnson et al., 1978).

Increased thermostability imparted to α-chymotrypsin by intramolecular cross-links has been related to the chain-length of such cross-links (Torchilin et al., 1978). Several enzymes glycosylated with HJO_4-oxidized dextran exhibited increased shelf-life (Reiner and Döring, 1978).

Patents

Several additives such als alkoxylated alkylamines (Henkel KGaA, DOS 2633601, 1978), polyethyleneglycol/tallow alcohol ethoxylate (Procter and Gamble, US 4090973, 1978) or polyalcohols and polyalkanolamines, respectively, in the presence of boron salts (Unilever N.V., DOS 2748211, 2748212, 1978) have been claimed to improve the storage stability of enzymes in liquid detergents.

p-Hydroxybenzoic ester and alcohols or glycine derivatives (Mitsubishi Chem. Ind., Jap. 52094482, 1977) or isopropanol and Mg-salts (CPC International Inc., Belg. 852593, 1977) have been found to increase the storage stability of glucose isomerase. Cholesterol oxidase was stabilized by inorganic salts (Boehringer Mannheim GmbH, DOS 2755799, 1978), glycerokinase by a protein fraction from yeast (Boehringer Mannheim GmbH, Brit. 1488988, 1977), and hexokinase, glucose-6-phosphate-dehydrogenase and creatine kinase by gelatine derivatives (Boehringer Mannheim GmbH, DOS 2648759, 1978). Addition of ethanol, glycerol, citrate, and NaCl led to increased storage stability of catalase

(Amano Pharmaceutical Co., Jap. 7824093, 1978). Horseradish peroxidase showed improved stability during freeze-drying after addition of albumin, mannitol, and Fe-salts (Akzo N.V., DOS 2742699, 1978). Urease and peroxidase preparations were stabilized by inorganic salts (Laboratories Normon SA, Span. 457490, 457491, 1978); a protease exhibited increased stability to orthostatic pressure in the presence of polyethyleneglycol and sorbitanmonostearate (Aikokuzoki Pharm., Jap. 50121485, 1975). Terrolitin protease was modified for biomedical applications by covalent bonding to a PVP-acroleyl-copolymer (Macromol. Compounds Institute, SU 522-198, 1976).

Literature

Ahmad, F., Bigelow, C.C.: Canad. J. Biochem. *57*, 1003 (1978)
Asakura, T., Adachi, K., Schwartz, E.: J. Biol. Chem. *253*, 6423 (1978)
Bock, P.E., Frieden, C.: Trends Biochem. Sci. *3*, 100 (1978)
Charm, S.E., Wong, B.L.: Biotechnol. Bioeng. *20*, 451 (1978)
Cockle, S.A., Epand, R.M.: J. Biol. Chem. *253*, 8019 (1978)
Delincee, H.: J. Food Biochem. *2*, 49 (1978a), *2*, 71 (1978b)
Dreyfus, M., Vandenbunder, B., Buc, H.: FEBS Lett. *95*, 185 (1978)
Drickamer, H.G.: High Temp. – High Pressures *9*, 505 (1977)
Hibbard, L.S., Tulinsky, A.: Biochemistry *17*, 5460 (1978)
Hilz, H., Fanick, W.: Hoppe-Seyler's Z. Physiol. Chem. *359*, 1447 (1978)
Hiramatsu, K., Aoki, K.: Chem. Abstr. *89*, 199440j (1978)
Johnson, R.E., Adams, P., Rupley, J.A.: Biochemistry *17*, 1479 (1978)
Kushner, D.J.: Microbial life in extreme environments. London, New York, San Francisco: Academic Press 1978
Lapanje, S.: Physicochemical aspects of protein denaturation. New York, Chichester, Brisbane, Toronto: Wiley-Interscience 1978
Martinek, K., Gol'dmakher, V.S., Torchilin, V.P., Mishin, A.A., Smirnov, V.N., Berezin, I.V.: Dokl. Akad. Nauk SSSR *239*, 227 (1978)
Martinek, K., Mozhaev, V.V., Berezin, I.V.: Dokl. Akad. Nauk SSSR *239*, 77 (1978a)
Martinek, K., Torchilin, V.P.: Chem. Abstr. *89*, 159210c (1978b)
Maurel, P.: J. Biol. Chem. *253*, 1677 (1978)
Nashef, A.S., Osuga, D.T., Lee, H.S., Ahmed, A.I., Whitaker, J.R., Feeney, R.E.: J. Agric. Food Chem. *25*, 245 (1977)
Oshima, T.: Properties of heat stable enzymes of extreme thermophiles. In: Enzyme engineering. Broun, G.B., Manecke, G., Wingard, L.B., Jr. (eds.), p. 41. New York, London: Plenum Press 1978
Reiner, R.R., Döring, H.: Surface modification of proteins. In: Enzyme engineering. Broun, G.B., Manecke, G., Wingard, L.B., Jr. (eds.), p. 111. New York, London: Plenum Press 1978
Roche, R.S., Voordouw, G.: The role of calcium as a conformational lock in the structure of thermostable extracellular proteases. In: Calcium-binding proteins calcium funct., Proc. Int. Symp., 2nd 1977. Wasserman, R.H. (ed.), p. 38. New York: Elsevier N. Holland 1977
Saito, M.: Chem. Abstr. *88*, 166079z (1978)
Schmid, G., Lüdemann, H.-D., Jaenicke, R.: Eur. J. Biochem. *86*, 219 (1978)
Siegel, S.M., Speitel, T.W.: Life Sci. Space Res. *15*, 77 (1977)
Tirrell, M., Middleman, S.: Biotechnol. Bioeng. *20*, 605 (1978)
Torchilin, V.P., Maksimenko, A.V., Smirnov, V.N., Berezin, I.V., Klibanov, A.M., Martinek, K.: Biochim. Biophys. Acta *522*, 277 (1978)
Ugarova, N.N.: Chem. Abstr. *89*, 159211d (1978)
Volini, M., Wang, S.-F.: Arch. Biochem. Biophys. *187*, 163 (1978)

The Use of Coenzymes in Biochemical Reactors

Shaw S. Wang, Chwan-Kong King
Department of Chemical and Biochemical Engineering
College of Engineering, Rutgers University
Piscataway, NJ 08854, U.S.A.

Coenzymes are classified according to their needs for regeneration in coenzyme requiring enzymatic reactions. The activities of various forms of immobilized NAD(H) are compared. Different coenzyme regeneration methods are discussed and their merits assessed. Reactors that are suitable for coenzyme requiring enzymatic conversion are presented.

1 Introduction

Coenzymes, as the word implies, are those chemicals nonprotein in nature, that act with enzymes in a cooperative fashion. Today, there are about two thousand enzymes known. Of these enzymes, approximately 40% require coenzymes. According to the International Union of Biochemistry, enzyme-catalyzed reactions have been classified and named into six general groups, namely those catalyzing:

1. oxidation-reduction reactions (oxidoreductase),
2. group transfer reactions (transferases),
3. hydrolytic reactions (hydrolases),
4. the addition of groups to double bonds or vice versa (lyases),
5. isomerizations (isomerases), and
6. the condensation of two molecules coupled with the cleavage of a pyrophosphate bond of ATP or similar triphosphates (ligases or synthetases).

Dehydrogenases, a sub-group of oxidoreductases, requires an electron acceptor to achieve their "dehydrogenation" (or oxidation) -catalytic activities. These electron acceptors, such as NAD^+/NADH are existing as free coenzymes. FAD/$FADH_2$, are tightly bound to the enzymes that require them as cofactors, and so FAD/$FADH_2$ are also referred to as prosthetic groups.

Group-transferring enzymes, such as transaminases and transketolases, also require coenzymes. In this case, they require pyridoxal phosphate and thiamin pyrophosphate, respectively. Decarboxylases, which belongs to the group of lyases, also require pyridoxal phosphate as coenzyme. Many of the ligases which catalyze biosynthetic reactions, require a nucleoside triphosphate (such as ATP) as coenzyme.

Vitamin B_{12}, or cyanocobalamin, is a coenzyme required in isomerization of methylmalonyl CoA to succinyl CoA as catalyzed by methyl malonyl CoA isomerase. The only group of enzymes, among the six groups, which does not include a coenzyme requiring enzyme is that of hydrolases.

At this time, most industrially important enzymes are various types of hydrolases, such as carbohydrases (i.e. α-amylases, β-amylase, γ-amylase, amyloglucosidase, maltase, and cellulases), proteases (i.e. papain, bromelain, ficin, rennin, and various microbial proteases), amidases (i.e. penicillin amidase), and aminoacylases (i.e. α-glutamyl amino acid acylase). These aforementioned hydrolytic enzymes are widely used in food or pharmaceutical industries. With the advent of the technology of immobilized enzymes, many enzymes are studied in terms of attaching them to solid supports and subsequently studied in selected reactor configurations for possible applications of the immobilized catalysts in continuous enzymatic conversion processes. Again, the most popular enzymes studied in this connection are hydrolases. This is partially due to the relative simplicity of the action of hydrolases. There are relatively few studies and reports on the applications of immobilized enzymes which require coenzyme(s). Successful applications of immobilized dehydrogenases, transferases, and ligases (synthetases) require simultaneous use of the required coenzymes (free or immobilized) with a coenzyme regenerating system to realize the advantage of using immobilized enzymes other than free enzymes. Thus, the application of immobilized coenzyme requiring enzymes depends, to a large extent, on the feasibility of regenerating the required coenzyme(s) in the system.

There are many review articles available on various aspects of coenzymes[1–8]. Recently, Yamada[133] review the productions of vitamins and coenzymes in Japan.

In this article the authors discuss the function of different coenzymes, the immobilizations, regenerations, and applications of some of these coenzymes.

2 Classifications of Coenzymes

Coenzymes are organic, non-proteinaceous cofactors which are used conventionally as a term to include inorganic metal ions (Mg^{+2}, Ca^{+2}, etc.) that are essential for enzymatic reactions. Prosthetic groups are organic, non-proteinaceous cofactors that are bound covalently or through secondary bonds) tightly to the protein moiety of the enzymes. To differentiate them from inorganic cofactors (i.e. metal ions), coenzymes and prosthetic groups can be called organic cofactors. Organic cofactors are classified into three groups according to their regeneration requirements. As shown in Table 1, group A includes those organic cofactors that are self-regenerating in the reactions they are involved. For example, pyridoxal phosphate is regenerated after a transamination reaction is completed. Mechanistically, pyridoxal phosphate forms a Schiff's base with the amino group of the donor amino acid and subsequently form pyridoxamine phosphate and the corresponding keto acid. Pyridoxamine phosphate then reacts with the acceptor keto acid to produce the product amino acid and in the mean time, pyridoxal phosphate is regenerated.

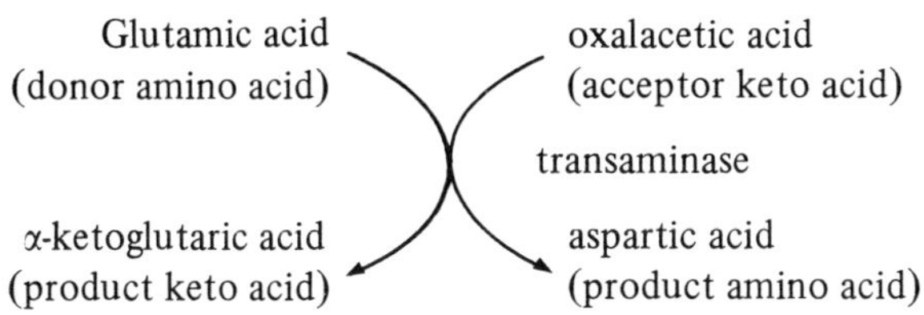

(1)

Group B includes those organic cofactors that can be regenerated by catalytic oxidation with molecular oxygen as the electron acceptor.

Table 1. Classifications of organic cofactors based on regeneration requirements

Group A	Group B	Group C
1. Biotin	1. FAD^+	1. Ascorbic acid (vitamin C)
2. Cyanocobalamin (vitamin B_{12}) (isomerization function)	2. FMN	2. Coenzyme A
3. Pyridoxal phosphate (vitamin B_6)	3. Lipoic acid	3. Coenzyme Q
4. Thiamine pyrophosphate (a derivative of vitamin B_1)	4. NAD^+	4. Cyanocobalamin (vitamin B_{12}) (transfer function)
	5. $NADP^+$	5. $FADH_2$
		6. $FMNH_2$
		7. Heme coenzymes (cytochromes)
		8. Glutathione
		9. NADH
		10. NADPH
		11. Nucleoside triphosphates (ATP, UTP, etc.)
		12. S-Adenosyl methionine
		13. THFA (tetrahydrofolic acid)

Group A: self-regenerating
Group B: can be regenerated by catalytic oxidation with molecular oxygen as the electron acceptor
Group C: require an appropriate substrate other than oxygen for regeneration (based on Baricos et al.[2] with modifications)

This group includes the disulfide-organic cofactors, lipoic acids, oxidized forms of flavin mono- and di-nucleotides, oxidized forms of nicotinamide adenine dinucleotides, and nicotinamide adenine dinucleotide phosphates. Mechanistically, the regeneration of group B organic cofactors can be examplified by schemes (2) and (3).

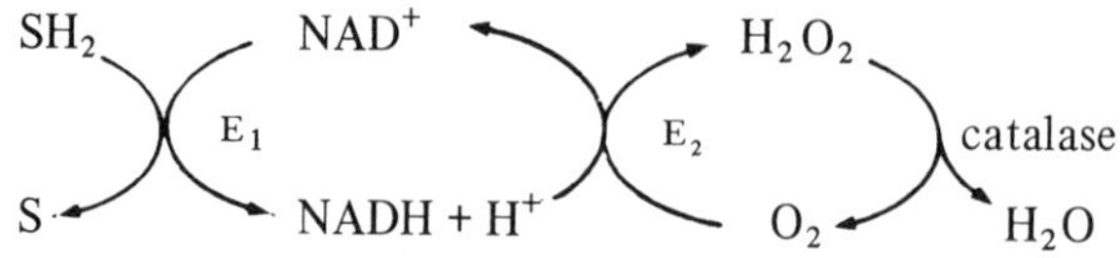

(2)

E_1 : a dehydrogenase, E_2 : diaphorase

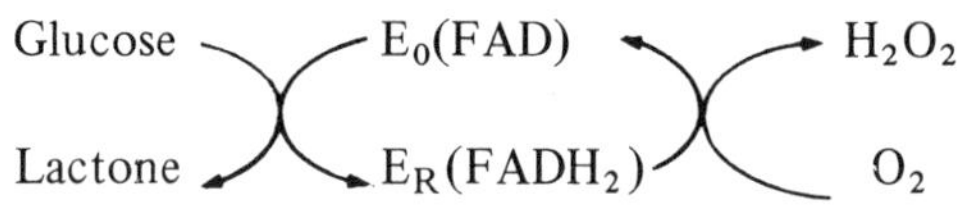

(3)

E_0 : oxidized form of glucose oxidase (FAD is the prosthetic group of the enzyme)

E_R : reduced form of glucose oxidase

In scheme (2) and (3), oxygen is used as the electron acceptor to regenerate the oxidized form of the organic cofactor involved.

According to such classifications, most of the organic cofactors require an appropriate substrate other than oxygen. Among the thirteen organic cofactors listed, NADH, NADPH, and ATP are the most important ones in terms of possible industrial applications involving cofactors in enzymatic conversions of substrates. Some of the other cofactors listed under group C have been studied in connection with applications in affinity chromatography for purification purposes (Table 2). Examples of the regeneration of organic cofactors of group C are illustrated in the following:

Regeneration of NADH (30):

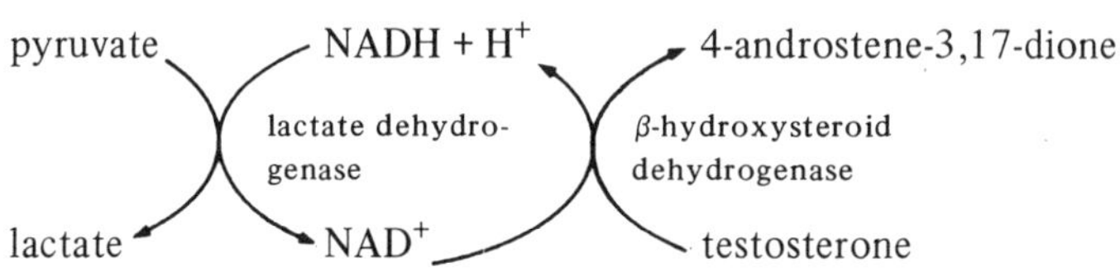

(4)

Regeneration of ATP from its "used" form AMP[31, 32]:

$$\text{AMP} + \text{ATP} \xrightleftharpoons{\text{adenylate kinase}} 2\ \text{ADP}$$

$$2\ \text{Acetate phosphate} + 2\ \text{ADP} \xrightleftharpoons[\text{Mg}^{++}]{\text{acetate kinase}} 2\ \text{ATP} + 2\ \text{acetate} \qquad (5)$$

Net reaction: $\text{AMP} + 2\ \text{acetate phosphate} \rightleftharpoons \text{ATP} + 2\ \text{acetate}$
(ATP is used in many biosynthetic processes as an energy donor)

Table 2. Classifications of cofactors (organic and inorganic) based on their need for immobilization in plausible industrial conversions

Group A (Does not need immobilization		*Group B* (Immobilization is helpful for its application)
Inorganic cofactors:		Free stoichiometric cofactors
Zn^{+2}, Mg^{+2}, Mn^{+2}, Fe^{+2}(Fe^{+3})		Ascorbic acid
Co^{+2}, Cu^{+2}(Cu^{+}), K^{+}, Na^{+}		* Biotin[43–45]
		* Coenzyme A[46]
Prostethic groups:		* Cyanocobalamin (vitamin B_{12})[47–49]
(bound stoichiometric organic cofactors)		* NAD^{+} (NADH) (Table 4)
1) In oxidoreductases	2) In transferases	* $NADP^{+}$ (NADPH)[50, 51]
FAD($FADH_2$)	** Pyridoxal phosphate (vitamin B_6)[37–39]	* Nucleoside mono-di- or triphosphates
** FMN($FMNH_2$)[33–35]		* ATP[22, 23, 52–55]
** Haem coenzymes[36]	** Pyridoxamine 5-phosphate[40, 41]	AMP[56–58]
		S-adenosyl methionine
	** Thiamine pyrophosphate (a derivative of vitamin B_1)[42]	* Folate[59]
		* Glutathione[60, 61]

** These cofactors were immobilized on carriers other than their own "native" carriers for their applications in affinity chromatography

* Immobilized forms have been reported

Immobilization of coenzymes are desirable in view of process enconomics. However, not all cofactors are needed to be immobilized. As classified in Table 2, inorganic cofactors are relatively inexpensive and there is no economical justification for their immobilizations. Prosthetic groups, such as flavin nucleotides, hematic organic cofactors, pyridoxal phosphate and thiamine pyridoxal phosphate and thiamine pyrophosphate are tightly bound to their protein moiety of the enzymes. They are immobilized if the enzymes containing them are. Immobilization is helpful for the applications of free stoichiometric organic cofactors such as those listed in Group B of Table 3. Again, the most important and most reported immobilized organic cofactors in this group are NAD^{+} (NADH), and a nucleoside triphosphate, ATP.

3 Immobilizations of Coenzymes

The purpose of immobilizations of coenzymes is twofold: (1) to extend immobilized enzymes technology from applications of immobilized hydrolases to that of immobilized oxidoreductases, ligases, and other groups of enzymes that also require coenzymes; (2) to use the immobilized coenzymes for enzyme purifications through affinity chromatography technique[62–64]. This section is devoted to the discussions of various methods of immobilization of NAD^{+}, an important co-enzyme which is required by most dehydrogenases for their catalytic activities.

Table 3. Methods of immobilization of NAD^+/NADH

Forms of coenzyme	Supports used	Methods of coupling	Position of attachment	Activities of the immobilized coenzyme	Ref.
1) NAD^+	Porous glass 40–60 mesh, 550 Å average pore diameter	Diazonium salt (azo linkage formation)	C_6 or C_8 on adenine moiety of NADH	About 10% of the activity of free NAD^+	65, 66
2) NAD^+	Alginic acid	a) 1-Cyclohexyl-3-(2-morpholino-ethyl)-carbodiimide metho-p-toluensulfonate as coupling agent	–	–	67
		b) 1,2,7,8-Diepoxyoctane	Through C-2 ribosyl ether linkage or C-6 amino group of the adenine moiety	–	68
3) Succinyl NAD^+	Polyethylene imine (M.W. 40,000–60,000)	Dicyclohexylcarbodiimide-coupling in aqueous pyridine	Possibly through 0-succinyl ribose ester moiety of succinyl NAD^+	K_M, ADH = 0.52×10^{-4} (at pH 6.5) T.N. = 0.458 s^{-1} (for ADH) (calculated by the authors) 20–45% active with YADH as compared to free NAD^+	69, 70
4) Succinyl NAD^+	Amino dextran (M.W. 500,000)	Carbodiimide-coupling	–	50% of the turn over rate of the free NAD^+ with YADH	71
5) (6)-Aminoethyl amino – NADH	Dextran T-40 (M.W. 40,000)	CNBr-coupling	C-6 of adenine moiety of NADH	Estimated maximum turn over number 5–15 s^{-1} (for three enzymes tested: LDH, YADH, and LADH)	72
6) 8-(6-aminohexyl)-Amino-adenine NAD^+	Sepharose 4 B	CNBr-coupling	C-8 of adenine moiety	100% of the turn over rate of the free NAD^+ with YADH	74
7) Nicotinamide-6-mercaptopurine	Bromoacetylated Sepharose 4 B	Spontaneous condensation of the thiol group of the mercaptopurine residue with the reactive bromoacetyl group of Sepharose 4 B derivative	C-6 mecapto-adenine moiety	–	75

8) Nicotinamide-8-(2-carboxyethylthio) adenine dinucleotide	(1) Polyethyleneimine (2) Polylysine (3) Aminohexyl-sepharose	Carbodiimide-coupling	C-8 position by an ω-carboxylic side chain	Efficiency for polyethyleneimine analogue: 47% of free NAD^+ with YADH	76
9) Nicotinamide-6-(2-hydroxy-3-carboxypropylamino) purine dinucleotide	(1) Polyethyleneimine (2) Polylysine (3) Aminohexyl-sepharose	Carbodiimide-coupling	N-6 position	Efficiency for polyethyleneimine analogue 2–7% of free NAD^+ with YADH	77
10) N^6 (N-(6-aminohexyl) Acetamide derivatives of NAD^+	(a) Sepharose 4B (b) Dextran	a) CNBr-coupling b) CNBr-coupling	a) N-6 position b) –	a) Rate of reduction with YADH in relation to free NAD^+: 0.7% b) Rate of reduction with YADH in relation to free NAD^+: 16%	78 79
11) N^6-(6-aminohexyl) Carbamoylmethyl) derivative of NAD^+	(a) Sepharose 4B (b) Dextran	CNBr-coupling CNBr-coupling	a) N-6 b) –		6
12) NAD^+	a) Sepharose 4B with ε-amino caproic acid as spacer	Dicyclohexyl carbodiimide	–	Rate of reduction with LADH in relation to free NAD^+: 0.2%	80
	b) Sepharose 4B with 6-amino-hexanoic acid as spacer	Dicyclohexyl carbodiimide	–	–	81
13) NAD^+	Cellulose	CNBr-coupling	–	–	82

Abbreviations: LADH: liver alcohol dehydrogenase
YADH: yeast alcohol dehydrogenase
LDH: lactate dehydrogenase

Apart from possible significant mass transfer resistances that are possibly developed in using immobilized enzymes and coenzymes as compared to using their free counterparts, the immobilization of enzymes and coenzymes, especially by covalent bond formation, may modify their specific activities from pure bond formation from pure catalytic reaction considerations. Some immobilized enzymes have been reported to have different K_m values, specific activities and even specificities from that of free enzymes. Although true intrinsic kinetic data are difficult to be obtained and are often disguised by mass transfer effect, specificity could be a good indication as to whether the intrinsic catalytic ability of the enzymes is modified through immobilization or not. The same argument can be applied to coenzymes. The immobilizations of coenzymes should be done in such a way that the modifications of the chemical structures of the coenzymes through immobilization does not change its reactivity appreciably.

The essential part of a NAD^+/NADH molecule for its coenzyme activity lies in the nicotinamide moiety. Based on the idea presented above, the immobilization of

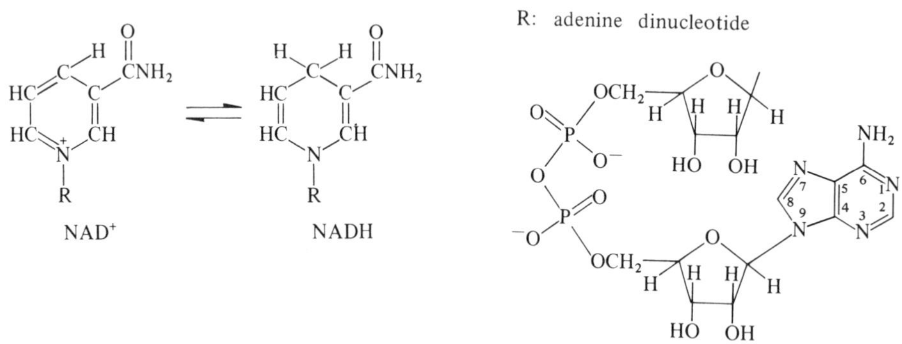

NAD^+/NADH can be achieved by linking the molecule to a support through the functionalities in the R group. The possible positions in the R group are ribosyl hydroxyl, C-6 amino group or its derivatives (different derivatives are used as spacers between the coenzyme and the support), and number 8 carbon of the adenine moiety. Table 3 illustrates the various methods of immobilization of NAD^+/NADH onto either soluble or insoluble supports. These methods all involve chemical coupling agents. Derivatization of the coenzymes involves creating a functional group and inserting a spacer. The purpose of including a spacer is to increase the availability of the coenzyme to the catalytic surface of the enzyme. The derivatized coenzymes are then coupled to a support. A typical procedure is illustrated in Fig. 1. The procedure of NAD^+ immobilization involves the following steps: 1) alkylation at N-1 position of the adenine moiety with iodoacetic acid and subsequent alkaline rearrangement to obtain N-6-carboxymethyl NAD^+, 2) the spacer is added by carbodiimide coupling of diaminohexane to the carboxyl group, and 3) the amino group on the spacer is then coupled to sepharose by cyanogen bromide activation.

The idea of using spacers is a good one in terms of reducing steric hindrance. However, the prerequisite of using spacers to derivatize NAD^+ is that the derivatives

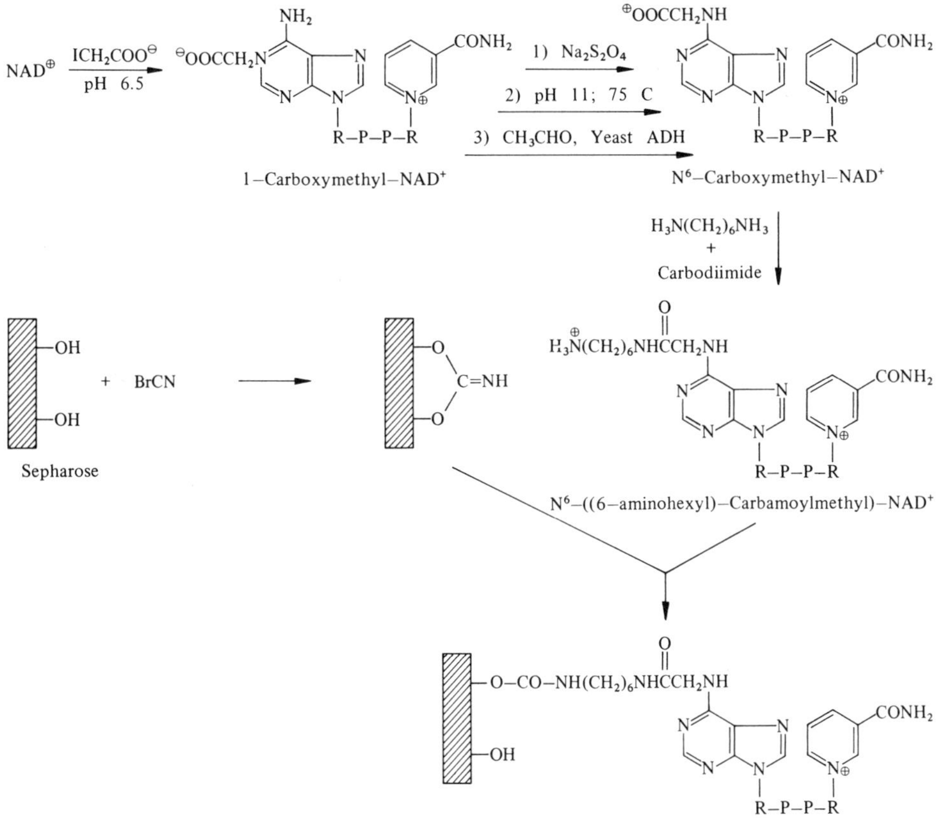

Fig. 1. A typical procedure for the immobilization of a NAD^+ analogue. ADH, alcohol dehydrogenase; R, ribose; P, phosphate[5]

obtained should retain the effectiveness of the coenzymes. Table 4 shows several derivatives of NAD^+ to have comparable K_m (Michaelis-Menten constant) and Vm (maximum catalytic activity) values for alcohol dehydrogenase.

As indicated in Table 3, cyanogen bromide and carbodiimide are the popular coupling agents for immobilization. Lee et al.[74] immobilized 8-(6-aminohexyl)-amino-adenine nucleotide to sepharose by cyanogen bromide coupling and found the turn over rate of immobilized coenzymes in YADH system to be as good as that of free NAD^+. The turn over rate of free NAD^+ in YADH system is calculated to be 28.15 s^{-1} (based on data in reference[83]). Weibel et al.[72] immobilized N^6-aminoethyl NADH on dextran T-40 by cyanogen bromide coupling and estimated the maximum turn over rates to be 5–15 s^{-1} for the three enzymes tested. So, the "specific activities" of the immobilized NAD^+ is about 18–54% of the free NAD^+. As shown in Table 3, the reported activities of the immobilized NAD^+ range from 20 to 100% of the free coenzyme. It is reasonable to assume that if insoluble supports are used for immobilization of the coenzymes,

Table 4. Effect of modification of NAD^+ on its coenzyme activity (system studied = ethanol oxidation catalyzed by alcohol dehydrogenase, pH 6.5)

Substrate	K_m (10^{-4}) M	V_m (μmole min^{-1} μg^{-1})
1) NAD^+	0.92	0.037
2) NAD^+ + PEI[a]	2.66	0.132
3) Succinyl-NAD^+	0.58	0.025
4) PEI-succinyl-NAD^+	0.53	0.036
5) NAD^+	5.5	1.00[b]
6) 8-Br-NAD^+	1.7	1.00[c]

1–4 from Wykes et al.[69)]
5 and 6 from Lee and Kaplan[74)]

[a] PEI: polyethyleneimine
[b] Molecular weight used in calculation is that of succinyl-NAD^+
[c] Relative activities

because of mass transfer resistances the measured turn over rates are expected to be smaller than that of the free counterparts. Based on this argument, coupling coenzyme derivatives to soluble supports such as dextran T-40[6, 72)], polyethyleneimine or polylysine[69, 70, 76, 77)] would give immobilized coenzymes with better turn over rates than that immobilized on insoluble supports such as glass beads and sepharose.

4 Regenerations of Coenzymes

From an industrial conversion point of view, for enzymatic catalysis requiring coenzymes to be technologically and economically feasible, the reactor systems used must include a convenient coenzyme regenerating scheme. For coenzymes that are self-regenerating (see group A, Table 1), such a scheme is built into the reaction system. So, one can look at these coenzymes, namely, biotin, cyanocobalamin, pyridoxal phosphate, and thiamine pyrophosphate, as true catalysts because they participated in the reaction and are totally regenerated after the catalytic reaction is completed. On the other hand, coenzymes classified in group B and C in Table 1 need some kind of coenzyme regenerating scheme.

If oxidized forms of flavin, nicotinamide, and lipoic coenzymes are used in the reactor system their regeneration can be effected by catalytic, enzymatic or non-enzymatic, oxidation with molecular oxygen as the elctron acceptor. However, if the reduced form of flavin and nicotinamide coenzymes are used, their regenerations require an appropriate substrate that can be catalyzed by another enzyme to serve as an electron and a proton donor. In such cases, two enzymes, two substrates, two products, and one coenzyme (both oxidized and reduced forms) can be found in the reactor system.

Other than oxidoreductases, which require electron carriers; (i.e. flavin coenzymes and nicotiamide coenzymes), ligases (or synthetases) require high energy donors, i.e. ATP. The regeneration of ATP from AMP can be achieved by using two enzymes as catalyst and acetate phosphate as the phosphate donor [see Reaction (5) in Sect. 2]. As illustrated by Reaction (5), the regeneration system requires two enzymes and an additional substrate. This represents one of the most complex systems that was devised to mimic *in vivo* biological reactions. In the field of biosynthesis it is difficult to compete with a biological machine (the *in vivo* system) which enjoys rather high efficiency in energy conservation. ATP regeneration with possible connection to *in vitro* gramicidin S synthesis represents a scheme that has been attempted in terms of possible use for industrial *in vitro* biosynthesis[84)].

Among the coenzymes listed (Table 1) efforts have been directed toward the regeneration of electron carrier coenzymes, especially NAD^+/NADH. Nicotinamide adenine dinucleotide is required as a coenzyme in many reactions catalyzed by various dehydrogenases. The regenerations of NAD^+/NADH can be done by one of the following methods.

1. Enzymatic Methods
1.1. Coupled enzymes method
1.1.1. Free coenzymes coupled with free enzymes
1.1.2. Free coenzymes coupled with immobilized enzymes
1.1.3. Immobilized coenzymes coupled with free enzymes
1.1.4. Immobilized coenzymes coupled with immobilized enzymes
1.2. Coupled substrate method
2. Nonenzymatic methods
2.1. Chemical regenerations
2.2. Electro-chemical regenerations
2.2.1. Electro-chemical regeneration of native NADH from NAD^+
2.2.2. Electrochemical regeneration of immobilized NADH from immobilized NAD^+.

These different methods are tabulated with reactions involved and references in Tables 5-8. However, quantitative comparisons of different regeneration schemes, such as by recycling rates, are difficult to obtain. Some qualitative conclusions can be drawn. Regenerations of coenzymes by coupled immobilized enzymes systems, are more efficient if coenzyme is in the free form or immobilized on soluble supports rather than immobilized on insoluble supports. This is apparent from severe diffusional resistances, or steric hindrance involved in the latter. The turn over rates of the coupled enzymes used should be compatible to each other for maximal performance by such a system. For the coupled substrates systems, the most efficient model would co-immobilize the enzyme and the coenzyme on the same support. Gestrelius et al.[95)] co-immobilized horse liver alcohol dehydrogenase and a NADH analogue, N^6-[(6-aminohexyl)carbamoyl methyl]-NADH on sepharose 4B and found a maximum recycling rate of 3400 cycles per hour for the immobilized NADH. The regeneration rates in this system is affected by the concentration driven redox reaction. The co-substrate used usually needs to be in high concentration. The co-substrate used is preferably a good substrate for the same enzyme, yet in the opposite direction to the accumulation of the desired product.

Table 5. Regenerations of NAD/NADH coenzyme in free coenzymes/free enzymes systems

Enzyme couples or triples	Substrates	Products	Ref.
Glucose dehydrogenase Yeast alcohol dehydrogenase	D-Glucose Acetaldehyde	D-Gluconolactone Ethyl alcohol	85
Lactate dehydrogenase Horse liver alcohol dehydrogenase	Pyruvic acid (deuterioglycollate) Ethylalcohol	Lactic acid Acetaldehyde	86
17-β-Hydroxysteroid dehydrogenase Lactate dehydrogenase	Testosterone Sodium pyruvic acid	4-Androstene-3,17-dione Sodium lactic acid	30
Aldolase, glyceraldehyde-3-dehydrogenase Horse liver alcohol dehydrogenase	Fructose-1,6-diphosphate Phosphate acetaldehyde	Phosphoenol pyruvic acid Glyceral 1,3-diphosphate Ethyl alcohol	87
Sorbitol dehydrogenase, diaphorase (with a redox agent)	Sorbitol O_2	Fructose H_2O	88

Table 6. Regenerations of NAD/NADH coenzymes in immobilized enzymes and/or immobilized coenzymes systems

Enzyme couples	Coenzyme form	Substrates	Products	Ref.
Immobilized sorbitol dehydrogenase	NAD^+/NADH	Sorbitol	Fructose	
Immobilized alcohol dehydrogenase		Acetaldehyde	Ethyl alcohol	89
Galactose dehydrogenase	Dextran-NAD^+/ dextran-NADH	Galactose	Galactonolactone	
Alanine dehydrogenase		Pyruvic acid NH_4^+	Alanine	90
Lactate dehydrogenase	Dextran NAD^+/ dextran NADH	Lactic acid	Pyruvic acid	
Alanine dehydrogenase		Pyruvic acid NH_4^+	Alanine	90
Alcohol dehydrogenase	Formylpolyethyleneimine-succinyl-NAD^+/ NADH	Ethyl alcohol	Acetaldehyde	
Lactate dehydrogenase		Pyruvic acid	Lactic acid	70
Alcohol dehydrogenase	Polyethyleneimine-NAD^+/NADH	Ethyl alcohol	Acetaldehyde	
Malic dehydrogenase		Oxaloacetic acid	Malic acid	91

Table 6 (continued)

Enzyme couples	Coenzyme form	Substrates	Products	Ref.
Alcohol dehydrogenase Aldehydrogenase	NAD^+/NADH on glass beads	Ethyl alcohol	Acetate	
20-β-Dehydrogenase		C-20 β-Ketogroup of a steroid	C-20 β-Alcohol	92

Table 7. Regenerations of coenzyme NAD^+/NADH with coupled substrate method

Enzyme	Coupled substrates	Products	Ref.
Horse liver alcohol	Trans-2-decalone	Trans-2-decalol	93
Dehydrogenase	2-Cyclohexanol (two-fold excess)	2-Cyclohexamone	
Horse liver alcohol	3-Methyl-cyclohexamone	3-Methyl cyclohexamol	94
Dehydrogenase	Ethanol (seven-fold excess)	Acetaldehyde	
Horse liver alcohol	Ethyl alcohol	Acetaldehyde	95[a]
Dehydrogenase (HLADH)	Lactaldehyde	Propanediol	96

[a] HLADH-NAD(H)-sepharose complex (enzyme-coenzyme coimmobilized on sepharose) was used instead of free coenzymes

Table 8. Regenerations of coenzymes – chemical methods

Enzymes	Redox couples	Reaction	Ref.
Horse liver alcohol dehydrogenase	$Na_2S_2O_3$, NAD^+ → $NaHSO_3$, NADH	Aldehyde or ketone reduction	97
Horse liver alcohol dehydrogenase	NADH, FMN, H_2O_2 → NAD^+, $FMNH_2$, O_2	Aldehyde or ketone reduction	98
Glucose-dehydrogenase	NADPH, Methylene blue (oxidized), H_2O_2 → $NADP^+$, Methylene blue (reduced), O_2	D-Glucose to Gluconolactone	99
Lactate dehydrogenase	NADH, Sepharose bound actiflavin (oxidized), H_2O_2 —hν→ NAD^+, Actiflavin (reduced), O_2	Lactate to Pyruvate	100

Note: Oxygen is the final electron acceptor in the redox couples. The hydrogen peroxide formed decomposes to water and oxygen slowly, and the reaction can be catalyzed by catalase

Theoretically, for oxidoreduction reactions, the coupled substrate approach is applicable to the production of either the oxidized or reduced product. However, for $C=O \rightleftharpoons CHOH$, equilibrium is heavily in favor of the formation of alcohol. For example, K_{eq} (25 C), equilibrium constant, of 9.72×10^{-12} and standard free energy change of +5.40 kcal $mole^{-1}$ were found for the following reaction catalyzed by alcohol dehydrogenase. $CH_3CH_2OH + NAD^+ \rightleftharpoons CH_3CHO + NADH + H^+$. (6) Consequently, the technique of using coupled substrates for the regeneration of coenzymes is restricted to the reduction of ketones and aldehydes. In summary, according to the following scheme [Reactions (7) and (8)],

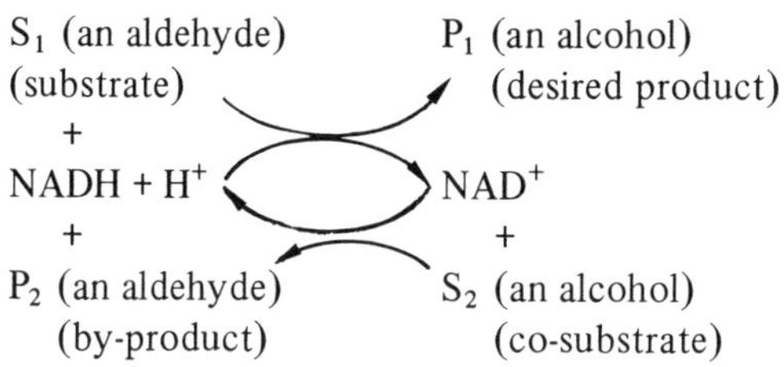

(7)

(8)

if P_1 is desired, high concentration of S_2, the co-substrate, is needed to counteract the unfavorable up-hill drive against equilibrium. Similarly, if a carbonyl product (either an aldehyde or a ketone) is desired, a very high concentration of the co-carbonyl substrate will have to be added to allow any significant accumulation of the product. Furthermore, the removal of the by-product P_2 [Reaction (8)] will promote the rate of reaction in this system. For example, if the by-product is acetaldehyde, it is rather volatile and it can be driven from the system by bubbling nitrogen through it. Volatilized acetaldehyde can then be trapped in a vessel containing semicarbazide to form acetaldehyde semicarbazone.

Chemical methods of regenerating coenzymes have received less attention than the enzyme-dependent processes. As shown on Table 8, sodium dithionite, flavin mononucleotide and its derivatives were studied by several researchers. Derivatives of pyridinium and dihydropyridine[102], phenazine methosulfate and 2,6-dichlorophenolindophenol[103] are also suitable for recycling of NAD^+/NADH. The advantages of using chemical catalysts are: (1) its stability, and (2) its low cost as compared to using enzymes. However, chemical catalysts are usually smaller in molecular weights as compared to that of enzymes. The separation of chemical catalysts from the reaction system is then a difficult job. One possible way of overcoming this difficulty is to polymerize the catalyst. As shown on Table 4, polymerized FMN was used by Chambers et al. and found to have a slightly higher rate in oxidizing NADH than when free FMN was used. This result should, however, be taken as a unique one instead of the general rule. The disadvantage of the chemical method as summarized by Jones et al.[8] is its low efficiency as compared to that for the best coupled-substrate or coupled enzyme systems. As shown in Table 9 the turnover rates of the chemical regeneration process range from 0.43 to 8.5 min^{-1}, or 0.007–0.14 s^{-1}. The enzyme catalyzed coenzyme turn over rates are higher (in the range of 0.5–1 s^{-1}). These are summarized in Table 10.

Table 9. Rates of catalytic oxidation of NADH[a] with O_2 as the ultimate electron acceptor. (After Chamber et al.[71)])

Catalysts	Rates (μM min^{-1})	Specific rates (min^{-1})
Diaphorase, 2.9 units/l[a]	1.7	0.43
Phenanzine methsulfate, 10 μM	84.8	8.5
Methylene blue 10 μM + light	48.3	4.8[b]
Flavomononucleotide 10 μM + light	36.7	3.7
Flavomononucleotide polymer 10 μM + light[c]	48.7	4.9

[a] Rates studies in 0.1 M phosphate buffer, pH 7.4 with 100 μM NADH
[b] 4.8 mg l^{-1}, equivalent to 10 μM FMN (flavinmononucleotide)
[c] Based on FMN

However, since less effort has been made to study the chemical method, its potential should not be overlooked.

Electrochemical methods have also been studied for the purpose of NAD^+/NADH regeneration[104–112)]. In general, the direct electrochemical reduction of NAD^+ could inactivate the coenzyme, and so it is not as good as the indirect electrochemical procedure[113)]. Electrochemical method of coenzyme regeneration can probably be used in analytical field as some immobilized enzyme electrodes have been. However, for possible industrial applications, based on coenzyme recycling rates, the coupled enzymes and coupled substrates regeneration methods are believed to be better than the electrochemical ones.

Table 10. Coenzyme turnover rates in enzymatic regeneration systems

Enzyme(s) (coenzymes)	Substrates	Products	Coenzyme turnover	Ref.
Alcohol dehydrogenase	Acetaldehyde	Ethylalcohol	0.25 s^{-1} [a]	89
Sorbitol dehydrogenase (co-immobilized on zerconium coated porous glass)	Sorbitol	Fructose		
Alcohol dehydrogenase and NAD co-immobilized on sepharose 4B	Ethylalcohol	Acetaldehyde	0.94[a] s^{-1}	95
	Lactaldehyde	1,2-Propanediol		

[a] Calculated by authors

5 Process Consideration, Reactor Design and Economics

5.1 General Considerations and Economics

Successful operation of enzymatic processes that require coenzymes has yet to be developed. In spite of the economic advantages introduced by using immobilized enzymes, the cost of the said processes may still be unacceptably high due to the coenzyme requirement. Therefore, one ought to ask the following questions in economic evaluations.

(a) How much will the cofactor cost contribute to the overall product cost
(b) How can the cost be reduced, and
(c) What are the alternatives, or competing processes?

For 90% pure β-NAD^+, the common NAD^+ for most dehydrogenases, the current price (Sigma price list, April, 1977) is about 1600 dollars per pound. For an oxidation reduction reaction (i.e. dehydrogenation) that requires stoichiometric NAD^+ as the coenzyme, the cost of NAD^+ per pound of product would be \$ 1600 x (molecular weight ratio of NAD^+ to that of product). For example, in the conversion of cortisone the calculated cost would be \$ 2932 per pound of cortisone produced. If the NAD^+ that is reduced to NADH in the reaction can be regenerated as discussed in Sect. 4, then the cost of NAD^+ can be reduced. The reduced cost would be 2930/N dollars, where N is the number of recycling of NAD^+. If the cost allowed for NAD^+ in the production of prednisolone is one dollar per pound of product, the N, the number of recycling of NAD^+, must be at least 2930. In other words, if N, the recycling rate is close to 2930, then a catalytic, instead of a stoichiometric, amount of NAD^+ is needed for the conversion.

Cremonesi et al.[114] using coupled enzymes system [Reactions (9) and (10)] was able to convert cortisone to 20-dihydrocortisone

$$\text{cortisone} + \text{NADH} + H^+ \underset{}{\overset{20\,\beta\text{-HSDH}}{\rightleftharpoons}} \text{20-dihydrocortisone} + NAD^+ \tag{9}$$

$$C_2H_5OH + NAD^+ \overset{\text{ADH}}{\rightleftharpoons} CH_3CHO + \text{NADH} + H^+ \tag{10}$$

where, 20 β-HSDH = 20 β-hydroxysteroid dehydrogenase ADH = alcohol dehydrogenase.

The reaction mixture contained 2.77 mmol of cortisone, 0.125 mmol of NAD^+, 15 ml of ethanol, (256 mmol) 2.98 mmol of semicarbazide, the enzymes and serum albumin in 500 ml butyl acetate and 500 ml of 0.05 M phosphate buffer, pH 7. They reported a rate of 100% conversion in four hours. The amount (moles) of NAD^+ used was 4.5% of cortisone (substrate 1) and 0.005% of ethanol (substrate 2). So NAD^+ is used in catalytic quantity, and its recycling rate is calculated to be 5.54 h^{-1} or a total number of 22 recycles of NAD^+ in four hours. This is about 1/8 of the cycling rate reported for ADH + LDH coupled system (see Table 11). The differences are probably due to the low solubility of steroids in water. Butyl acetate was used to promote the coupled reaction in the two phases-systems. Semicarbazide was used to form a carbazone

Table 11. Comparisons of rates of reaction of NAD^+ and its derivatives as coenzymes to least alcohol dehydrogenase (ADH) and/or muscle lactate dehydrogenase (LDH) (coupled or separated)

Cofactor	Reduction rate with ADH		Oxidation rate with LDH		Cycling rate with ADH + LDH			Ref.
	$\mu M\ min^{-1}$	%	$\mu M\ min^{-1}$	%	$\mu M\ min^{-1}$	Cycles h	%	
NAD(H)	28	100	16	100	69	41	100	a
NAD(H) + blank dextran	29	104	15	95	–	–		a
N^6-R-NAD(H)	17	61	12	76	42	25	61	a
N^6-R-NAD(H) + blank dextran	16	57	11	70	–	–	–	a
Dextran-NAD(H) (T-40)	4.3	16	2.3	14	12	7.4	18	a
Sepharose-NAD(H) (4B)	0.2	0.7	0.02	0.1	0.3	0.2	0.5	a
N^6 R′-NAD(H)	–	55[d]	–	65[e]				b
N^6 R″-NAD(H)	–	61	–	50[e]				b
Poly-EI-NAD/1	–	3	–	24				c
Poly-EI-NAD/2	–	100	–	27				c
Poly-EI-NAD/4	–	–	–	40				c
Poly-EI-NAD/5	–	100	–	44				c
Poly-lys NAD	–	27	–	25				c

[a] Larsson and Mosbach[79)]
[b] Mosbach et al.[6)]
[c] Marconi et al.[115)]
[d] For liver alcohol dehydrogenase, rather than yeast ADH
[e] For beef heart lactate dehydrogenase, rather than muscle LDH

R: N-(6-aminohexyl)-acetamide)
R′: carboxymethyl.
R″: N-(6-aminohexyl) carbamoylmethyl
Poly-EI: polyethyleneimine
Poly-lys: polylysine

with acetaldehyde to promote the oxidation of ethanol which is not thermodynamically favored according to the equilibrium of the reaction. The cost of NAD^+ in this process is calculated to be 133 dollars (2930/22) per pound of 20-dihydrocortisone produced.

5.2 Alternatives - Whole Cells?

The alternatives to the above-described processes are: (1) using immobilized enzymes and immobilized coenzymes to make the reuse of them easy and so further reduce the cost, and (2) the use of microbial processes and/or whole cells systems where regeneration of coenzymes are naturally found in the intra-cellular environment. Such processes are beyond the scope of this article. It is sufficent to say that the cultivation of cells and induction of enzymes are only part of the economical consideration. Since a complex medium is used for such type of experiments, the separation of the product from the final reaction mixture is usually quite expensive. The use of whole cells for enzymatic conversions requiring coenzymes can be considered to be an extension of the traditional biological process to include a secondary process which is of biochemical nature, since in this latter process cells are not expected to reproduce. However, since regenerations of coenzymes are needed in the process, the immobilized whole cells are expected to still retain the ability of maintaining a substantial degree of redox potential in terms of NADH and NADPH productions. How does one produce this type of cells? Can one add chemicals such as azaserine in the cultivation media during the process to produce cells which cannot reproduce well, and because of their leaky membrane can one also have higher rate of transport of substrates and products? There are many questions to be answered if one wants to use whole cells.

The rate of hydrocortisone production by 11 β-hydroxylation of Reichstein's Compound S (11 β-deoxy-17-hydrocorticosterone) using polyacrylamide gel entrapped *Curvularia lunata* was found to be slow[116] (5 g wet gel in 20 ml solution produce 2.0–2.9 mmoles of cortisone per hour). The slow rate could be due to mass transfer resistance and/or limitation of coenzyme regeneration by the intracellular enzymes. However, it is encouraging to know that immobilized whole cells have been used to produce other coenzymes. Shimizu et al.[26] reported the use of patothenic acid, cysteine and ATP as substrates to produce coenzyme A repeatedly (four times reuse; relative activities are 100, 183, 57, and 21 for the four times). They used polyacrylamide gel-entrapped *Brevibacterium ammoniagenes* IFO 1207. Recently, at the 4th Enzyme Engineering Conference, Samejima et al.[24] reported (1) synthesis of ATP from AMP by immobilized yeast cells, and (2) synthesis of coenzyme A from pantothenate, cysteine and ATP by co-immobilized *Brevibacterium ammoniagenes,* and yeast cells. They were making use of the glycolytic activity of yeast cells for ATP regeneration in such a system.

After the idea of immobilized whole cells was found to have some advantages over immobilized enzymes in some cases[117, 118] immobilized whole cells have been used just as a source for the immobilized enzyme in systems where the coenzyme still has to be added as a separate entity. Many oxidoreductases are membrane bound enzymes. The stability of these enzymes may be higher in the original organelle environment than in the isolated enzyme state. So in the use of whole cells, cell disruption, enzyme isolation, and purification and handling of a rather unstable enzyme are nonexisting problems. As reported by Larsson and Mosbach[119], 1 g of *A simplex* cells when entrapped in gel showed a Δ^1-dehydrogenase activity of 100 μmoles per hour, whereas the isolated enzyme from the same amount of cells and subsequently entrapped yielded about one-tenth the activity, namely 10 μmoles/h. Whole cells of *Corynebacterium simples* were also immobilized in a collagen matrix[120]. The use of whole cells as enzyme and coenzyme source may have such advantages as high specific activity and stability of enzyme, nevertheless one should also be aware of the possible enhancement of mass transfer resistances due to the whole cell structure.

In the following, attentions are directed toward the regenerations of coenzyme NAD^+/NADH not by using whole cells, but by simple enzymatic coupled enzymes or coupled substrates method. Since these methods are generally accepted to be more efficient than that of chemical, electrochemical, or other methods, we will center our discussions on the process and reactor design using the said preferred regeneration methods.

5.3 Coupled Enzymes and Coupled Substrates

In possible continuous enzymatic processes requiring NADH regeneration by coupled enzymes or coupled substrates method (see Sect. 4) there are two things which deserve general considerations. These are, (1) retention of coenzymes and (2) continuous removal of products to promote the coupled reaction rates which usually have equilibrium constants around unity or not much away from it. This is illustrated in the following two examples,

$$\text{pyruvate} + \text{NADH} + \text{H}^+ \xrightleftharpoons{K_{eq} = 0.34 \times 10^{12}\ (M^{-1})} \text{lactate} + \text{NAD+},$$

$$\text{ethanol} + \text{NAD}^+ \xrightleftharpoons{K_{eq} = 9.72 \times 10^{-12}\ (M)} \text{acetaldehyde} + \text{NADH} + \text{H}^+ .$$

So the overall equilibrium constant for the coupled reaction is $(0.34 \times 10^{12})(9.72 \times 10^{-12}) = 3.30$[121, 122], or

$$\text{pyruvate} + \text{ethanol} \xrightleftharpoons[(pH = 7.2)]{K_{eq} = 3.30} \text{lactate} + \text{acetaldehyde}, \qquad (11)$$

$$\text{sorbitol} + \text{NAD}^+ \xrightleftharpoons{K_{eq} = 0.11} \text{fructose} + \text{NADH},$$

$$\text{acetaldehyde} + \text{NADH} \xrightleftharpoons{K_{eq} = 1250} \text{ethanol} + \text{NAD}^+ .$$

So the overall equilibrium constant for the coupled reaction is (1250) (0.11) = 137.5[89], or

$$\text{sorbitol} + \text{acetaldehyde} \xrightleftharpoons[(pH = 8,\ 25\ ^\circ C)]{K_{eq} = 137.5} \text{fructose} + \text{ethanol} . \qquad (12)$$

As shown in Reaction (11) and (12), with proper selection of regeneration system (preferably with large K_{eq} and a high negative free energy change) thermodynamically unfavorable reactions can be coupled to form products through "up hill" driving by maintaining high (active/spent)-cofactor ratios.

In simple enzymatic, coupled enzymes-(or coupled substrates) processes of NADH regeneration, NADH or NAD^+ is supplied in catalytic quantity instead of stoichiometric quantity. Regeneration enables the reuse of the coenzyme. The practical problem involved in reusing the coenzyme is to have an efficient and selective method of separating it from the substrate and product stream. The molecular weight of NAD^+/NADH is 663/664, the substrates and products involved in reactions catalyzed by dehydrogenases are usually smaller than this. Based on this difference in molecular weights hollow fibers with proper molecular weight cut-off) can be used to retain NAD^+ but not substrate(s) and product(s). Chambers et al.[71] reported the use of this system for the oxidation of ethanol to acetate as catalyzed sequentially by alcohol dehydrogenase (ADH) and acetaldehyde dehydrogenase (ALDDH). Diaphorase (DIA) was used to regenerate NAD^+, and catalase was used to scavenge hydrogen peroxide produced. They reported an optimum enzyme ratio of ADH to ALDDH to DIA as 2:1:2.4 units for such a system using maximum rate of ethanol conversion as the criterion.

Molecular weight cut off of an ultrafiltration membrane is an average number[123]. Clear cut separation of molecules with close molecular weights are not easily obtained. In 1974, Rony[124] disclosed a patent application on the preparation and use of selective hollow fibers. The fibers were impregnated with suitable liquids, usually hydrophobic ones to make "tubular liquid membranes". Rony cited a reaction system involving a three-enzyme system containing alcohol dehydrogenase, diaphorase and catalase, plus NAD^+, and a substrate, a low molecular weight alcohol. This system, housed in a hollow fiber reactor, was used to produce aldehyde. The wall of the hollow fibers was impregnated with a fluorocarbon liquid membrane. The selectivity of the membrane let the product, an aldehyde, pass through but retained the rest of the materials in the reaction including the coenzyme and the substrate. If the liquid-membrane-hollow-fiber reactor is used, one can design a simple system as depicted in Fig. 2 to continuously produce an oxidation-reduction product such as an aldehyde, a ketone, an alcohol or an acid by enzymatic methods. The economical feasibility of such a process has yet to be studied.

One of the large expenses in the operation of a hollow fiber type reactor is the power requirement. If, for example, semicarbazone formation is used to remove the aldehyde formed, the only other requirement is to retain the coenzyme NAD^+/NADH which enables a continuous enzymatic oxidation of an alcohol. In such a case, in order to reduce the power expenditure in the ultrafiltration process, one can immobilize NAD^+/NADH on Dextran T-40, alginic acid, polyethylene imine, polylysine or other soluble high molecular weight carriers, but not high molecular weight insoluble carriers such as Sephadex which would cause plugging in the hollow fibers.

NAD^+, and many other coenzymes are charged compounds, an alternative for their separations is the use of ion selective membranes, because substrates and products are usually uncharged, poor or non electrolytes. Although this type of technique is rather common in separations of inorganic ions, its application to separations of coenzymes has not been reported. Gardner[126] described the use of an electric potential gradient across a composite membrane to remove acetate selectively from enzymatic ATP regeneration system [see Reaction (5), Sect. 2] by a combination of charge and size separation processes.

A potential problem in using a hollow fiber reactor is the inactivation of enzymes due to the shear generated in the circulation of enzyme solutions through small tubes[127]. Most immobilized enzymes show better storage and thermal stability than

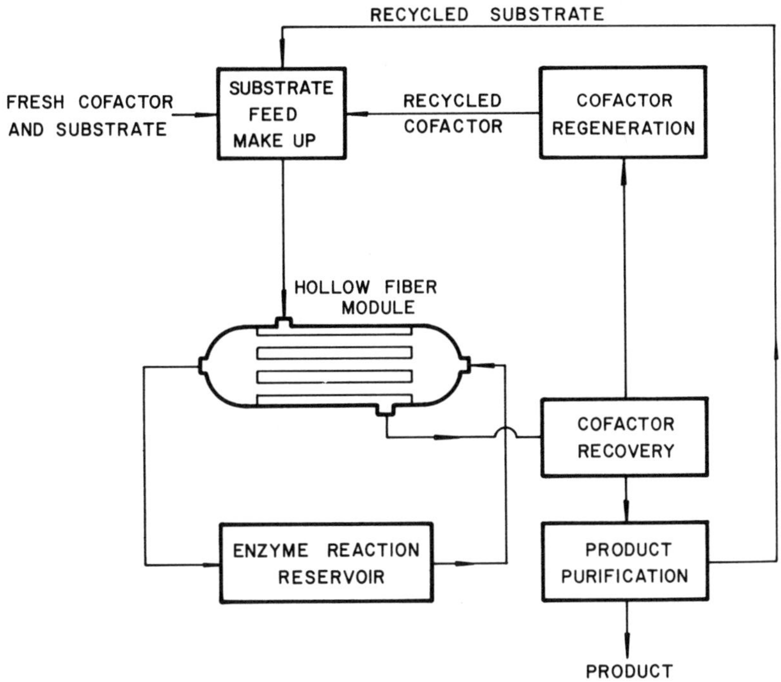

Fig. 2. Hollow fiber – Enzyme reactor. The hollow fiber can be coated with liquid membrane to achieve selective separation of products from the reaction stream. (After Fink and Rodwell[125])

their free enzyme counterpart. If one can extrapolate such a phenomenon to include the stability toward shear rate, then it is desirable to use immobilized enzymes in the hollow fiber system, provided that enzymes are immobilized on soluble macromolecular carriers.

Gestrelius et al.[95] co-immobilized NAD^+ and alcohol dehydrogenase on Sephadex 4B and found that the turnover rate of NAD^+ was quite high (0.94 s^{-1} for ethanol and lactaldehyde coupled system, Table 10). They reported that a molar ratio of one to one for the amount of bound nucleotide and bound enzyme was sufficient to obtain high activities in the absence of free NAD^+. Furthermore, immobilized ADH was found to have higher thermal and storage stability as compared to free ADH.

Table 12 shows the K_m values of different substrates for ADH. Since K_m for ethanol (0.35 mM) is higher than that for acetaldehyde (0.18 mM), kinetically speaking acetaldehyde is a better substrate for ADH than ethanol. However, addition of NAD^+ to the system, and removal of acetaldehyde will drive the reaction in the direction of oxidation of ethanol. Since lactaldehyde has a much lower K_m (1.1 mM) than that of propandiol (50.0 mM), from kinetic point of view the reduction of lactaldehyde is preferred, and NAD^+ is regenerated in the meantime. On the other hand, glyceraldehyde to glycerol can also be catalyzed by ADH but is not suitable to be used here. In such a system, retention of coenzyme is automatic and coenzyme loss is nonexistent. The key is the se-

Table 12. Kinetic parameters of different alcohols and aldehydes for horse liver alcohol dehydrogenase catalyzed reactions[128)]

Substrate	K_m (mM)	K_m -RCH_2OH / K_m -RCHO
Ethanol	0.35	1.9
Acetaldehyde	0.18	
Glycerol	13.5	0.66
Glyceraldehyde	20.4	
Propandiol	54.0	49
Lactaldehyde	1.1	

lection of a suitable co-substrate. If a good selection of the co-substrate is made, this system is most desirable in terms of coenzyme retention and regeneration.

If single immobilized coenzyme requiring enzyme is used in a packed bed reactor, or a stirred tank reactor, the coenzyme used in such a system needs to be separated, regenerated and concentrated before it can be recycled. In such a case half of the initial cofactor used in the system will be lost in the process of reuse after only 69 cycles if the combined efficiency of coenzyme separation, regeneration, and concentration is 99% [$(0.99)^{69} = 0.5$], or after only 14 cycles if the efficiency is 95% [$(0.95)^{14} = 0.49$]. With coupled enzymes and coupled substrates systems such problems can be partially avoided. Because in this case only catalytic quantity, instead of stoichiometric quantity, of coenzyme is needed, and regeneration is internal. Chambers et al.[89)] reported the performance of such a system. Using a small column reactor (Fig. 3) containing sorbitol dehydrogenase and alcohol dehydrogenase co-immobilized on amino-hexyl sepharose they were able to operate continuously at a steady state conversion of 99%. This steady state was achieved at a space time of 210 min and a coenzyme turnover rate of 10,000/210 min, or 0.8 s^{-1}. This turnover rate is higher than that reported by the same authors using a hollow fiber reactor[71)] (see Table 10). The turnover of cofactor in the column can be increased to 1,000,000 if 99% of the coenzyme NAD^+ in the product stream can be separated and returned to the feed stream. Gardner et al.[31)] have reported a 99% recovery of another coenzyme, ATP, by combined active carbon adsorption and ethanol elution. Fink and Rodwell[125)] also reported quantitative removal of NAD^+ from product stream by anion exchange resin.

Due to the similarity in chemical structures of the oxidized and the reduced forms of coenzymes such as NAD^+/NADH, they are known to be potent competitive inhibitors for each other in the oxidation reduction reactions. For instance, the K_i for NADH in the oxidation of ethanol by yeast alcohol dehydrogenase (YADH) equals to 1.8×10^{-5} M at pH 7.2, whereas the K_m for NAD^+–NADH is 7.4×10^{-5} M at pH 7.2[83)]. With an internal regeneration system such as that shown on Fig. 3, the build-up of the spent coenzymes is minimized, and thus allows maximal conversion throughout the course of the process.

Another form of reactor that can be used to regenerate and retain coenzymes is that of microencapsulated coupled enzymes systems. Campbell and Chang[91)] prepared semi-

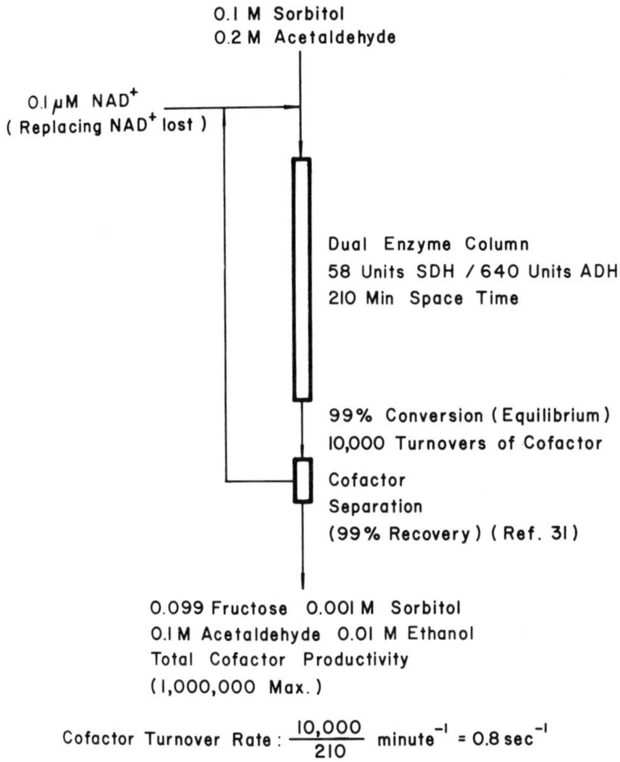

Fig. 3. Schematic diagram of a high cofactor turn over system. (Modified from Chambers et al.[89])

permeable aqueous collodion microcapsules containing alcohol dehydrogenase, malic dehydrogenase, and a soluble derivatized NAD^+, polyethyleneimine-NAD^+. Ethanol and oxaloacetate were used as substrates. The products are acetaldehyde and malate. This partially self-sufficient system functioned well but not as efficient as when free NAD^+ was used. This is apparently due to diffusional limitations within the capsules. These microcapsules can be used in packed bed reactors with suitable filler as spacers among capsules, or in stirred tank reactors.

Another important coenzyme, namely ATP, has been studied extensively in terms of this regeneration and separation for the purpose of its reuse in possible biosynthetic processes. Several excellent reviews have been published. These include the articles by Langer et al.[129, 130] and Archer et al.[131].

6 Concluding Remarks

The research interests on immobilized coenzymes practically follow the footsteps of immobilized enzymes. However, its possible industrial application is still hampered by economical considerations, whereas several immobilized enzyme processes have been in industrial operations now. The most immediate applications of immobilized coenzymes is in affinity chromatography, clinical work such as the correction of a metabolic imbalance, analytical chemistry, and mechanistic studies of enzymatic catalysis rather than in industrial enzymatic conversions.

For an enzymatic process involving coenzymes to be attractive, the process will have to be competitive enough or better than the conventional biological processes involving cultivation of cells, or the up and coming practice of immobilized whole cells. Process efficiency and economics are two major considerations for the selection of a process. The key factors that determine the degree of desirability of immobilized coenzyme processes are efficiency and economics of coenzyme retention and regeneration.

Immobilization of coenzymes makes their retention relatively easy. However, some immobilized coenzymes have been shown to have less activity than their free counterparts. Based on activities, it is, in general, better to immobilized coenzyme on soluble macromolecules than on particulate matrices. Dextran and polyethyleneimine have been studied in this connection. Such macromolecular immobilized coenzymes usually have good activity rating and can be retained by using ultrafiltration membranes. More research in this direction is needed.

Regeneration of the immobilized coenzymes is probably the most important and difficult task involved in large scale applications of coenzymes requiring immobilized enzymes for biosynthesis or oxidation reduction reactions. ATP regeneration has been studied and demonstrated to be technologically feasible and possibly economically feasible[132)] However, the regeneration of NAD(P)H has not received equal attention. At this time, the preferred methods of NAD(P)H regeneration seems to be coupled enzyme and coupled substrate methods. Nevertheless, the practicality of these regeneration schemes remains to be demonstrated.

7 Nomenclature

K_M: Michaelis-Menten constant in

$$v = \frac{V_M S}{K_M + S},$$

where v: rate of enzyme catalyzed reaction; V_M: maximum rate of enzyme catalyzed reaction; S: substrate concentration; K_i: competitive inhibition constant in

$$v = \frac{V_M S}{S + K_M (1 + i/K_i)},$$

where i: concentration of the competitive inhibitor; $FAD/FADH_2$: flavin adenine dinucleotide oxidized/reduced form; NAD/NADH: nicotinamide adenine dinucleotide oxidized/reduced form; YADH: yeast alcohol dehydrogenase.

8 References

1. Baricos, W.H., Chambers, R.P., Cohen, W.: Enzyme Technol. Dig. *4*, 39 (1975)
2. Baricos, W.H., Chambers, R.P., Cohen, W.: Anal. Lett. *9*, 257 (1976)
3. Bright, H.J.: In: Industrial application of immobilized enzyme reactors. Messing, R.A. (ed.), p. 137. New York: Academic Press 1975
4. Goldstein, L., Katchalski-Katzir, E.: In: Applied biochemistry and bioengineering, Vol. 1, p. 1. New York: Academic Press 1976
5. Mosbach, K.: In: Applications of biochemical systems in organic chemistry. Jones, J.B., Shih, C.J., Perlman, D. (eds.), p. 969. New York: Wiley-Interscience 1976
6. Mosbach, K., Larrson, P.-O., Lowe, C.: In: Methods in enzymology. Mosbach, K. (ed.), Vol. XLIV, p. 859. New York: Academic Press 1976
7. Skinner, K.J.: Chem. and Eng. News *53*, 22 (1975)
8. Jones, J.B., Beck, J.F.: In: Applications of biochemical systems in organic chemistry. Jones, J.B., Shih, C.J., Perlman, E. (eds.), p. 107. New York: Wiley-Interscience 1976
9. Diem, K., Lentner, C. (eds.): Geigy scientific tables, 7th ed., Ciba-Geigy Limited, Basle, Switzerland 1970
10. Conn, E.E., Stumpf, P.K.: Outlines of biochemistry, 3rd ed. New York: Wiley and Sons 1972
11. Lehninger, A.L.: Biochemistry, 2nd ed. Worth Publishers, Inc. 1975
12. Guilbault, G.G.: In: Enzymatic methods of analysis, p. 176. New York: Pergamon Press 1970
13. Bergmeyer, H.-U. (ed.): Methods of enzymatic analysis. New York: Academic Press 1965
14. Wagner, A.F., Folkers, K.: Vitamins and coenzymes. New York: Wiley-Interscience 1964
15. Dawson, R.M.C., Elliott, D.C., Elliott, W.H., Jones, K.M. (eds.): Data for biochemical research. Oxford: Oxford University Press 1969
16. Nakayama, K., Sato, Z., Tanaka, H., Kinoshita, S.: Agr. Biol. Chem. *37*, 1041 (1973)
17. Sakai, T., Uchida, T., Chibata, I.: Agr. Biol. Chem. *37*, 1049 (1973)
18. Watanake, T., Uchida, T., Kato, J., Chibata, I.: Appl. Microbiol. *27*, 531 (1974)
19. Sakai, T., Watanake, T., Chibata, I.: Agr. Biol. Chem. *37*, 849 (1973)
20. Tanaka, H., Sato, Z., Nakayama, K., Kinoshita, S.: Agr. Biol. Chem. *32*, 721 (1968)
21. Tanaka, A., Hironaka, J.: Agr. Biol. Chem. *36*, 867 (1972)
22. Whitesides, G.M., Lamotte, A., Adalsteinsson, O., Colton, C.K.: In: Methods in enzymology. Mosbach, K. (ed.), Vol. XLIV, p. 887. New York: Academic Press 1976
23. Marshall, D.L.: Biotechnol. Bioeng. *15*, 447 (1973)
24. Samejima, H., Kimura, K., Ado, K., Suzuki, Y., Tadokoro, T.: Research summary presented at the 4th enzyme engineering conference, P.T-20 Bad Neuenahr (F. R. Germany) 1977
25. Ogata, K., Shimizu, S., Taini, Y.: Agr. Biol. Chem. *36*, 84 (1972)
26. Shimizu, S., Morioko, H., Taini, Y., Ogata, K.: J. Ferm. Technol. *53*, 77 (1975)
27. Shimizu, S., Miyata, K., Taini, Y., Ogata, K.: Agr. Biol. Chem. *37*, 607 (1973)
28. Skeggs, H.R.: In: Industrial microbiology. Miller, B.M., Litaky, W. (eds.), p. 47. New York: McGraw-Hill 1977
29. Sigma price list. Sigma Co., St. Louis, Missouri, USA, April, 1977
30. Cremonesi, P., Carrea, G., Ferrara, L., Antonini, F.: Eur. J. Biochem. *44*, 401 (1974)
31. Gardner, C.R., Colton, C.K., Langer, R.S., Hamilton, B.K., Archer, M.C., Whitesides, G.M.: In: Enzyme engineering. Pye, E.K., Wingard, L.B., Jr. (eds.), Vol. 2, p. 209. New York: Plenum Press 1974
32. Whitesides, G.M., Siegel, M., Garrett, P.: J. Org. Chem. *40*, 2516 (1975)

33. Arsenis, C.A., McCormick, D.B.: J. Biol. Chem. *241*, 330 (1966)
34. Waters, C.A., Murphy, J.R., Hasting, J.W.: Biochem. Biophys. Res. Comm. *57*, 1152 (1974)
35. Blankenhorn, G., Osuga, D.T., Lee, H.S., Feeney, R.E.: Biochem. Biophys. Acta *386*, 470 (1975)
36. Conway, T.P., Muller-Eberhard, U.: Fed. Proc. *32*, 1382 (1973)
37. Fukui, S., Ikeda, S.-I., Fujimura, M., Yamada, H., Kumagai, H.: Eur. J. Biochem. *51*, 155 (1975)
38. Ikeda, S., Fukui, S.: Biochem. Biophys. Res. Comm. *52*, 482 (1973)
39. Ikeda, S., Hara, H., Sugimoto, S., Fukui, S.: FEBS Letts. *56*, 307 (1975)
40. Collier, R., Kohlhan, G.: Anal. Biochem. *42*, 48 (1971)
41. Miller, J.B., Cuatrecasas, P., Thompson, E.B.: Biochem. Biophys. Acta *276*, 407 (1972)
42. Matsuura, A., Iwashima, A., Nose, Y.: Biochem. Biophys. Res. Commun. *51*, 241 (1973)
43. McCormick, D.B.: Anal. Biochem. *13*, 194 (1965)
44. Cuatrecasas, Wilchek, M.: Biochem. Biophys. Res. Commun. *33*, 235 (1968)
45. Wolpert, J.S., Ernst-Fonberg, M.L.: Anal. Biochem. *52*, 111 (1973)
46. Chibata, I., Tosa, T., Matuo, Y.: In: Enzyme engineering. Pye, E.K., Wingard, L.B., Jr. (eds.), Vol. 2, p. 229. New York: Plenum Press 1974
47. Allen, R.H., Majerus, P.W.: J. Biol. Chem. *247*, 7695 (1972)
48. Olesen, H., Hippe, E., Haber, E.: Biochem. Biophys. Acta *243*, 66 (1971)
49. Yamada, R.H., Hogenkamp, H.P.C.: J. Biol. Chem. *247*, 6266 (1972)
50. Lowe, C.R., Mosbach, K.: Eur. J. Biochem. *49*, 511 (1974)
51. Hy, Margit, Reeves, H.C.: Biochem. Biophys. Acta *445*, 280 (1976)
52. Lindberg, M., Mosbach, K.: Eur. J. Biochem. *53*, 481 (1975)
53. Lamed, R., Oplatka, A.: Biochemistry *13*, 3137 (1974)
54. Oplatka, A., Mahlrad, A., Lamed, R.: J. Biol. Chem. *251*, 3972 (1976)
55. Berglund, O., Eckstein, F.: Eur. J. Biochem. *28*, 492 (1972)
56. Hojeberg, B., Brodelius, P., Rydstrom, J., Mosbach, K.: Eur. J. Biochem. *66*, 467 (1976)
57. Harvey, M.J., Lowe, C.R., Dean, P.D.G.: Eur. J. Biochem. *41*, 353 (1974)
58. Brodelius, P., Mosbach, K.: Anal. Biochem. *72*, 629 (1976)
59. Salter, D.N., Ford, J.E., Scott, K.J., Andrews, P.: FEBS Lett. *20*, 302 (1972)
60. Harding, J.J.: J. Chromatog. *77*, 191 (1973)
61. Danner, J., Lenhoff, H.M., Heagy, W.: J. Solid Phase Biochem. *1*, 177 (1976)
62. Lowe, C.R., Dean, P.D.G.: In: Affinity chromatography, p. 90. New York: Wiley and Sons 1974
63. Baum, G., Wrobel, S.J.: In: Immobilized enzymes, antigens, antibodies, and peptides. Weetall, H.H. (ed.), p. 419. New York: Marcel Dekker, Inc. 1975
64. Turkova, J.: J. Chromatog. *91*, 267 (1974)
65. Weibel, M.K., Weetall, H.H., Bright, H.T.: Biochem. Biophys. Res. Commun. *44*, 347 (1971)
66. Weibel, M.K., Doyle, E.R., Humphrey, A.E., Bright, H.J.: In: Enzyme engineering. Wingard, L.B., Jr. (ed.), p. 167. New York: Wiley and Sons 1972
67. Aizawa, M., Coughlin, R.W., Charles, M.: Biotechnol. Bioeng. *18*, 209 (1976)
68. Coughlin, R., Aizawa, M., Charles, M.: Biotechnol. Bioeng. *18*, 199 (1976)
69. Wyke, J.R., Dunill, P., Lilly, M.D.: Biochem. Biophys. Acta *286*, 260 (1972)
70. Wyke, J.R., Dunill, P., Lilly, M.D.: Biotechnol. Bioeng. *17*, 51 (1975)
71. Chambers, R.P., Ford, J.R., Allender, J.H.: In: Enzyme engineering. Pye, E.K., Wingard, L.B., Jr. (eds.), Vol. 2, p. 195. New York: Wiley 1974
72. Weibel, M.K., Fuller, C.W., Stadel, J.M., Buckmann, A.F.E.P., Doyle, T., Bright, H.J.: In: Enzyme engineering. Pye, E.K., Wingard, L.B., Jr. (eds.), Vol. 2, p. 203. New York: Wiley 1974
73. Pye, E.K., Humphrey, A.E., Bright, H.J., Graves, D.J., Kallen, R.G., Quinn, J., Weibel, M.K.: In: Enzyme technology grantees – users conference. Pye, E.K. (Org.), p. 9. Philadelphia, Pa.: University of Pennsylvania 1975
74. Lee, C.Y., Kaplan, N.: Arch. Biochem. Biophys. *168*, 665 (1975)

75. Barry, S., O'Carra, P.: FEBS Lett. *37*, 134 (1973)
76. Zappelli, P., Rossodivita, A., Prosperi, G., Pappa, R., Re, L.: Eur. J. Biochem. *62*, 211 (1976)
77. Zappelli, P., Rossodivita, A., Re, L.: Eur. J. Biochem. *54*, 475 (1975)
78. Lindberg, M., Larsson, P.O., Mosbach, K.: Eur. J. Biochem. *40*, 187 (1973)
79. Larrson, P.-O., Mosbach, K.: FEBS Lett. *46*, 119 (1974)
80. Larrson, P.-O., Mosbach, K.: Biotechnol. Bioeng. *13*, 393 (1971)
81. Mosbach, K.: In: Methods in enzymology. Jakoby, W.B. (ed.), Vol. XXXIV, p. 229. New York: Academic Press 1974
82. Lowe, C.R., Dean, P.D.G.: FEBS Lett. *14*, 313 (1971)
83. Barman, T.E. (ed.): Enzyme handbook. Berlin, Heidelberg, New York: Springer 1969
84. Hamilton, B.K., Montgometry, J.P., Wang, D.I.C.: In: Enzyme engineering. Pye, E.K., Wingard, L.B., Jr. (eds.), Vol. 2, p. 153. New York: Plenum Press 1974
85. Levy, H.R., Loewus, F.A., Vennesland, B.: J. Am. Chem. Soc. *79*, 2949 (1957)
86. Johnson, C.K., Gabe, E.J., Taylor, M.R., Rose, I.A.: J. Amer. Chem. Soc. *87*, 1802 (1965)
87. Helmchen-Zeier, R.E.: Über die enzymatische Reduktion von 2-, 3- und 4-Alkyl-Cyclohexanonen. Ph.D. thesis 4991, ETH, Zürich 1973
88. Chambers, R.P., McElrath, K.O., Cohen, W.: Research summary presented in 4th enzyme engineering conference, P.M-9, Bad Neuenahr (F. R. Germany) 1977
89. Chambers, R.P., Baricos, W.H., Cohen, W.: In: Enzyme technology grantees – users conference. Pye E.K. (Org.), p. 26. New York: Wiley 1975
90. Davies, P., Mosbach, K.: Biochim. Biophys. Acta *370*, 329 (1974)
91. Campbell, J., Chang, T.M.S.: Biochim. Biophys. Res. Commun. *69*, 562 (1976)
92. Anonymous: Chem. & Eng. News *52*, 19 (1974)
93. Mislin, R.: Enzymatische Reduktionen von Dekalonen-(1) and -(2). Ph.D. Thesis 4169, ETH, Zürich 1968
94. Retey, J.: Über die Stereospezifizität der enzymatischen Reduktion von Carbonylverbindungen und Oxydation von Alkoholen. Ph.D. thesis 3409. ETH, Zürich 1963
95. Gestrelius, S., Masson, M.-O., Mosbach, K.: Eur. J. Biochem. *57*, 529 (1975)
96. Gupta, N.K., Robinson, W.G.: Biochim. Biophys. Acta *118*, 431 (1966)
97. Jones, J.B., Sneddon, D.W., Higgins, W., Lewis, A.J.: Chem. Commun. *856* (1972)
98. Taylor, K.E.: Studies related to nicotinamide coenzyme recycling in preparative alcohol dehydrogenase-mediated reactions. Ph.D. thesis, University of Toronto, Toronto 1973
99. Wagner, D.-C.F., Bernt, E., Nelbock, M.: Angeu. Chem. Int. Ed., *3*, 587 (1964)
100. Måsson, M.O., Mattiasson, B., Grestrelius, S., Mosbach, K.: Biotechnol. Bioeng. *18*, 1145 (1976)
101. Lowry, O.H., Passonneau, J.V.: A flexible system of enzymatic analysis, p. 129. New York: Academic Press 1972
102. Jones, J.B., Taylor, K.E.: Chem. Commun. *205* (1973)
103. Pinder, S., Clark, J.B., Greenbaum, A.L.: Methods in enzymology. McCormick, D.B., Wright, L.D. (eds.), Vol. XVIII, B, p. 20. New York: Academic Press 1971
104. Christian, G.D.: In: Ion and enzyme electrodes in biology and medicine. Kessler, M., Clark, L.C., Jr., Lubbers, D.W., Silver, I.A., Simon, W. (eds.), p. 173. Baltimore: University Park Press 1976
105. Aizawa, M., Coughlin, R.W., Charles, M.: Biochim. Biophys. Acta *385*, 362 (1975)
106. Aizawa, M., Coughlin, R.W., Charles, M.: Biochim. Biophys. Acta *440*, 233 (1976)
107. Aizawa, M., Coughlin, R.W., Charles, M.: Biotechnol. Bioeng. *18*, 209 (1976)
108. Coughlin, R.W., Aizawa, M., Alexander, B.F.: Biotechnol. Bioeng. *17*, 515 (1975)
109. Aizawa, M., Ikariyama, Y., Suzuki, S.: J. Solid. phase Biochem. *1*, 249 (1976)
110. Wallace, T.C., Leh, M.B., Coughlin, R.W.: Biotechnol. Bioeng. *19*, 901 (1977)
111. Wallace, T.C., Coughlin, R.W.: Anal. Biochem. *80*, 133 (1977)
112. Braun, R.D., Santhanam, K.S.V., Elving, P.T.: J. Am. Chem. Soc. *97*, 2591 (1975)
113. Schmarkel, C.O., Santhanam, K.S.V., Elving, P.T.: J. Electrochem. Soc. *21*, 345 (1974)
114. Gremonesi, P., Carrea, G., Ferrara, L., Antonini, E.: Biotechnol. Bioeng. *17*, 1101 (1975)
115. Marconi, W., Prosperi, G., Giovenco, S., Morisi, F.: J. Mol. Catal. *1*, 111 (1975)

116. Mosbach, K., Larrson, P.-O.: Biotechnol. Bioeng. *10*, 19 (1970)
117. Vieth, W.R., Wang, S.S., Saini, R.: Biotechnol. Bioeng. *15*, 565 (1973)
118. Chibata, I., Tosa, T., Sato, T.: In: Methods in enzymology. Mosbach, K. (ed.), Vol. XLIV, p. 739. New York: Academic Press 1976
119. Larsson, P.-O., Mosbach, K.: In: Methods in enzymology. Mosbach, K. (ed.), Vol. XLIV, p. 183. New York: Academic Press 1976
120. Venkatasubrumanian, K., Vieth, W.R., Constantinides, A.: In: 75th Ann. Meet. Am. Soc. Microbiol. Session 116, 1975
121. Stinson, R.A., Holbrook, J.J.: Biochem. J. *131*, 719 (1973)
122. Wratten, C.C., Cleland, W.W.: Biochem. *2*, 935 (1963)
123. Zabin, B.A.: In: Methods of protein separation. Castsimpoolas, N. (ed.), Vol. 1, p. 239. New York: Plenum Press 1975
124. Rony, P.R.: Chem. & Eng. News, 21, Sept. 16, 1974
125. Fink, D.J., Rodwell, V.W.: Biotechnol. Bioeng. *17*, 1029 (1975)
126. Gardner, C.R.: Research summary presented at the 4th enzyme engineering conference, p. M19, Bad Neuenahr (F. R. Germany) 1977
127. Charm, S.E., Wong, B.L.: Biotechnol. Bioeng. *12*, 1103 (1970)
128. Woodley, C.L., Gupta, N.K.: Arch. Biochem. Biophys. *148*, 238 (1972)
129. Langer, R.S., Hamilton, B.K., Gardner, C.R., Archer, M.C., Colton, C.K.: AIChE J. *22*, 1079 (1976)
130. Langer, R.S., Gardner, C.R., Hamilton, B.K., Colton, C.K.: AIChE J. *23*, 1 (1977)
131. Archer, M.C., Colton, C.K., Cooney, C.L., Demain, A.L., Wang, D.I.C., Whitesides, G.M.: Enzyme technology grantees – users conference. Pye, E.K. (Org.), p. 1. Philadelphia, Pa.: University of Pennsylvania 1975
132. Whitesides, G.M.: In: Applications of biochemical systems in organic chemistry. Jones, J.B., Shih, C.J., Perlman, D. (eds.), p. 901. New York: Wiley 1976
133. Yamada, K.: Bioengineering report. Recent Adv. in Industrial Fermentation in Japan, Biotechnol. and Bioeng. *19*, 1563 (1977)

Process Development and Economic Aspects in Enzyme Engineering. Acylase L-methionine System

Christian Wandrey
Institut für Biotechnologie II
Kernforschungsanlage Jülich
D-5170 Jülich, West Germany

Erwin Flaschel
Institut de Génie Chimique III, Ecole Polytechnique Fédérale de Lausanne
Ecublens, CH-1015 Lausanne, Switzerland

An integral study of chemical reaction engineering and process development including economic aspects was carried out using the optical resolution of aminoacids by means of native and carrier-fixed acylase as example. Continuous operation is not only possible with carrier-fixed enzymes but also with native enzymes using ultrafiltration devices for catalyst retention. It could experimentally be proved on a 1 kg L-methionine/d-scale that the use of soluble acylase in membrane-reactors is economically superior to the carrier-fixed type in the tube-reactors. In all cases where immobilization of enzymes cannot be achieved with high activity yield or remarkable increase in stability, the homogeneous catalysis with soluble enzymes in continuously operating membrane-reactors might be a very promising alternative. – These studies were performed at the "Institut für Technische Chemie der Technischen Universität Hannover", West-Germany.

1 Introduction

The development of immobilized enzymes has introduced the possibility of continuous operation with biocatalysts. A great variety of methods for enzyme immobilization can be found in the literature[1]. Most of the papers dealing with immobilization techniques are also concerned with the evaluation of kinetic data and modelling of mass transfer phenomena. Fewer papers deal with chemical reaction engineering problems arising in the field of enzyme engineering. There are only a few examples where economic aspects are also taken into account (e.g.[2]). But one has to be aware of the situation that probably the most promising examples of such calculations are very unlikely to be published.

It is the aim of the authors to deal with the peculiarities of process development for reusable enzymatic catalysts. Continuous operation can be established not only by immobilization, but also by means of ultrafiltration devices for enzyme retention (for a general survey see[65–67]). These two main alternatives will be dealt with under chemical reaction engineering as well as in economic terms, using the optical resolution of amino-acids by means of acylase as example.

1.1 Biocatalysts

Enzymes are biopolymers, which accelerate special biochemical reactions. Most of them act highly selectively, with great activity, at room temperature and normal pressure. In comparison with inorganic catalysts their temperature- and pH-stability is low. Due to the fact that enzymes have to be isolated from biological material they seem to be expensive. For example 1 mol platinum (distributed on an aluminium oxide carrier) catalyses at 75 °C the production of 14 mol/min ethane from ethylene[3].

1 mol chymotrypsin catalyses at 25 °C the hydrolysis of 1.335 mol phenylalanine-methylester/min[4]. If the molecular weight (platinum: 195 g/mol, chymotrypsin: 24,800 g/mol) and the prices (platinum: 13.15 DM/g[a] at a content of 99.99%, chymotrypsin 37.84 DM/g[b] at a content of 85%) are taken into account, for 1 DM 5.5 mmol ethane/min or 1.2 mmol L-phenylalanine/min can be produced with the catalytic activity available. These values show that for the installation of a certain capacity one has to be aware of higher cost of catalyst per unit weight of product when biocatalysts are used instead of inorganic types. Nevertheless, enzymatically catalysed reactions are of great economic interest, because the increment value in most cases is comparatively high.

[a] Westdeutsche Edelmetalle, 1976
[b] Boehringer Mannheim, Pricelist, 1976

For continuous operation, the catalyst must be separated from the reactants. This does not mean that only heterogeneous catalysis can be applied. The difference in molecular weight between enzymes and their substrates is in most cases high enough that ultrafiltration devices can be used for retention of soluble enzymes. In the past, carrier fixed enzymes have been developed in order to use simple separation units like glass frits for enzyme retention in a continuously operating reactor. Meanwhile, membrane technology has been developed so far that ultrafiltration devices are available for enzyme retention on an industrial scale.

If the molecular weight of an enzyme is comparatively small, its molecular weight can be enlarged artificially by copolymerisation, for instance, with glutardialdehyde. Thus, soluble enzyme oligomers can be obtained[5)] which may be retained by ultrafiltration membranes of higher flux per unit area of separation. The alternatives discussed above are illustrated in Fig. 1. The main advantages of continuous operation with soluble enzymes are: homogeneous catalysis, approximately quantitative use of native activity and no dependency on special immobilization know-how. The main advantages of carrier fixed enzymes are: retention by simple separation units, simply applicable in tube reactors and often increased stability.

To compare different types of catalysts, a clear definition of catalyst activity is needed. The most precise term is the turn-over number which is defined as number of mols substrate converted per mol catalyst and unit time. For this definition the molecular weight and the purity of the enzyme must be known. When the number of active sites per unit weight of catalyst can be obtained – which is e.g., possible for chymotrypsin[6–9)] – this definition should be used.

If the content of active enzyme protein is not known, the activity per unit weight of dry catalyst should be used. The unit microcatal – the activity which converts one micromol substrate per second – is recommended by "Commission Biochemical Nomenclature"[10)], but seldom used in the literature.

Here the mass specific activity (A_m) in mmol · g^{-1} · min^{-1} and the volume specific activity (A_v) in mmol · l^{-1} · min^{-1} are used. The activity itself (A) will be stated in mmol · min^{-1} or μmol · min^{-1}. As abbreviation for the latter unit the letter U is often used in literature on the subject. Activity specifications are only meaningful when the temperature, the substrate concentration, the pH value and the concentration of possible activators are stated, too. Initial rates should be measured in order to avoid product inhibition influence.

Activity and integral productivity (PR) should be distinguished. The productivity is defined as the number of kilogram product produced per gram dry catalyst and day at a given operating condition, whereas the term activity is defined for initial rate conditions, the term productivity depends on the degree of conversion.

If carrier fixed enzymes are used, the particle size (d_K) and the particle Reynolds number (Re_K) should be specified. In order to calculate volumetric activities from mass specific activities the catalyst concentration (E) must be known. Here the unit g · l^{-1} is used. For carrier-fixed enzymes the void volume (V_f) has to be used.

All values of stability should relate to continuous operation. If the deactivation can be described in terms of a 1st order process it is sensible to use a deactivation constant (k_{de}) in the unit d^{-1}.

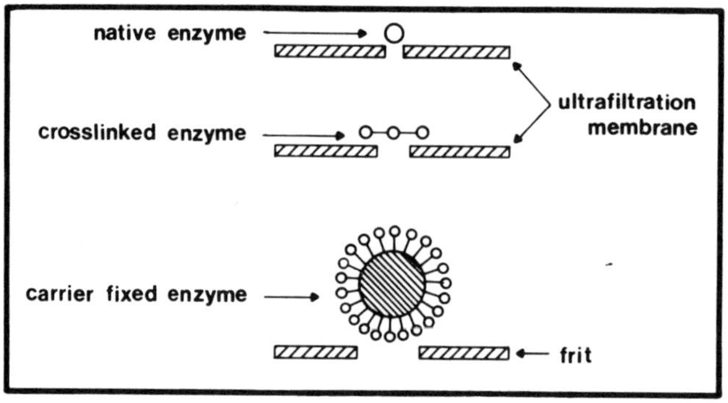

Fig. 1. Rentention of various catalyst types

1.2 Enzyme Reaction Engineering

There are some characteristic peculiarities in enzyme reaction engineering. In the literature dealing with many cases, kinetic data can only be found for a concentration range which is not interesting for technical applications. This is because most of the experiments are carried out at physiological concentrations. The proposed kinetic models can seldom be applied for higher concentrations. Similarly very little information is given on statistical significance of kinetic parameters. Besides kinetic data, thermodynamic data and some physical properties of the reaction mixture must be known for process development purposes.

As there is a great number of parameters which influence the whole process performance, in the past enzyme reactors have been optimized empirically. Mathematical modelling of enzyme reactors has been known for quite some time. But only since about 1970 there are examples that such models have been checked at technically interesting conditions[11–18].

1.3 Economic Aspects

To evaluate economically optimal operating points, the optimal use of raw material, of catalyst activity and not least of energy for unit operations has to be taken into account. This will result in a rather complicate economic model containing the reactor model only as one part. Such models are still meaningful if the number of parameters is kept at a reasonable level, and if the parameters used have been measured with sufficient precision. A biocatalyst can be regarded as raw material which is slowly consumed. Therefore, precise information on catalyst effectivity and stability must be available.

2 Process Development

The stereospecific hydrolysis of N-acetyl-D, L-methionine by means of acylase for the production of L-methionine is thought to be the first example where an immobilized enzyme has been used[19] on an industrial scale. The main emphasis seems to have been given to the catalyst development, while the reactor has been more or less empirically optimized.

The reaction system is shown in Fig. 2. Out of the racemic substrate mixture only the L-enantiomer is deacylated. The D-enantiomer can be used after reracemization in order to improve the yield based on raw material. As catalyst hog kidney acylase or microbial acylase may be used. Typical reaction temperatures are between 37 and 50 °C. The pH-optimum of the reaction is at 7.0. As metal effector Co^{2+}-ions are used at a concentration of 0.5 mmol · l^{-1}. Product recovery is achieved by crystallization, because L-menthionine is much less soluble than the substrate.

2.1 Continuous Analysis

For studying continuous operation, a continuous method of analysis should be applied. Thus for kinetic evaluation, quick data processing is possible. Reactors for production purposes can be controlled continuously owing to the rapid analysis.

In principle, spectroscopic methods are useful for continuous analysis. Mitz and Schlüter[20] developed a differential spectroscopic method to monitor the enzymatically catalysed hydrolysis of N-acetyl-L-methionine. The concentrations used at that time – up to 100 mmol · l^{-1} were, however, too high. This can be seen from Table 1. Even modern photometers do not amplify the remaining light intensity of less than 0.1% precisely. Thus using a cuvette of 1 cm layer thickness, only substrate solutions up to 1.5 mmol · l^{-1} can be monitored.

Figure 3 shows calibrating measurements using a substrate concentration of 1 mmol · l^{-1} and a layer thickness of 1 cm for 0, 20, 40, 60, and 80% conversion. Measurements were carried out against a product concentration of 1 mmol · l^{-1}. Since the molecular extinction coefficient for the substrate is higher than that for the product, the extinction decreases with conversion. If cuvettes with a layer thickness of 1 mm are used, continuous analysis is possible up to substrate concentrations of 15 mmol · l^{-1}.

Table 1. Spectroscopic data for N-acetyl-L-methionine (S_L) and L-methionine (P_L), measured against water

Compound \ Parameter	C mol · cm^{-3}	λ nm	pH –	T °C	d cm	E –	ϵ_M $cm^2 \cdot mol^{-1}$
S_L	1×10^{-6}	214	7.0	25	1	1.80	1.80×10^6
P_L	1×10^{-6}	214	7.0	25	1	0.84	0.84×10^6

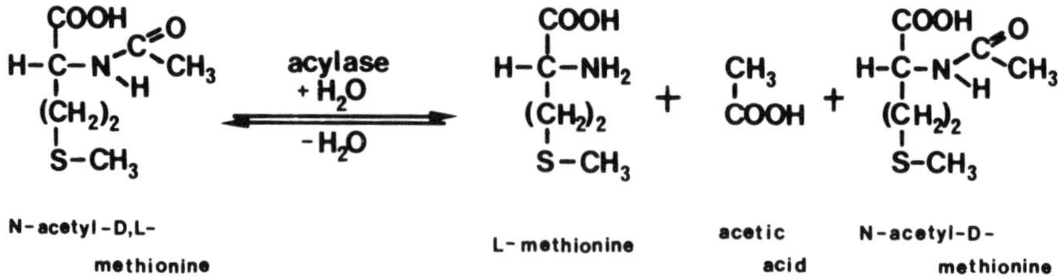

Fig. 2. Reaction system

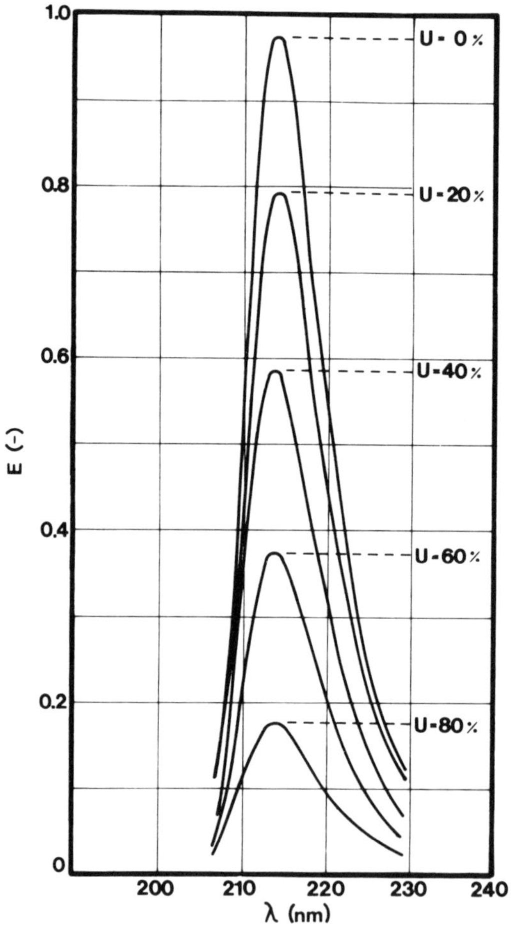

Fig. 3. Measurements for differential spectrophotometric calibration (1 mmol · l^{-1} L-substrate against 1 mmol · l^{-1} L-product, path length 1 cm, wave length 214 nm)

Like in many enzymatically catalysed reactions the environment of an asymmetric carbon atom is changed. Therefore polarimetry may be used as analysing method. An automatic polarimeter with analog and digital output has been used[a]. The digital sensitivity was 1 millidegree and the recorder sensitivity 2.5 mm · millidegree^{-1}. The reproducibility was ±1 millidegree. Using a flow-through cuvette with a volume of 1 ml and tubes of small diameter, the dead-time can be limited to about 2 s. In principle the signal is available without time delay. In practice after a sudden change of the optical rotation value of 1 degree there is a response time (90%-time) of about 3 s due to the mechanically driven analyser.

A review on spectro-polarimetry as an analytical method is given by Potapow[21], and there are some recent papers on the application of polarimetry for enzyme activity determination and continuous analysis of enzymatically catalysed reactions[22–26].

For reaction engineering measurements the molecular rotation $[(M)_\lambda^T]$ is more suitable than the specific rotation. The molecular rotation is defined as the degree of rotation using a concentration of 1 mol · l^{-1} and a layer thickness of 1 m. Temperature (T) and wave length (λ) must be specified. For calibrating measurements it is useful to apply the same pH-value, concentration level and amount of optically inactive components present in the reaction mixture.

The application of polarimetry to the stereospecific hydrolysis of N-acetyl-D, L-methionine is illustrated in Table 2. Acylase only catalyses the hydrolysis of the L-substrate (L-S) to the L-product (L-P) and acetate. The constitution of the D-substrate (D-S) remains unchanged. In practice, racemic substrate mixtures are used, so that the rotation value is zero at the starting point of the reaction.

Table 2. Application of polarimetry for the analysis of the acylase – methionine system (T = 37 °C, λ = 365 nm, extrapolated to zero concentration)

L-S (L-substrate) $(M_{L\text{-}S})_\lambda^T$	$\xrightarrow{\text{acylase}}$	L-P (L-product) $(M_{L\text{-}P})_\lambda^T$
D-S (D-substrate) $(M_{D\text{-}S})_\lambda^T$	– – – –→	D-S (D-substrate) $(M_{D\text{-}S})_\lambda^T$
for U = 0% $(M_U) = 0$		for U = 100% $(M_U) = (M_{L\text{-}P}) + (M_{D\text{-}S})$

$(M_{L\text{-}P})_{365}^{37} = -21.0\ (^\circ \cdot l \cdot mol^{-1} \cdot m^{-1})$

$(M_{D\text{-}S})_{365}^{37} = -64.5\ (^\circ \cdot l \cdot mol^{-1} \cdot m^{-1})$

for U = 100%

$(M_U)_{365}^{37} = -85.5\ (^\circ \cdot l \cdot mol^{-1} \cdot m^{-1})$

[a] Perkin Elmer, Überlingen, FRG, model 11241

For quantitative conversion the sum of the molecular rotation values of L-product and D-substrate results. In this case it is favourable that the rotation values for substrate and product of each enantiomer are of opposite sign. The data given in Table 2 refer to highly diluted solutions. With increasing concentration level the molecular rotation decreases due to the intermolecular interactions. As concentration level the starting concentration of L-substrate (S_{oL}) in mol · l^{-1} is used.

The layer thickness (d) is 0.1 m in all cases. Using the symbols of Table 2 Eq. (1) results, where the effective rotation value $(\alpha_{eff})_\lambda^T$ is given as a function of conversion (U) and initial substrate concentration (S_{oL}).

$$(\alpha_{eff})_\lambda^T = [(M_{D\text{-}S})_\lambda^T + (1\text{-}U)(M_{L\text{-}S})_\lambda^T + U(M_{L\text{-}P})_\lambda^T]\, S_{oL} \cdot d \,. \tag{1}$$

Taking into account that

$$(M_{D\text{-}S})_\lambda^T = -(M_{L\text{-}S})_\lambda^T \tag{2}$$

it follows:

$$\left[\left(\frac{\alpha_{eff}}{S_{oL} \cdot d}\right)\right]_\lambda^T = [-(M_{L\text{-}S})_\lambda^T + (M_{L\text{-}P})_\lambda^T]\; U = (M_U)_\lambda^T \cdot U \,, \tag{3}$$

where $(M_U)_\lambda^T$ is the symbol for the molecular rotation at total conversion. For sufficiently dilute solutions Eq. (4) holds:

$$U = \frac{1}{(M_U)_\lambda^T} \cdot \left[\left(\frac{\alpha_{eff}}{S_{oL} \cdot d}\right)\right]_\lambda^T . \tag{4}$$

All calibration measurements were carried out at a wave length $\lambda = 365$ nm at pH = 7.0. To determine the temperature dependence of the molecular rotation for substrate and product, measurements at a concentration level of 0.2 m · l^{-1} were carried out in a temperature range between 20 and 50 °C. The results are shown in Fig. 4. Using a basis of 37 °C the following two regression formulae can be used:

$$S{:}\ [M]_{365}^T = +\,59.0 + 1.66\,(T\text{-}37) - 0.009\,(T\text{-}37)^2 \,, \tag{5}$$

$$P{:}\ [M]_{365}^T = -\,16.0 + 0.34\,(T\text{-}37) - 0.003\,(T\text{-}37)^2 \,. \tag{6}$$

At 37 °C the alteration of the rotation value for the substrate is +2.8% · K^{-1} and +2.1% · K^{-1} for the product respectively. Obviously the polarimetric cuvette must be thermostated precisely.

For practical applications it has to be considered that the reaction mixture always contains the D-substrate. Even the acetate ions, which are produced during the reaction, influence the molecular rotation values due to molecular interactions. To get a precise calibration, all the solutions contained 0.5 mmol · l^{-1} Co^{2+} as did those used during all further investigations.

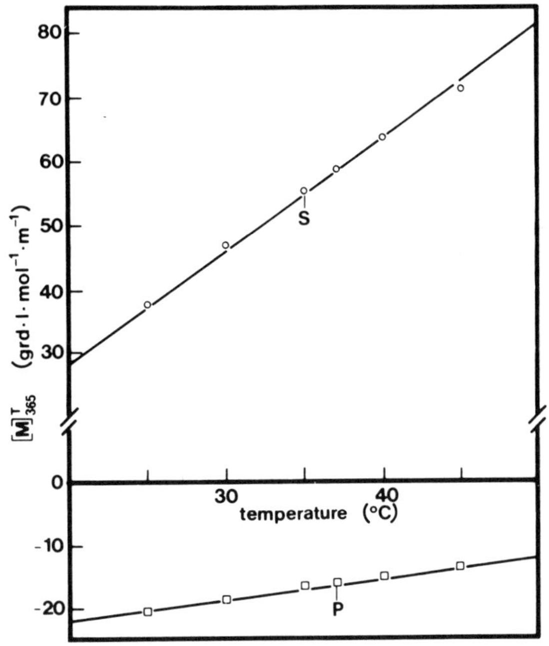

Fig. 4. Temperature dependence of molecular rotation for substrate (S) and product (P)

Since there is a non-linear relation between the molecular rotation and the concentration level, the non-linear Eq. (7) is more accurate than the linear approximation of Eq. (4).

$$U = A \cdot \left[\left(\frac{\alpha_{eff}}{S_{oL} \cdot d}\right)\right]_{\lambda}^{T} + B \cdot \left[\left(\frac{\alpha_{eff}}{S_{oL} \cdot d}\right)\right]_{\lambda}^{T^2} . \tag{7}$$

In order to use Eq. (7) for more than one concentration level, the dependence of the coefficients A and B on the concentration level has to be evaluated. For this purpose calibration measurements at 0, 25, 50, 75, and 100% conversion were carried out up to a concentration level of 1.0 mol $\cdot l^{-1}$ at 365 nm and pH 7. All measurements were carried out at 25, 37, and 50 °C. Table 3 summarizes the results of a non-linear regression analysis for the coefficients A and B and their dependence on the concentration level.

Using Eq. (7) the conversion can be calculated for a given concentration level from measured effective rotation values.

The results for all concentration levels are shown graphically in Fig. 5 where the abscissa is given in molecular rotation. The parameter of the resulting family of curves is the concentration level. One realises that the molecular rotation for a given conversion decreases with increasing concentration level.

When Eq. (7) is solved for α_{eff}, Table 4 results for 37 °C. Since the reproducibility is ±1 millidegree, polarimetry proves to be a very precise method. It is an additional advantage that this method can be used up to the saturation point without a pre-programmed dilution step.

Table 3. Regression analysis of polarimetric calibration-function

Concentration-range mol · l^{-1}	Conversion range %	T °C	Parameters of quadratic regression
0–1 (D,L)	0–100	25	$A = 0.0146 + 0.0138\ S_{oL} - 0.0286\ S_{oL}^2$ $B = + 0.000112\ S_{oL} + 0.000224\ S_{oL}^2$
0–1 (D,L)	0–100	37	$A = 0.0117 + 0.0114\ S_{oL} - 0.0144\ S_{oL}^2$ $B = + 0.000059\ S_{oL} + 0.000037\ S_{oL}^2$
0–1 (D,L)	0–100	50	$A = 0.0100 + 0.0069\ S_{oL} - 0.0103\ S_{oL}^2$ $B = + 0.000038\ S_{oL} + 0.000055\ S_{oL}^2$

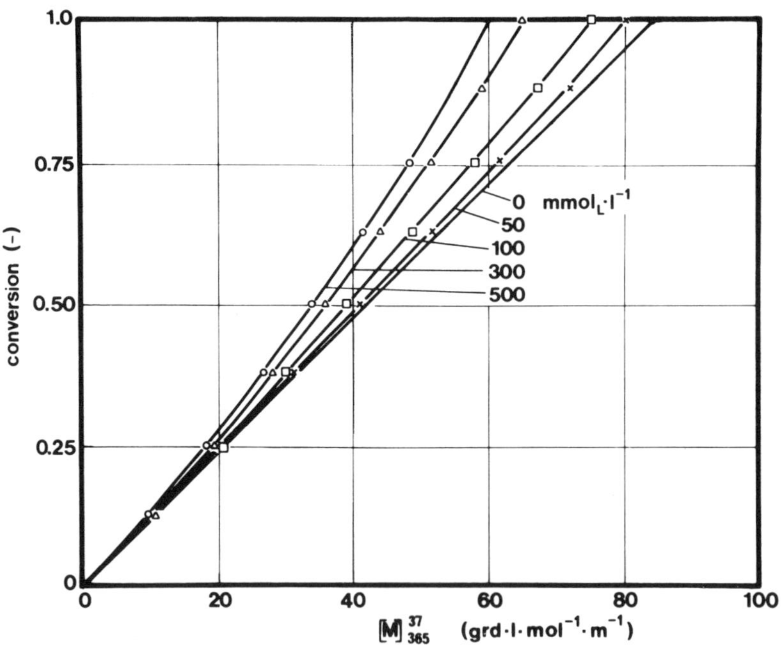

Fig. 5. Polarimeter calibration of methionine-system (standard condition, parameter: concentration of L-substrate in racemate)

Table 4. Effective optical rotation $(\alpha_{eff})^{37}_{365}$ in millidegree as function of initial substrate concentration and conversion

	Initial L-substrate concentration (in racemic mixture) (mmol · l^{-1})									
U (%)	50	100	150	200	250	300	350	400	450	500
0.	0.	0.	0.	0.	0.	0.	0.	0.	0.	0.
2.	8.	16.	23.	30.	37.	43.	50.	57.	64.	72.
4.	16.	31.	46.	60.	73.	86.	100.	114.	128.	144.
6.	24.	47.	69.	89.	109.	129.	150.	170.	192.	215.
8.	33.	63.	91.	119.	145.	172.	199.	227.	255.	285.
10.	41.	78.	114.	148.	181.	215.	248.	282.	318.	355.
12.	49.	94.	137.	177.	217.	257.	297.	338.	380.	425.
14.	57.	110.	159.	207.	253.	299.	346.	393.	442.	493.
16.	65.	125.	182.	236.	289.	341.	394.	448.	504.	562.
18.	73.	141.	204.	265.	324.	383.	442.	502.	565.	630.
20.	81.	156.	227.	294.	360.	425.	490.	557.	625.	697.
22.	90.	172.	249.	323.	395.	466.	538.	611.	686.	764.
24.	98.	187.	271.	352.	430.	508.	585.	664.	746.	831.
26.	106.	203.	294.	381.	465.	549.	633.	718.	805.	897.
28.	114.	218.	316.	409.	500.	590.	680.	771.	865.	962.
30.	122.	234.	338.	438.	535.	631.	727.	824.	924.	1027.
32.	130.	249.	360.	467.	570.	671.	773.	876.	982.	1092.
34.	138.	264.	382.	495.	604.	712.	820.	929.	1041.	1157.
36.	146.	280.	405.	524.	639.	752.	866.	981.	1098.	1220.
38.	154.	295.	427.	552.	673.	793.	912.	1033.	1156.	1284.
40.	162.	310.	449.	580.	707.	833.	958.	1084.	1213.	1347.
42.	170.	326.	470.	608.	742.	873.	1003.	1135.	1270.	1410.
44.	178.	341.	492.	636.	776.	913.	1049.	1186.	1327.	1472.
46.	186.	356.	514.	664.	810.	952.	1094.	1237.	1383.	1534.
48.	194.	371.	536.	692.	843.	992.	1139.	1288.	1439.	1596.
50.	202.	386.	558.	720.	877.	1031.	1184.	1338.	1495.	1657.
52.	210.	402.	579.	748.	911.	1070.	1229.	1388.	1551.	1718.
54.	218.	417.	601.	776.	944.	1109.	1273.	1438.	1606.	1779.
56.	226.	432.	623.	804.	978.	1148.	1318.	1488.	1661.	1839.
58.	234.	447.	644.	831.	1011.	1187.	1362.	1537.	1715.	1899.
60.	242.	462.	666.	859.	1044.	1226.	1406.	1586.	1770.	1958.
62.	250.	477.	687.	886.	1077.	1264.	1449.	1635.	1824.	2017.
64.	258.	492.	709.	914.	1110.	1303.	1493.	1684.	1878.	2076.
66.	266.	507.	730.	941.	1143.	1341.	1537.	1733.	1931.	2135.
68.	274.	522.	752.	969.	1176.	1379.	1580.	1781.	1985.	2193.
70.	282.	537.	773.	995.	1209.	1417.	1623.	1829.	2038.	2251.
72.	290.	552.	794.	1022.	1241.	1455.	1666.	1877.	2091.	2309.
74.	298.	567.	815.	1049.	1274.	1493.	1709.	1925.	2143.	2366.
76.	306.	582.	837.	1076.	1306.	1530.	1751.	1972.	2195.	2423.
78.	314.	597.	858.	1103.	1339.	1568.	1794.	2020.	2248.	2480.
80.	322.	612.	879.	1130.	1371.	1605.	1836.	2067.	2299.	2537.
82.	330.	626.	900.	1157.	1403.	1642.	1878.	2114.	2351.	2593.
84.	338.	641.	921.	1184.	1435.	1680.	1920.	2160.	2402.	2649.
86.	346.	656.	942.	1211.	1467.	1717.	1962.	2207.	2454.	2705.
88.	353.	671.	963.	1237.	1499.	1754.	2004.	2253.	2505.	2760.
90.	361.	686.	984.	1264.	1531.	1790.	2045.	2300.	2555.	2815.
92.	369.	700.	1005.	1290.	1563.	1827.	2087.	2346.	2606.	2870.
94.	377.	715.	1026.	1317.	1594.	1864.	2128.	2391.	2656.	2925.
96.	385.	730.	1046.	1343.	1626.	1900.	2169.	2437.	2706.	2979.
98.	393.	744.	1067.	1369.	1657.	1936.	2210.	2483.	2756.	3033.
100.	401.	759.	1088.	1395.	1689.	1973.	2251.	2528.	2806.	3087.

2.2 Reaction System

It is typical for enzymatically catalysed reactions that the activity depends on pH. Therefore the pH should be kept at the optimal value during the whole conversion range. And it should be investigated if a buffer or pH-control by means of a pH-stat is needed. The pH-optimum for acylase is at pH 7 (refer to Part 2.3.2). For methionine and N-acetyl-methionine, respectively, there are two pK values. At pH 7 it is not necessary to take into account the pK-values of the carboxyl group, because the carboxyl group is nearly quantitatively deprotonized. The pK-values of the nitrogen group (pK_2) were determined from titration curves. The values for acetic acid and water were taken from the literature[27, 28]. All the data needed for a quantitative calculation are summarized in Table 5.

Table 5. pK-values of the methionine-system

T		(°C)	25	37
N-acetyl-methionine	pK_2	(–)	3.52	3.60
Methionine	pK_2	(–)	9.21	9.10
Acetic acid	pK_S	(–)	4.75	4.67
Water	pK_W	(–)	14.00	13.60

The pH-change caused by adding a strong acid can be calculated by means of the buffer capacity (β)[4, 28]. In this case no acid is added, but protons are generated during the reaction. These are partly intercepted by the acid base equilibrium of the reaction mixture, therefore only a part of them will give rise to a pH change, which can be calculated by means of a proton balance:

$$\begin{array}{l} \text{free (accessory)} \\ \text{protons} \end{array} = \begin{array}{l} \text{protons generated} \\ \text{by reaction} \end{array} - \begin{array}{l} \text{protons intercepted} \\ \text{by the buffer system} \end{array} .$$

The substrate is partly protonized at the amino group (SH^+) and partly non-protonized (S). Similar expressions are used for the protonized form (PH^+) and non-protonized form (P) of the product L-methionine, acetic acid (HAc), and acetate (Ac^-) respectively.

$$\begin{array}{ccccccc} \text{N-acetyl-methionine} & & \text{water} & & \text{acetate} & & \text{methionine} \\ H_3C\text{-}CO\text{-}\underset{\overset{|}{H^+}}{NH}\text{-}R & + & H_2O & \rightleftharpoons & H_3C\text{-}COOH & + & \underset{\overset{|}{H^+}}{NH_2}\text{-}R \end{array}$$

$$\text{I} \quad \boxed{\begin{array}{ccccc} (SH^+) & \longrightarrow & (HAc) & + & (PH^+) \\ & & \downarrow\uparrow & & \downarrow\uparrow \\ & & (Ac^-) + H^+ & & (P) + H^+ \end{array}} \; , \qquad (8a)$$

$$H_3C-CO-NH-R + H_2O \rightleftharpoons H_3C-COOH + NH_2-R$$

$$\text{II} \quad \boxed{\begin{array}{ccccc} (S) & \longrightarrow & (HAc) & + & (P) \\ & & \downarrow\uparrow & & +H^+ \downarrow\uparrow -H^+ \\ & & (Ac^-) + H^+ & & (PH^+) \end{array}} \;. \tag{9a}$$

The protonized form of the substrate (SH^+) generates acetic acid (HAc) and the protonized form (PH^+) of the product. The non-protonized substrate (S) is converted to acetic acid (HAc) and the non-protonized form of the product (P). The proton balance in differential form is represented by Eqs. (8b) and (9b):

$$dC_{H^+}\ (I) = \left(\frac{S_0 - (AcH)}{S_0}\right) \cdot (SH^+) \cdot dU + \left(\frac{S_0 - (PH^+)}{S_0}\right) (SH^+) \cdot dU\,, \tag{8b}$$

$$dC_{H^+}\ (II) = \left(\frac{S_0 - (AcH)}{S_0}\right) \cdot (S) \cdot dU - \left(\frac{(PH^+)}{S_0}\right) \cdot (S) \cdot dU\,. \tag{9b}$$

For the total substrate concentration (S_0) Eq. (10) holds:

$$S_0 = (SH^+) + (S)\,. \tag{10}$$

Using Eq. (10) in combination with Eqs. (8b) and (9b), Eq. (11) results:

$$dC_{H^+} = \left(1 + \frac{(SH^+)}{S_0} - \frac{(AcH)}{S_0} - \frac{(PH^+)}{S_0}\right) \cdot S_0 \cdot dU\,. \tag{11}$$

If the pK values of the corresponding acid-base pairs are used from Eq. (11), Eq. (12) can be developed:

$$dC_{H^+} = \left(1 + \frac{1}{1 + 10^{pH - pK_{Sub}}} - \frac{1}{1 + 10^{pH - pK_{Pro}}} - \frac{1}{1 + 10^{pH - pK_{HAc}}}\right) \cdot S_0 \cdot dU\,. \tag{12}$$

In Eq. (12) the expression in brackets represents the factor by which the product $S_0 \cdot dU$ has to be multiplied to get the effective change in the concentration of free protons. The different terms within the expression in brackets can be interpreted as follows:

In principle, one proton is formed during the reaction. Additionally, there is another proton from the protonized form of the substrate. Part of these protons is intercepted by the acetate ions (formation of (HAc) and by the product forming (PH^+). If the symbol F_{H^+} is used for the expression in brackets in Eq. (12), Eq. (13) results:

$$dC_{H^+} = (F_{H^+}) \cdot S_0 \cdot dU\,. \tag{13}$$

The expression F_{H^+} as a function of the pH value is given in Fig. 6. One realizes that the reaction system with initial pH values > 6.9 generates protons and using initial pH-values < 6.9 consumes protons. The concentration of accessory (i.e. consumed) protons (C_{H^+}) may be considered by analogy with the conditions of an acid base titration as the concentration of a strong acid. Using the buffer capacity (β) of the system Eq. (14) holds:

$$dpH = - dC_{H^+}/\beta \, . \tag{14}$$

The buffer capacity (β) is a function of the total concentration of the acid base pairs and of the pK-values of these systems (7):

$$\beta = \ln 10 \left[10^{-pH} + 10^{pH-pK_W} + \frac{S_0(1-U)}{(1+10^{pK_{Sub}-pH})(1+10^{pH-pK_{Sub}})} \right.$$
$$+ \frac{S_0 \cdot U}{(1+10^{pK_{Pro}-pH})(1+10^{pH-pK_{Pro}})}$$
$$\left. + \frac{S_0 \cdot U}{(1+10^{pK_{HAc}-pH})(1+10^{pH-pK_{HAc}})} \right] . \tag{15}$$

Equations (13) and (14) represent a linked differential equation system with the initial conditions $pH = pH_0$, $S = S_0$, and $U = 0$. Since in practice racemic mixtures are used, the integration of the system is only meaningful for conversions up to $U = 0.5$. The function

$$U = f(pH)_{pH_0} \tag{16}$$

can be obtained by numerical integration. After the initial conditions have been fixed with Eq. (13), a differential conversion step is assumed to calculate the concentration of accessory (i.e. consumed) protons using the last calculated pH-value. The algorithm is started with the initial pH value. Using the pH-value at the beginning of each differential step as well as the total concentration of substrate and product, the buffer capacity following Eq. (15) is calculated. Thus with Eq. (14) a differential pH change results and the pH-value for the next differential step is fixed.

In Fig. 7 the change of pH during the reaction is shown graphically. Theoretical results are compared with measurements at 37 °C and a racemate concentration of $0.4 \text{ mol}_{DL} \cdot l^{-1}$ represented by small circles. As can be seen from the graph, the model excellently fits the data. At a starting pH-value of $pH_0 = 7$ the pH-shift with increasing conversion is so small that no buffer is needed. At $pH_0 = 6.88$ no protons are generated or consumed. Using a starting pH of $pH_0 = 8$, protons are generated, because acetate is no longer an effective buffer at this pH value. At $pH_0 = 6$ methionine intercepts more than 99.9% of the generated protons. Moreover, there is still some buffer capacity of acetate, so that in all a consumption of protons results. Obviously the system is self-stabilizing to pH 6.88.

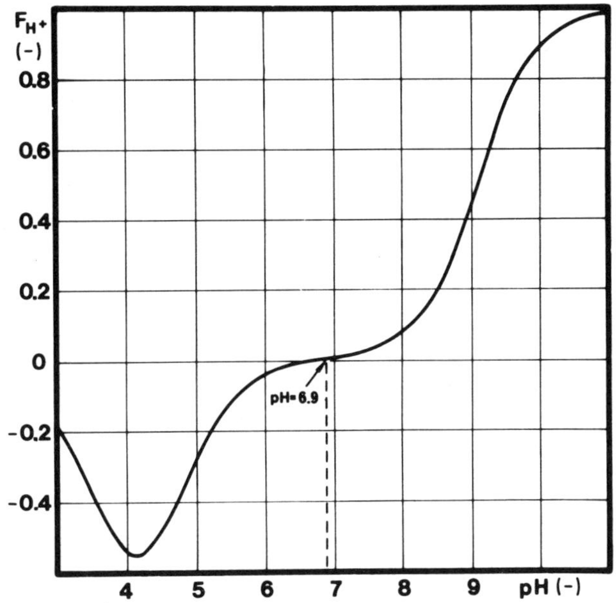

Fig. 6. pH dependence of the free fraction of hydrogen ions formed

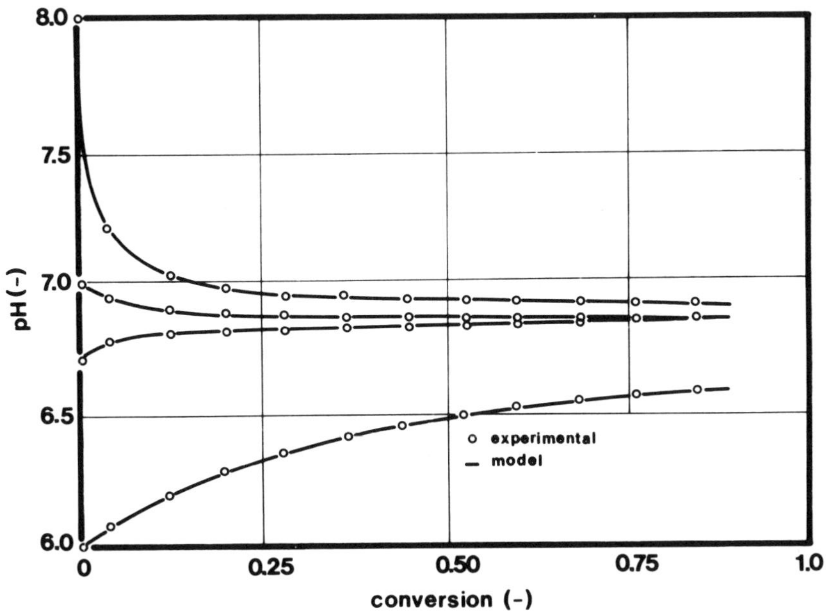

Fig. 7. pH value as function of conversion (parameter: initial pH-value)

As far as thermodynamic data are concerned, the state of equilibrium is especially important. It can be found in literature that the reaction catalysed by acylase is a quantitative one[29–38]. Other authors presumed that the deacylation of aminoacids cannot be achieved quantitatively. These contradictory statements may be explained by the limited accuracy of analysis as well as by the fact that at low substrate concentrations the state of equilibrium lies near quantitative conversion. At higher concentrations it could simply be proved by means of the polarimetric analysis that an equilibrium is established. Both the forward and reverse reaction could be monitored at pH 7 and 37 °C. The forward reaction was started using a substrate concentration of 0.4 $mol_L \cdot l^{-1}$. The reverse reaction was carried out using 0.4 mol L-methionine $\cdot l^{-1}$ and the corresponding acetate concentration. Figure 8 shows the conversion for both reactions as a function of time.

To calculate the equilibrium constant, the state of equilibrium was measured as a function of the concentration level. The results are given in Fig. 9. The line represents the result of a fit using an equilibrium constant defined by Eq. (17).

$$K_{37\,°C,\,pH\,=\,7} = \frac{[Ac] \cdot [L\text{-}Met]}{[N\text{-}Ac\text{-}L\text{-}Met]} = 2.75\ (mol \cdot l^{-1})\,. \tag{17}$$

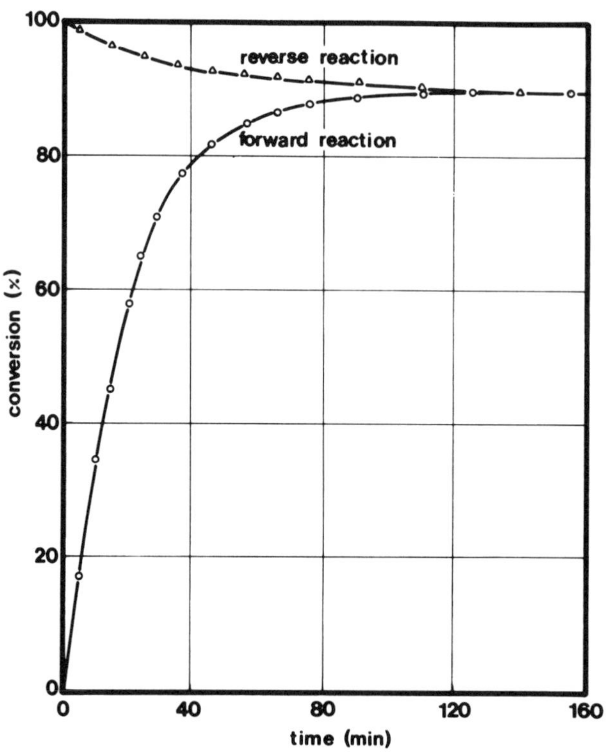

Fig. 8. Establishment of equilibrium at standard condition

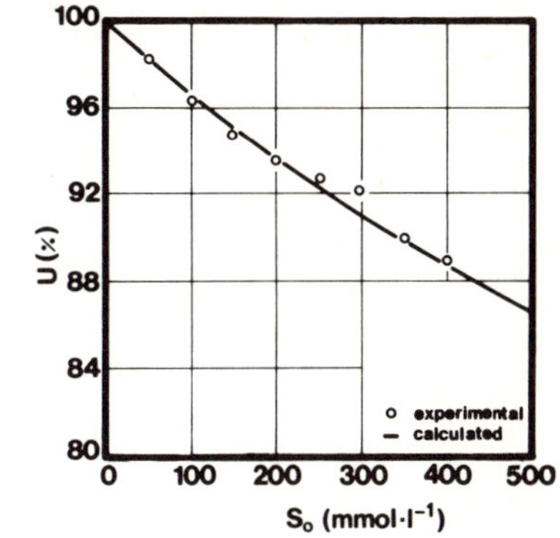

Fig. 9. Equilibrium conversion as function of initial substrate concentration (pH 7.0, 37 °C)

At pH 7 and a concentration level of 0.3 $\text{mol}_L \cdot l^{-1}$ the temperature dependence of the state of equilibrium was measured. The results are given in the form of an Arrhenius-plot in Fig. 10. The achievable conversion decreases with increasing temperature. From the slope of the straight line in Fig. 10 the reaction enthalpy can be calculated following Eq. (18):

$$\ln K_2 - \ln K_1 = -\frac{\Delta H}{R}\left(\frac{1}{T_2} - \frac{1}{T_1}\right). \tag{18}$$

A reaction enthalpy of 1.9 kcal · mol^{-1}, 7.9 kg mol^{-1} respectively, results.

More physical-chemical data of the reaction system can be found elsewhere[39].

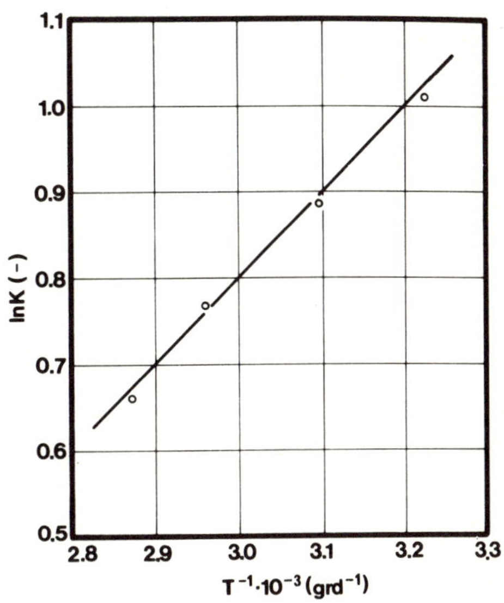

Fig. 10. Temperature dependence of the equilibrium constant

2.3 Characterization of the Catalysts

The specific properties of the catalysts will be described in this part. For kinetic measurements refer to Part 2.4.

2.3.1 Physical Properties

Soluble hog kidney acylase (EC 3.4.1.14)[a] was used. The mass specific activity using N-acetyl-L-methionine as substrate at 25 °C and pH 7 is reported to be 15 mmol · g^{-1} · min^{-1}. It is a lyophilized powder. At 25 °C about 20 g · l^{-1} can be dissolved in water. Data for the molecular weight of acylase vary from 76,600 to 119,000[40, 41]. Commercial hog kidney acylase was purified by means of gel-chromatography. It was proved that pure fractions could be achieved by testing with disc gel-electrophoresis. This fraction was used for molecular weight determination in an analytical ultracentrifuge by means of the sedimentation-equilibrium-method[42]. As the basis for calculations the well-known Eq. (19) was used:

$$\ln \frac{C}{C_0} = \frac{\frac{1}{2} \cdot M\,(1-\bar{v} \cdot \rho) \cdot \omega^2 \cdot r^2}{RT} \tag{19}$$

where

C	concentration	mol · l^{-1}
M	molecular weight	g · mol^{-1}
ω	angular velocity	rad · s^{-1}
R	gas constant	erg · K^{-1} · mol^{-1}
T	temperature	K
$\bar{v}$	partial specific volume	cm^3 · g^{-1}
ρ	density	g · cm^{-3} .

For most proteins in water a value for the partial specific volume of 0.74 cm^3 · g^{-1} can be used[43]. If, for constant rpm, the logarithm of concentration is plotted against the second power of the distance from the rotation axis, a straight line should result with a slope from which the molecular weight can be calculated. From Eq. (19) can be seen that for half the value of rpm the slope is four times smaller. This can clearly be seen in Fig. 11. If Eq. (19) is solved for the molecular weight M, Eq. (20) results:

$$M = \frac{2\,RT \ln (C_2/C_1)}{(1-\bar{v} \cdot \rho) \cdot \omega^2\,(r_2^2 - r_1^2)} \,. \tag{20}$$

From different runs, the molecular weight was calculated to be M = 91,000 ± 6000 g · mol^{-1}.

This value is in good agreement with 86,000 g · mol^{-1} recently published by Kördel and Schneider[44].

---

[a] Serva, Heidelberg, FRG, cat. no. 10,720

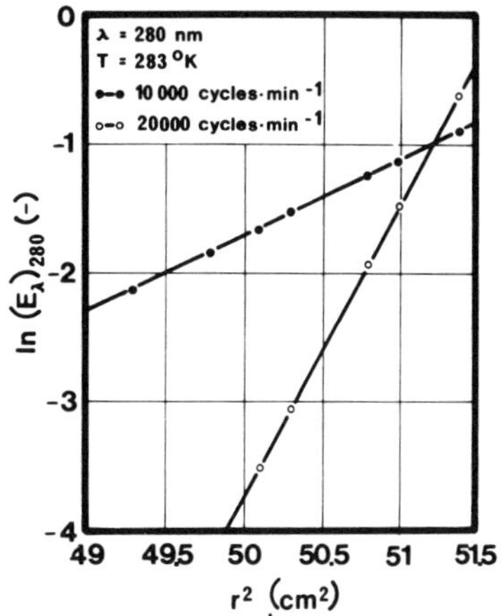

Fig. 11. Determination of the molecular weight of acylase (sedimentation equilibrium method by ultracentrifugation)

Operating with soluble enzymes, the molecular weight determines, which ultrafiltration device should be chosen (refer to Part 2.5.1).

The physical properties of insoluble enzymes are mainly determined by the carrier. The carrier fixed acylase described here was obtained by copolymerisation of modified acylase with acrylamide in the pores of an aluminium silicate carrier[45]. The following data are valid for the particle fraction 250–315 μ as summarized in Table 6.

Table 6. Characterization of acylase carrier fixed on porous clay (sieve fraction 250–315 μ)

True density	2.090 ($g \cdot cm^{-3}$)	Specific volume	0.478 ($cm^3 \cdot g^{-1}$)
Rel. void volume (V_{free}/V_{total})	0.828 (–)	Specific free volume	2.308 ($cm^3 \cdot g^{-1}$)
Porosity (V_{pore}/V_{grain})	0.004 (–)	Pore volume	0.002 ($cm^3 \cdot g^{-1}$)
Settled apparent density (in H_2O)	0.359 ($g \cdot cm^{-3}$)	Space requirement (in H_2O)	2.788 ($cm^3 \cdot g^{-1}$)
Swelling 4.8%, maximal catalyst concentration 433 $g \cdot l_{free}^{-1}$			

2.3.2 Chemical Properties

The chemical properties of the catalyst cannot be investigated strictly separated from the reaction system. This chapter will deal with the catalyst itself. Kinetic investigations will be described in Part 2.4.

The activity of soluble hog kidney acylase was measured at the following standard conditions:

- concentration (N-ac-D,L-met) 0.4 $mol_{D,L} \cdot l^{-1}$
- temperature 37 °C

- pH 7.0
- buffer –
- effector (Co^{2+}) 0.5 mmol · l^{-1} .

The polarimetric cuvette itself could be used as a batch reactor. At standard conditions an activity of 27.3 mmol · g^{-1} · min^{-1} was found.

For studying the pH dependence of activity, the reaction mixture was kept at constant pH in a batch reactor by means of a pH-stat using a flow-through polarimetric cuvette in a by-pass connection. The flowthrough cuvette is shown in Fig. (12). The sample stream is heated electrically (a) and passes through a bubble trap (b). Only a part of the stream reaches the light path across a glass frit, so that a pure solution can be measured between the quartz windows (d). The light path is surrounded by a thermostatic jacket (e). The temperature in the thermostatic jacket is compared with the temperature of the sample stream reaching the light path. The sample stream is heated until both temperatures are identical. By means of a needle valve (g) it is possible to adjust the sample stream.

The obtained pH-optimum is shown in Fig. 13.

Furthermore the temperature dependence of the initial rate was measured. The result is shown in Fig. 14. From an Arrhenius-plot an activation energy of 6.9 kcal · mol^{-1} (28.9 kJ · mol^{-1}) can be calculated.

Small concentrations of metal ions can influence the activity remarkably in a positive (effectors) or negative (inhibitor) way. In Fig. 15 the initial rate in relation to the maximal obtainable rate is plotted as function of the effector concentration (logarithmic scale). Besides cobalt, several alkaline earth metals and transition metals of the fourth period were tested at a concentration of 0.5 mmol · l^{-1}. The results are shown in Fig. 16, where the initial rate is referred to the maximal obtainable value using cobalt as effector.

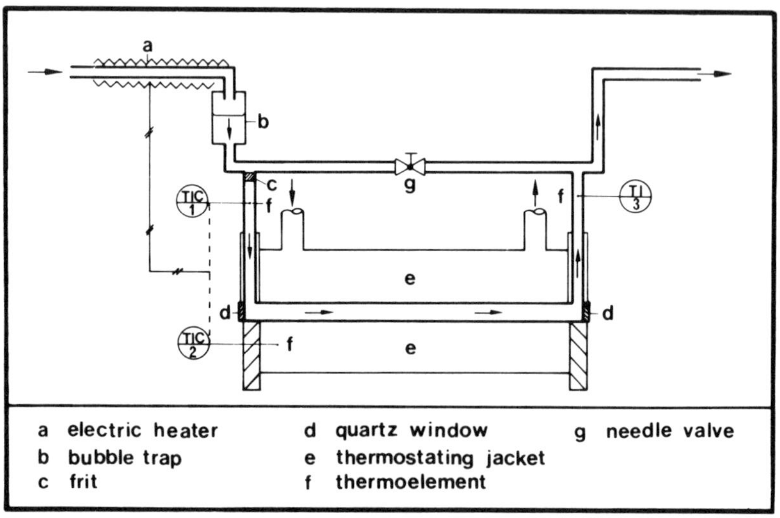

Fig. 12. Flow-through polarimeter cell

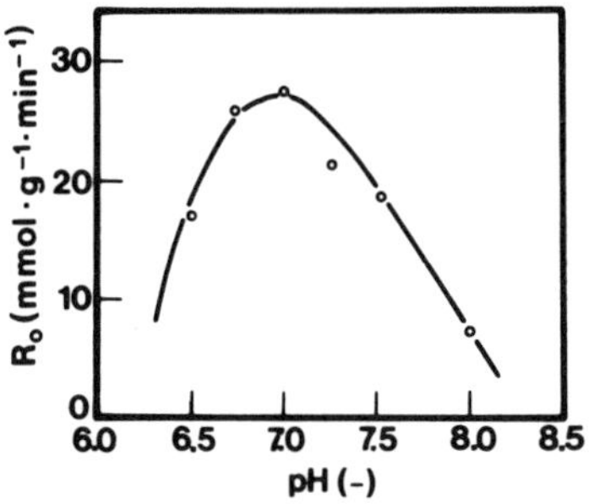

Fig. 13. pH dependence of the initial reaction rate for native acylase (standard condition)

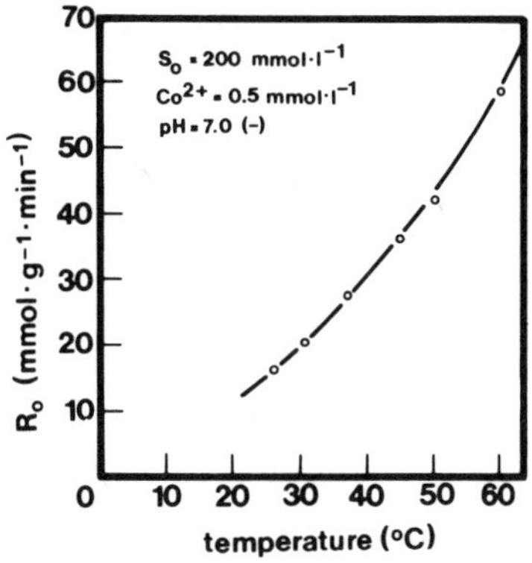

Fig. 14. Temperature dependence of the initial reaction rate for native acylase (standard condition)

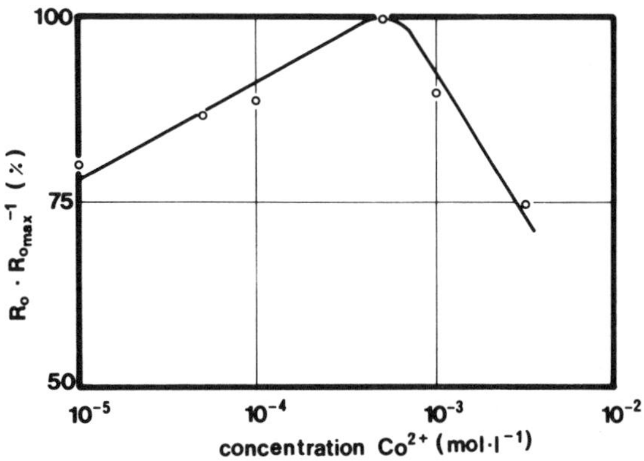

Fig. 15. Relative reaction rate in dependence of effector concentration (native acylase)

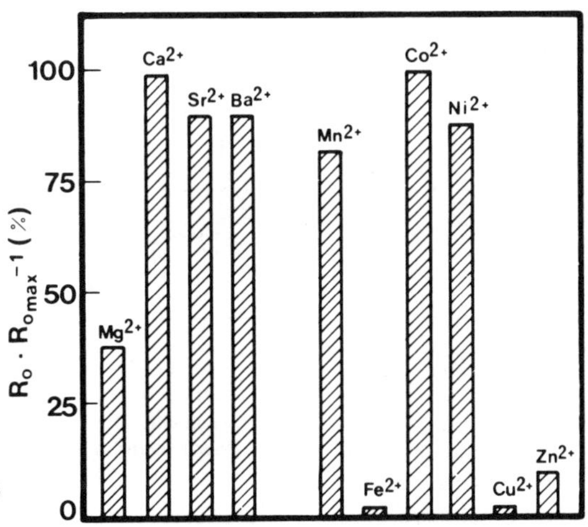

Fig. 16. Comparison of metal ion effectors for native acylase

The high activity using calcium as effector is remarkable. When different calcium concentrations were applied, only a little influence on concentration was found, so that the influence of calcium was perhaps more a case of a non-inhibitor than a case of an effector.

The membrane reactor shown in Fig. 17 was used for further investigations. A dosing pump conveys a substrate stream into a flat cylindrical vessel. The catalyst is retained by means of a membrane (e) on top of the reaction vessel. This membrane is permeable to substrate and product molecules (c) but not to enzyme molecules (d). Mixing is carried out by means of a magnetic stirrer (f), the distance between the magnetic stirrer and the membrane being so small that a turbulent flow pattern is reached near the membrane. The product solution leaves the reactor through a perforated plate (b) at the top of the reactor. The product stream is analysed continuously by means of polarimetry (g). Using this set-up, a small reactor volume ($15.3\ cm^3$) with a comparatively large separation area ($19.6\ cm^2$) can be established. An ultrafiltration membrane with a nominal cut-off of 10,000 was used under standard reaction conditions with a volumetric flow of $1.2\ cm^3 \cdot min^{-1}$ (average residence time = 12.75 min). The results are shown in Fig. 18. The conversion is given as function of operation time referred to the average residence time and alternatively in hours on top of the graph. At first no effector was present. During the first seven residence times the conversion decreases due to the bleeding of cobalt from the catalyst. From the beginning of the 8th residence time the substrate solution contained ethylenediaminetetraacetate at a concentration of $100\ mmol \cdot l^{-1}$ additionally as a complex forming agent for cobalt. Thus the catalyst could completely be deactivated and the conversion dropped to zero. After 30 residence times the substrate solution was changed to another one without a complex forming agent, but containing calcium ions at a concentration of $10\ mmol \cdot l^{-1}$. Reactivation by means of calcium can be observed. The catalyst reaches 56% of the activity achievable with cobalt as effector. That cobalt is the more powerful effector can be seen when, after about 130 residence times, the effector is changed from calcium to cobalt at a concentration of $0.5\ mmol \cdot l^{-1}$.

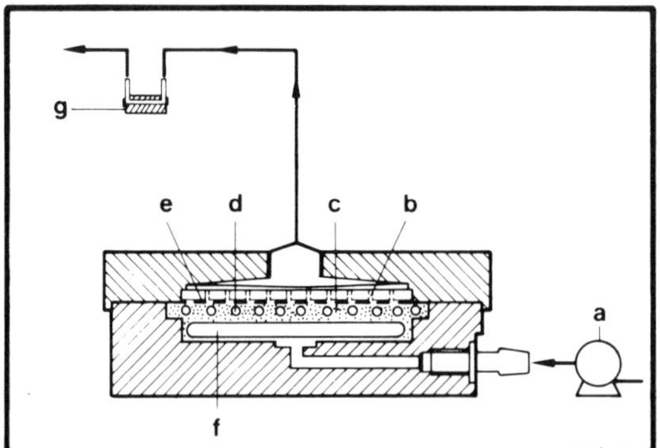

a pump
b bore
c reaction mixture
d enzyme
e membrane
f magnetic stirrer
g flow through cell

Fig. 17. Reactor with flat membrane for kinetic measurements

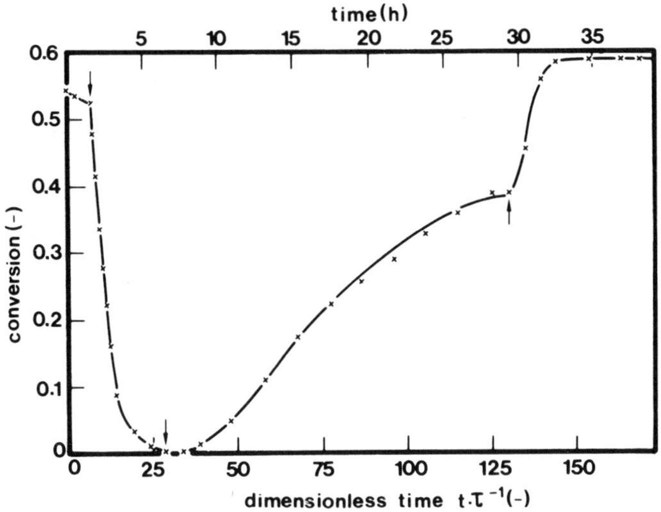

Fig. 18. Conversion in a reactor with flat membrane as a function of time and effector (described in the text)

A further quick activation occurs in spite of a 20 times lower concentration of the new effector.

The results of the effector experiments may be summarized as follows:

- Using ethylenediaminetetraacetate as complex forming agent acylase can be reversibly deactivated.
- The highest activity is achieved at a cobalt effector concentration of 0.5 mmol $\cdot l^{-1}$.
- If a substrate solution is used without any effector, there is bleeding of cobalt from the enzyme.
- Cobalt is able to displace calcium from the catalyst.
- With a physiologically harmless concentration of calcium (10 mmol $\cdot l^{-1}$) 56% of the maximal activity obtainable with cobalt can be reached.

The use of an enzyme for continuous operation is strongly influenced by its stability. Enzyme protein may be deactivated by unsuitable temperatures, pH-values, solvents, surfactants, interfaces, proteases, or powerful radiation.

To characterize the stability of an enzyme, a well defined deactivation rate should be used. Assuming a first order deactivating mechanism, the deactivation constant k_{de} is defined as the relative decrease of the maximal reaction rate constant per unit time. Deactivation constants should be measured in continuous operation. It may lead to great mistakes if the decrease in conversion is interpreted as decrease in activity[46]. A decrease in conversion from 99.9 to 99% (i.e. about 1%) may be caused, depending on the kinetic conditions, by an activity loss of more than 50%. Thus in graphs where the conversion is plotted against operating time no very meaningful turning points may occur. If the remaining activity is used instead of conversion, or better, if the logarithm of the remaining activity referred to the initial activity is used, it is quite simple to determine deactivation constants. The situation is summarized in Fig. 19.

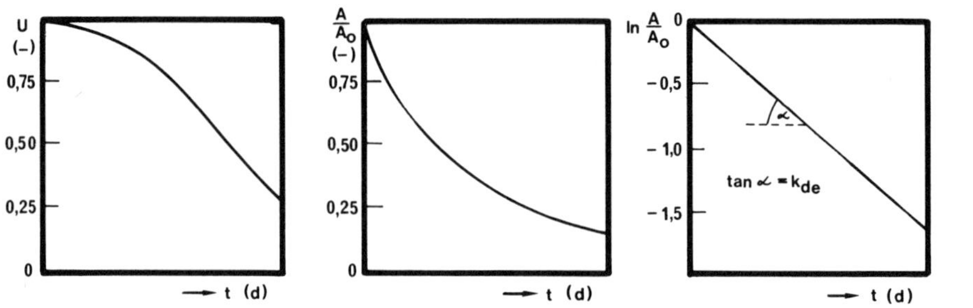

Fig. 19. Schematic plots of conversion, activity, logarithm of activity as functions of operation time

In most cases only the temperature stability is investigated. Instead of so-called temperature optima or short-time temperature stability it is more meaningful to follow the deactivation process in continuous operation. Figure 20 shows results for soluble acylase at standard reaction conditions using different temperatures as parameters. Since a logarithmic scale is used as ordinate, deactivation constants assuming a first order process can be calculated from the slope. Moreover, from the variation of the slope with the temperature a deactivation energy can be calculated. A value of 43 kcal · mol^{-1} (180 kJ · mol^{-1}) results, indicating that deactivation of enzymes is a rather complicated process.

In Table 7 the corresponding deactivation constants and half-life times respectively, are summarized.

Table 7. Deactivation constant and half-life of native acylase as function of temperature

T (°C)	k_{de} (d^{-1})	$t_{1/2}$ (d)
37	0.02	34.7
50	0.28	2.50
60	2.54	0.27
65	4.82	0.14
70	15.8	0.044
75	37.0	0.019

The chemical properties of carrier fixed acylase are not only determined by the enzyme protein, but also by the carrier material and the comonomer acrylamide. The same standard reaction conditions were applied as used during the investigations with soluble acylase. A particle fraction of 250–315 μ was chosen. Kinetic measurements were carried out in a loop reactor shown in Fig. 21. By means of a four way valve (a) the apparatus is first flushed with fresh substrate solution. Constant temperatuie in the loop reactor is achieved by means of the heat exchanger (b). The circulation pump (c) conveys

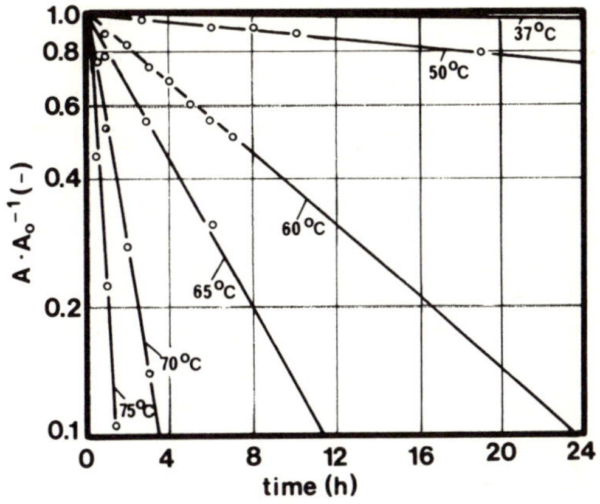

Fig. 20. Activity decay as function of operation time and temperature

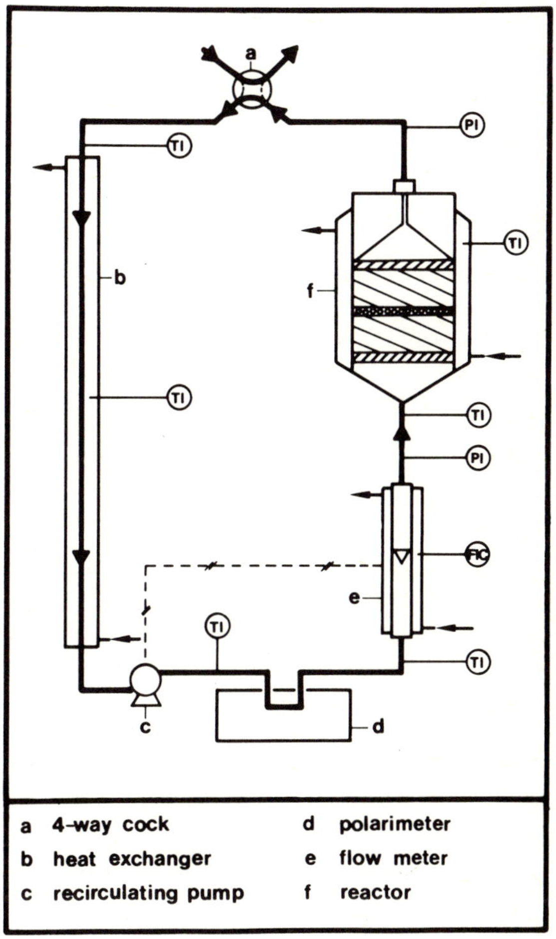

Fig. 21. Loop reactor for studying the kinetics of carrier fixed enzymes

the reaction solution through the polarimetric cuvette (d) and through a thermostatted flow-meter (e) to the differential fixed-bed reactor (f). In the thermostatted reactor of 25 mm inner diameter on top of a glass frit there is a layer of inactive carrier of 5 mm height. Then a layer of active carrier fixed acylase follows with a height of about 1 mm. On top of this, another layer of inactive carrier – 5 mm high – follows. The entire fixed bed is closed on top by means of a glass frit in a Teflon fitting. With this set-up the fixed bed can be flushed upwards by the fluid so that gas bubbles can easily be removed from the reactor. After flushing the set-up with new substrate solution the four-way valve is turned for batch-wise operation, achieving well-defined initial conditions. The apparatus can also be operated continuously, using two T-shaped pipes, one before the pump and another behind the reactor to enable connection to a dosing pump and a product vessel. Continuous operation in a loop reactor is especially interesting, because at constant average residence times the relative velocity of fluid to catalyst can independently be changed by means of the circulation pump. From such experiments it was found that the initial rate increases only by 2% when the particle Reynolds number (Re_K) is changed from 0.7 to 4.2. Therefore all activity measurements were carried out in this range of particle Reynolds numbers.

$$Re_K = \frac{u \cdot d_K}{\nu} \tag{21}$$

where u = fluid velocity (referred to the empty tube) $cm \cdot s^{-1}$
d_K = particle diameter cm
ν = kinematic viscosity $cm^2 \cdot s^{-1}$.

At standard reaction conditions the mass specific activity was found to be:

$A_m = 0.172\ mmol \cdot g^{-1} \cdot min^{-1}$ [a]

$(d_K = 250 - 315\ \mu)$

$(Re_K = 1.5)$.

With this catalyst, volume specific activities of more than 100 mmol $\cdot\, l^{-1} \cdot min^{-1}$ can be achieved (refer to Part 2.5.2).

The dependence of initial rate on pH and temperature was measured. No substantial difference in the pH optimum or temperature dependence in comparison with the soluble acylase could be found.

The stability of carrier fixed acylase at standard reaction conditions was investigated using a tube reactor. The slowly decreasing conversion was continuously monitored at constant flow by means of polarimetric analysis. The change of conversion with operat-

[a] Referring to Jaworek[47)] 0.07 g hog kidney acylase are used per gram carrier fixed enzyme. The mass specific activity of the native enzyme used is 42.6 mmol $\cdot\, g^{-1} \cdot min^{-1}$ at standard reaction conditions. Thus an activity yield of 5.77% results

ing time was converted to a change of activity by means of Eq. (22)

$$\frac{E \cdot \tau}{S_0} = \int_0^U \frac{1}{R}\, dU\,. \tag{22}$$

If the reciprocal value of the reaction rate, calculated from kinetic measurements (refer to Part 2.4) is plotted against the conversion, as shown in Fig. 22, the area under the curve up to a certain conversion represents the expression on the left side[64a)] of Eq. (22).

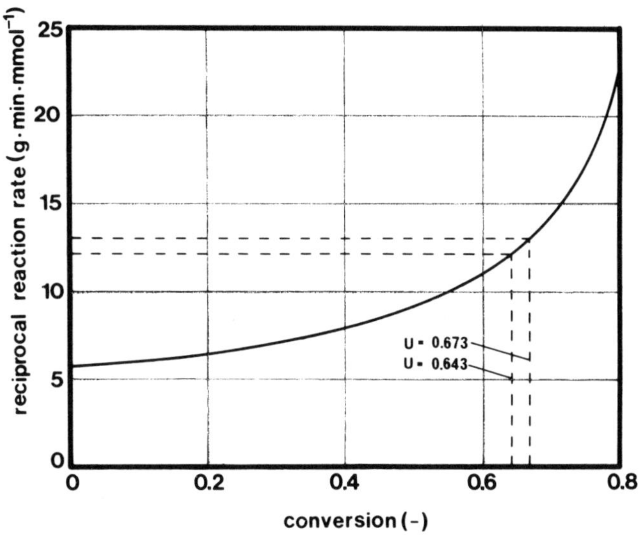

Fig. 22. Plot of reciprocal reaction rate versus conversion to estimate activities numerically (carrier fixed acylase, standard condition)

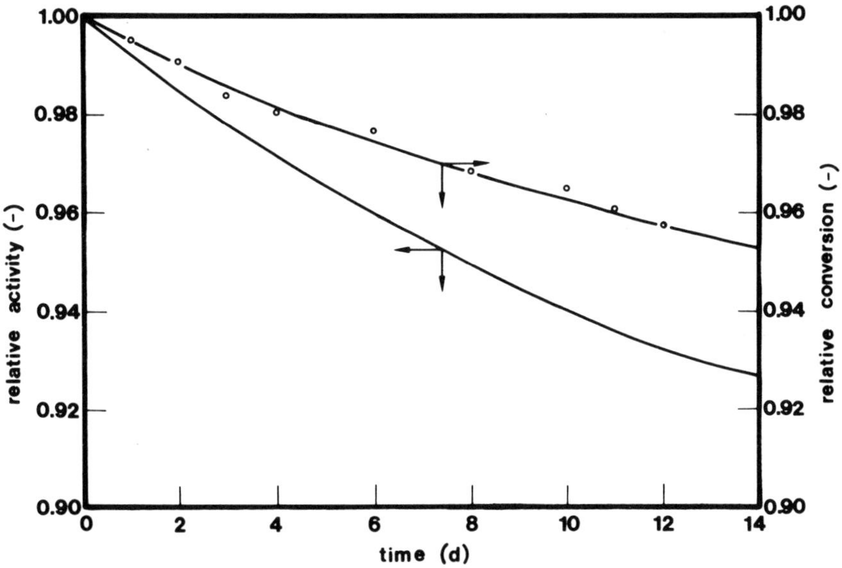

Fig. 23. Relative conversion and activity as functions of operating time (carrier fixed acylase)

Since the residence time (τ) and the initial substrate concentration (S_0) are kept constant, the decrease of the conversion seems to be caused by a decrease in enzyme concentration. Since the protein concentration was kept constant, this effect has to be interpreted as decrease of activity. The conversion decreased within 12.7 days from 67.3 to 64.3% as marked by broken lines in Fig. 23. The decrease of activity is represented by the area under the curve in Fig. 23 up to the conversion concerned. Integration [refer to Eq. (22)] was carried out numerically. The result is shown in comparison to the experimentally determined decrease of conversion in Fig. 23. It can be realized that after two weeks of continuous operation 95.3% of the initial conversion reached corresponding to a residual activity of 92.5%. These differences will increase with increasing initial conversion. From a plot of remaining activity (on a logarithmic scale) against the operating time the deactivation constant of carrier fixed acylase at standard reaction conditions can be calculated to be $k_{de} = 0.0056\ d^{-1}$.

2.4 Kinetic Investigations

2.4.1 Microkinetics

In this part, microkinetic investigations are described using the polarimetric cuvette as batch reactor and the membrane reactor (Fig. 17) as reactor for continuous operation.

The aim of the investigations was to determine the parameters of a handable kinetic model with sufficient precision.

Starting point for the modelling was the well-known equation of Michaelis and Menten[48] and Briggs and Haldane[49] respectively:

$$R = E \cdot \frac{V_{max} \cdot S}{k_m + S} \tag{23}$$

where

R	= reaction rate	$(\text{mmol} \cdot l^{-1} \cdot \text{min}^{-1})$
E	= enzyme concentration	$(g \cdot l^{-1})$
V_{max}	= reaction rate constant	$(\text{mmol} \cdot g^{-1} \cdot \text{min}^{-1})$
k_m	= Michaelis-Menten constant (reciprocal value of the adsorption equilibrium constant for substrate S)	$(\text{mmol} \cdot l^{-1})$
S	= actual substrate concentration	$(\text{mmol} \cdot l^{-1})$.

The expression of Michaelis and Menten is practically identical with a formulation of Hougen and Watson[50] following an expression of Langmuir[51] for the reaction rate of a heterogeneously catalysed reaction. In both cases (enzyme catalysis and heterogeneous catalysis) a quick adsorption equilibrium for the substrate at the catalyst is assumed followed by a comparatively slow chemical reaction.

The reaction rate constant (V_{max}) is at first a model parameter and has to be distinguished from the mass specific activity (A_m) of a catalyst which is only defined for certain experimental conditions. In the case of the simple model following Michaelis and Menten Eq. (24) holds:

$$\lim_{S \to \infty} A_m = V_{max} \,. \qquad (24)$$

The kinetic parameters were determined by methods of non-linear regression according to Hooke and Jeeves[52], Nelder and Mead[53], and Marquardt[54]. Graphical methods according to Lineweaver and Burke[55], Eddie[56] or Hofstee[57] were not used, since linearization of the kinetic equation would lead to a distortion of the error distribution.

The kinetic measurements were carried out in a batch reactor. A photometer cuvette and a polarimeter cuvette served for this purpose. The reaction mixture was thermostatted in a separate container. At time zero a certain amount of soluble enzyme was added. After dissolution an aliquot was quickly transferred to the cuvette, which was kept at the same temperature as the container. The whole procedure took approximately 90 s. Initial rates were measured with a catalyst concentration of 20 mg lyophylised enzyme per liter. Integral conversion data were measured using a tenfold catalyst concentration. From the calibration function of optical rotation vs. concentration and conversion, it was possible to calculate the concentration vs. time curves. By numerical differentiation of these curves, reaction rate vs. concentration curves were obtained. Because data were collected every 5 s, accurate determination of reaction rates was possible. In order to avoid giving too much weight to the data at low rates, only equidistant data points on the rate vs. concentration curve were taken. Due to the high sensitivity of the differential photometric analysis, this method was applied to measure initial rates in the substrate concentration range from 0.5 to 5 $mmol_L \cdot l^{-1}$. The results are shown in Fig. 24. The

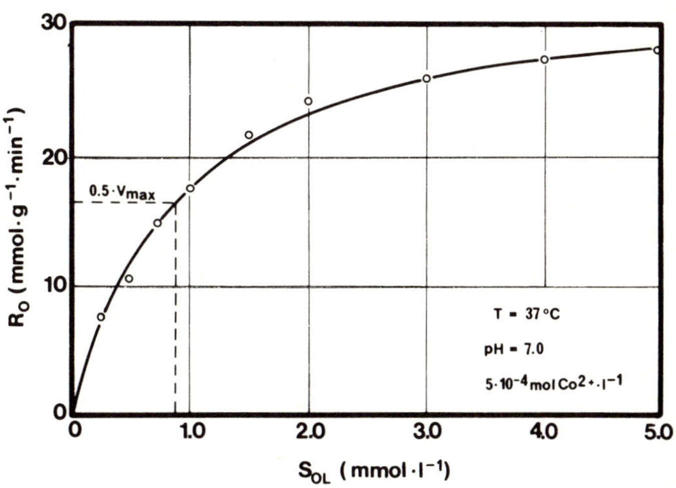

Fig. 24. Initial rate kinetics of native acylase at low substrate concentrations

following kinetic parameters were determined by non-linear regression analysis:

$$V_{max} = 33.7 \text{ mmol} \cdot g^{-1} \cdot \text{min}^{-1} \qquad k_m = 0.91 \text{ mmol} \cdot l^{-1}$$

(concentration range: 0 – 5 $\text{mmol}_L \cdot l^{-1}$) .

A comparison of the k_m values published in literature[58] shows that this parameter increases with increasing concentration range applied in the investigations. This indicates that not always true initial rates have been estimated.

Owing to strong product inhibition (see below), unduly high k_m values were found. In this investigation initial rates were measured up to substrate concentrations of 1000 mmol · l^{-1}. In Fig. 25 the measured values are plotted as a function of the concentration of L-enantiomer. The investigations were carried out with pure L-substrate as well as with racemic mixtures. In both cases a maximum occurs due to substrate inhibition. This inhibition is increased in the presence of the D-enantiomer (compare the curve for pure L-substrate and the curve for DL-substrate). Thus the maximal initial rate is attained at comparatively low substrate concentrations. The inhibition is obviously due to the overall concentration level. By plotting the initial rate as a function of the overall substrate concentration both curves coincide at high concentrations.

To describe the concentration level-dependent inhibition, the following model proved to be useful:

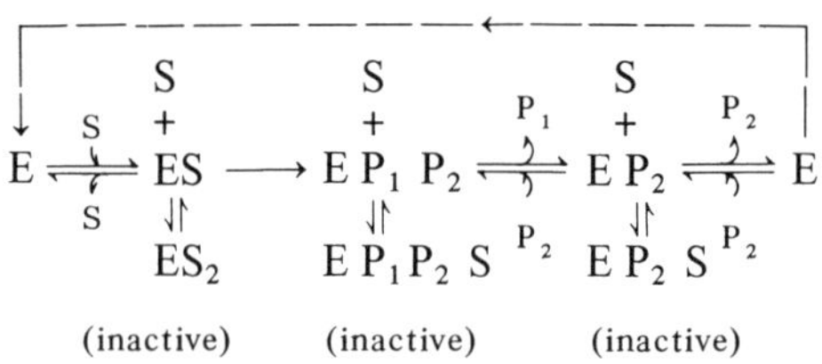

(25)

Theoretically three inactive intermediate complexes (ES_2, EP_1P_2S, and EP_2S) may be formulated. The symbols P_1 and P_2 represent the two products (acetate and L-aminoacid). The model described by Eq. (25) leads to an extension of Eq. (23) where the reaction rate constant (V_{max}) is decreased according to the substrate concentration level. This is formulated in Eq. (26):

$$R_o = E \cdot \frac{V_{max}}{\left(1 + \dfrac{S_{oL}\ (-) + S_{oD}}{k_{in}}\right)} \cdot \frac{S_{oL}}{k_m + S_{oL}}\,, \qquad (26)$$

where k_{in} inhibition constant for substrate (non-competitive) (mmol · l^{-1})

o initial condition

L L-enantiomer

D D-enantiomer .

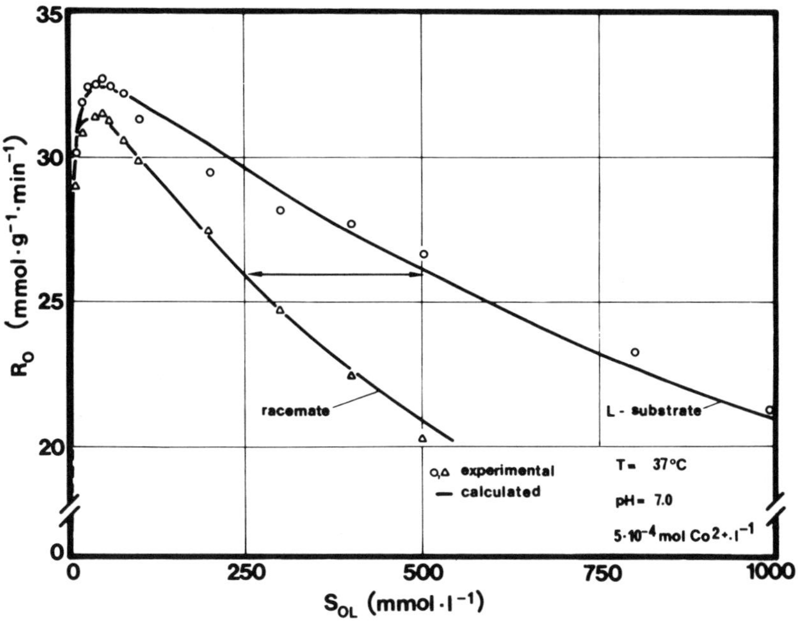

Fig. 25. Initial rate kinetics of native acylase at high substrate concentrations

As long as only initial rates are considered, the influence of the state of equilibrium or of the product do not have to be taken into account. Equation (26) is valid for the L-substrate as well as for racemic mixtures. The kinetic parameters were obtained by non-linear regression. The common fit for all measurements, shown in Fig. 25, led to the following parameters:

$$V_{max} = 34.5 \text{ mmol} \cdot g^{-1} \cdot \text{min}^{-1} \qquad k_m = 1.60 \text{ mmol} \cdot l^{-1}$$

$$k_{in} = 1566 \text{ mmol} \cdot l^{-1} .$$

The continuous lines in Fig. 25 were calculated with these parameters. One realizes that Eq. (26) describes the situation for the L-substrate as well as for the racemic mixtures satisfactorily. That the model is of practical importance can be seen by the fact that in spite of a 200-fold higher concentration level in comparison with the first measurements, very similar values for the model parameters V_{max} and k_m result.

The substrate inhibition, as described above, diminishes with increasing temperature as can be seen from Fig. 26. In this figure the initial reaction rates are plotted as a function of the substrate (racemate) concentration for three different temperatures. The result of the regression analysis for all three temperatures are summarized in Table 8. The continuous lines in Fig. 26 were calculated with these parameters. According to a decreasing substrate inhibition with increasing temperature the corresponding inhibitor constant increases because this constant is used in the model in a reciprocal form. An

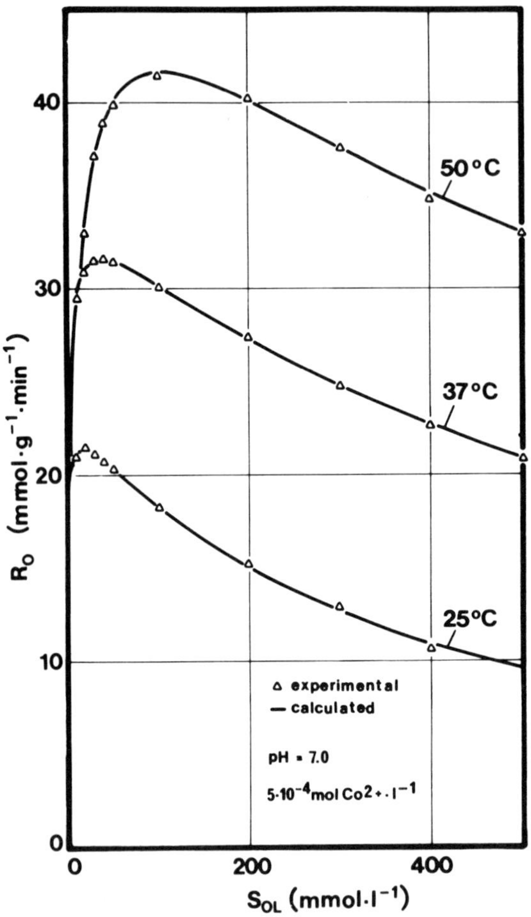

Fig. 26. Initial rate kinetics of native acylase at different temperatures

Table 8. Parameters of initial rate kinetic model as function of temperature

T	(°C)	25	37	50
V_{max}	$(mmol \cdot g^{-1} \cdot min^{-1})$	24.1	34.5	50.2
k_m	$(mmol \cdot l^{-1})$	1.2	1.6	9.8
k_{in}	$(mmol \cdot l^{-1})$	682	1566	2045

increase of the k_m value with temperature can be understood analogous to a decreasing adsorption equilibrium constant.

To investigate inhibition by product, the reaction rates were measured in the presence of different amounts of L-methionine or acetate as shown in Fig. 27. To describe these measurements, the kinetic model has to be extended once more:

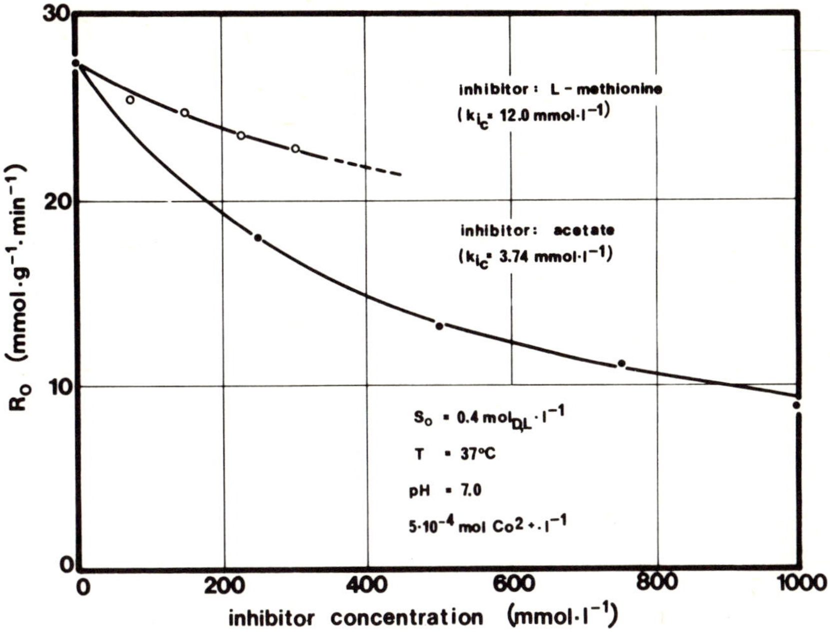

Fig. 27. Inhibition by products

$$R_0 = E \cdot \frac{V_{max}}{\left(1 + \frac{S_{oL} + S_{oD}}{k_{in}}\right)} \cdot \frac{S_{oL}}{k_m \left(1 + \frac{I}{k_{ic}}\right) + S_{oL}} \tag{27}$$

where I = inhibitor (product) concentration ($mmol \cdot l^{-1}$)
k_{ic} = inhibitor constant for the product (competitive) ($mmol \cdot l^{-1}$).

Using the model parameters determined before, the new parameters describing the product inhibition were found to be:

$$k_{ic}(\text{L-met}) = 12.0 \text{ mmol} \cdot l^{-1} \qquad k_{ic}(\text{Ac}) = 3.74 \text{ mmol} \cdot l^{-1}.$$

If it is considered that L-methionine and acetate are formed during the reaction in equimolar amounts, a common inhibition constant should be used. This constant can be calculated to be:

$$k_{ic}(\text{common}) = 2.85 \text{ mmol} \cdot l^{-1}.$$

For practical applications the reaction rate must be known as a function of conversion. For this purpose the kinetic model must be extended once more:

$$R = E \cdot \frac{V_{max}}{\left(1 + \frac{S_L + S_D}{k_{in}}\right)} \cdot \frac{S - \frac{(S_{oL} - S_L)^2}{K}}{k_m \left(1 + \frac{S_{oL} - S_L}{k_{ic}}\right) + S_L} \tag{28}$$

where K (equilibrium constant) = 2750 mmol $\cdot$ l^{-1} .

Now the reduction of the driving force due to the reverse reaction is taken into account. To estimate the model parameters valid for integral kinetics, measurements were carried out in a batch reactor up to the equilibrium state with different initial substrate concentrations: 0.1, 0.2, 0.4, 0.6, and 0.8 mol racemate $\cdot$ l^{-1}.

In Fig. 28 the experimental data are plotted as a function of the dimensionless concentration of substrate S_L/S_{oL} with different initial concentrations as parameter. From this figure the influence of substrate inhibition on the reaction rate can clearly be recognized: The reaction rate decreases with increasing initial substrate concentration.

The expected maximum of the reaction rate as a function of conversion – due to the substrate inhibition – is overlapped by the product inhibition. Therefore no maximum appears as a function of conversion. Model parameters were estimated using Eq. (28). The continuous lines in Fig. 28 were calculated using these parameters which are summarized in Table 9 together with the corresponding parameters evaluated from the initial rate experiments. The good agreement between these parameters obtained from initial rate experiments as well as from integral runs, indicate that the model is of physicochemical significance.

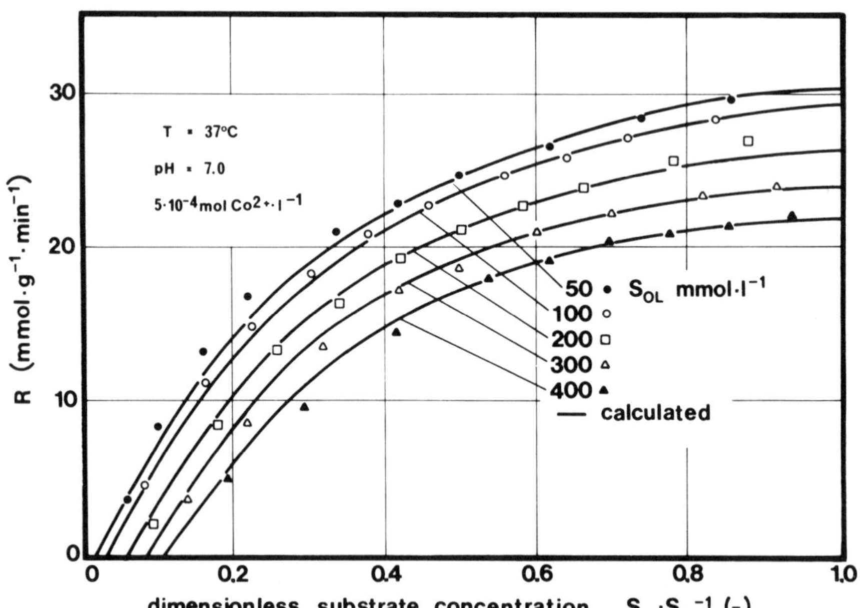

Fig. 28. Integral kinetics of native acylase

Table 9. Comparison of parameter-sets estimated out of initial-rate data and integral measurements

Parameter	V_{max} mmol · g^{-1} · min^{-1}	k_m	k_{in}	k_{ic}
		mmol · l^{-1}		
Initial-rate	34.5	0.91	1566	2.85
Integral runs	34.0	0.77	1450	2.80

2.4.2 Macrokinetics

In this part kinetic investigations with carrier fixed acylase are described. The experiments were carried out in the loop reactor (Fig. 21). The influence of the relative velocity of fluid to catalyst was investigated in the continuously operated loop reactor by means of varying the recycle stream at constant average residence time. The results are shown in Fig. 29. The initial rate obtained at a particle Reynolds number Re_K = 1.5 is shown as well. At two average residence times (6.1 and 11.8 min) the particle Reynolds number was varied between 0.7 and 4.2. It can be seen that the reaction rate is slightly increased with increasing Reynolds number, which is related to an increased conversion at constant residence time. Also the kinetic investigations were carried out at Reynolds number of 1.5 or even more, thus film diffusion effects could be neglected.

Initial rates as a function of the concentration level are shown in Fig. 30 for different particle sizes. The increase of rate with decreasing particle diameter is not only due to

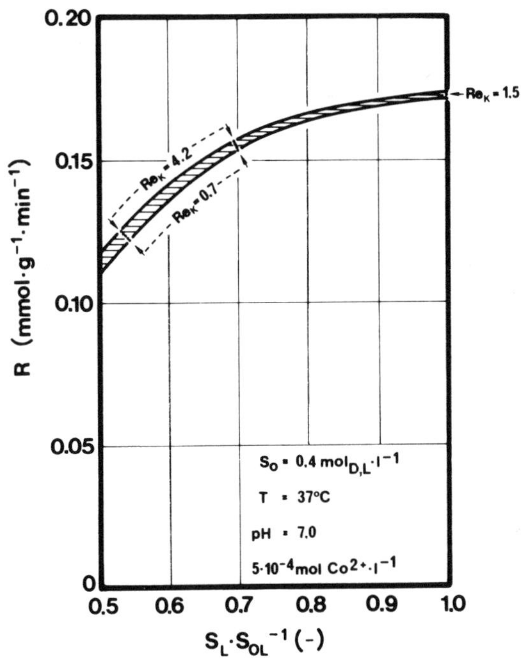

Fig. 29. Reaction rate as function of dimensionless substrate concentration (parameter: Reynolds number)

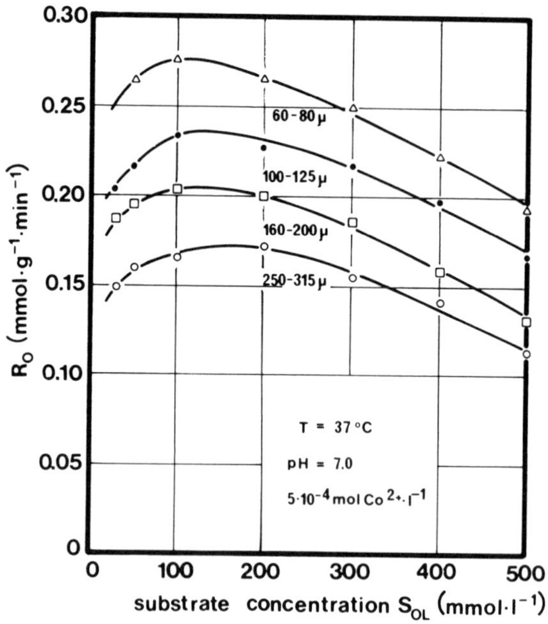

Fig. 30. Initial rate kinetics of carrier fixed acylase with different particle sizes

less severe pore diffusion influence but also to the method of catalyst preparation. Using smaller particles, the relative fraction penetrated by the polymerisation-solution is higher than if larger particles are used. All further investigations were carried out with a particle fraction of 250–315 μ, because only with this fraction the pressure drop in a fixed bed reactor could be kept at a reasonable level. It would theoretically be possible to decrease pressure drop by using small relative velocities of catalyst to fluid. In this case the film diffusion influence would increase, so that a simple transfer of kinetic results from the loop reactor to fixed bed reactors would not be possible.

In order to investigate the rates as a function of conversion, batch experiments were carried out in the loop reactor at concentration levels of 0.1, 0.2, 0.4, 0.6, and 0.8 mmol$_{D,L}$ · l^{-1}. The results are shown in Fig. 31. The dashed lines in Fig. 31 represent only an interpolation between the experimental points. At high conversion one realizes a similar dependenc of the rates as was found with native acylase. At low conversion the influence of pore diffusion limitation can be realized when low substrate concentrations are used. With decreasing concentration level the substrate inhibition decreases at first, but later on the substrate saturation of catalyst is obviously no longer quantitative.

If the same type of model as for native acylase is used for the carrier fixed type of catalyst, this model can obviously be only a simple approximation. In Table 10 the results of parameter estimation are compared with those obtained for the native catalyst. For reactor design purposes the parameter set is a very useful approximation, when a concentration of more than 0.2 mol$_{D,L}$ · l^{-1} is used. For lower concentration levels or better accuracy, Eq. (22) can be used:

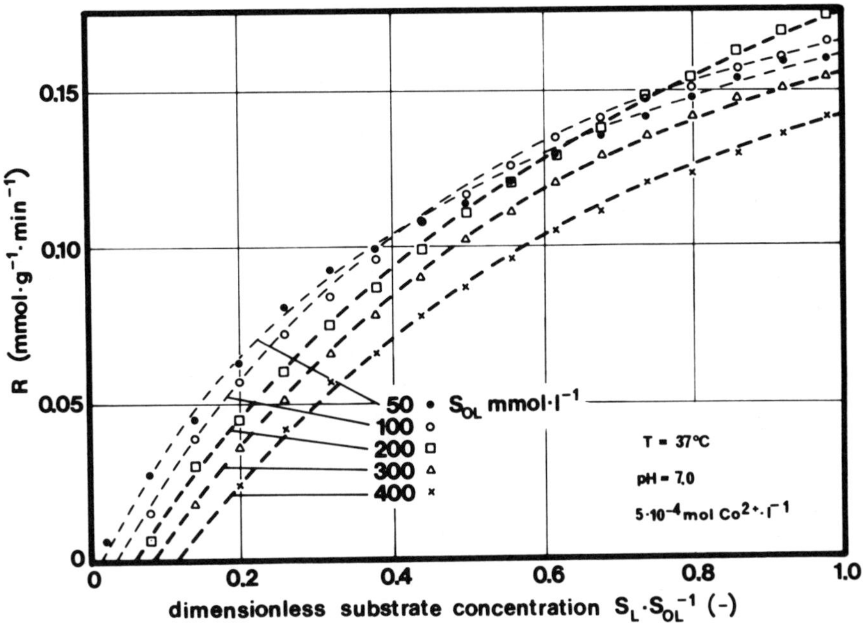

Fig. 31. Intergral kinetics of carrier fixed acylase

Table 10. Comparison of kinetic parameters for native and carrier fixed acylase

Parameter	V_{max} mmol · g^{-1} · min^{-1}	k_m	k_{in}	k_{ic}
		mmol · l^{-1}		
Native acylase	34.0	0.77	1450	2.80
Carrier fixed acylase (250–315 μ)	0.182	1.00	3000	2.20

$$\frac{E \cdot \tau}{S_0} = \int_0^U \frac{1}{R} dU . \quad (22)$$

Design tasks can be solved by numerical integration using the reciprocal value of the rate as a function of conversion.

2.5 Reactor Performance

In this part the experimental investigations on reactors for production purposes are described. The results of further modelling and cost estimation will be given in Part 3.

2.5.1 Reactors for Soluble Enzymes

As the physical characterization (molecular weight) and the chemical characterization (stability) of soluble acylase has shown, this enzyme may be used for continuous homogeneous catalysis. Different membrane materials were tested for their enzyme-compatibility. Cellulose acetate proved to be the most favourable material. A bunch of hollow fibers was used as separation unit. The properties are summarized in Fig. 32.

A flowsheet of the apparatus for the continuous production of L-methionine is given in Fig. 33[59]. The substrate is conveyed by means of a dosing pump (a) and a safety electrovalve controlling the pressure in front of two sterile filters (c) with a cut off of 0.2 μ by means of an electric manometer. The substrate passes an additional safety-valve and a thermostat (d) to reach the reactor (e) itself. The reactor-loop with the separation unit (g), the circulation pump (f) and another thermostat is, in principle, a CSTR. Product and unconverted substrate leave the separation unit through the inner core of the hollow fibers. The product solution passes the flow-through cuvette of the polar-

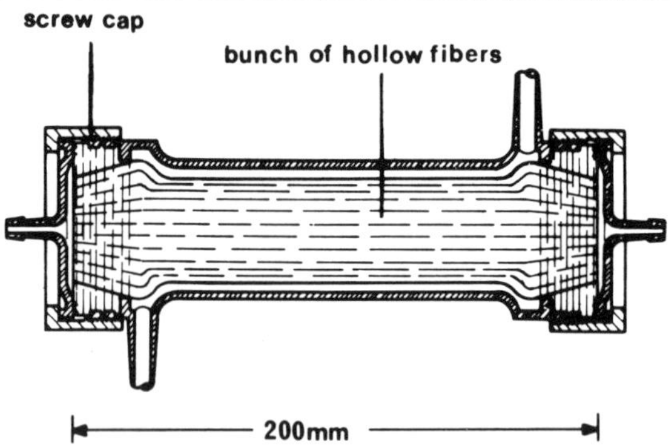

type	Dow Miniplant, c/HFU – 10
membrane material	celluloseacetate
membrane structure	symmetric
molecular cut-off	MG 70 000
membrane area	1.0 (m^2)
permeation rate 25 °C	100 ($ml \cdot min^{-1} \cdot bar^{-1} \cdot m^{-2}$)
pressure stability	2 (bar)
temperature stability	50 (°C)
pH – stability	4 – 9
outer volume	100 (ml)
fiber volume	75 (ml)

Fig. 32. Hollow-fiber separation unit

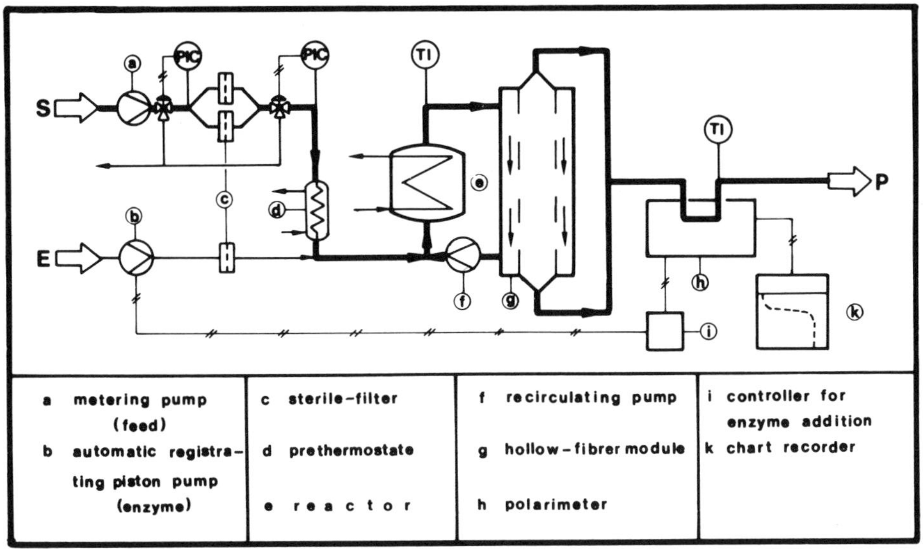

Fig. 33. Flow chart of a plant for continuous production of L-methionine using native acylase

imeter. The analog signal of the polarimeter (h) is recorded (k). The same signal is used in a control system (i) to operate an automatic burette (b). Thus catalyst solution is added, when the product concentration drops owing to catalyst deactivation. The flux from the automatic burette is recorded continuously, allowing the amount of catalyst needed per unit weight of product to be easily calculated. The recirculation pump is operated at high rates so that concentration polarization or adsorption of enzyme at the membrane can be kept low. Using a flux of about $1\,l \cdot h^{-1}$, enzyme retention was quantitative after about 25 residence times, due to an additional filter effect of a gel layer of enzyme protein.

The catalyst cannot be used quantitatively in a homogeneous way, because adsorption at the membrane cannot be completely avoided. It is favourable to operate the reactor at low enzyme concentrations, especially when it is considered that the protein concentration in the reactor gradually increases owing to the control system described above. In practice, enzyme concentrations from 0.5 to $2.0\,g \cdot l^{-1}$ were used. If necessary, it would not be limiting to operate the system even at lower enzyme concentration, because the flux capacity of the separation unit proved to be the true limiting feature. It is advisable to operate the separation unit at less than the maximal possible flux, in order to keep the enzyme retention practically quantitative.

The effectiveness of acylase in a membrane reactor was measured in continuous operation. For this purpose the membrane reactor was operated at different flow rates. The resulting reaction rates were compared with the corresponding rates of "free" soluble enzyme at the same conversion and concentration level. Figure 34 shows the data. The reaction rate (left ordinate) is plotted as function of fraction (S_o/S_{oL}). The ratio of reaction rates is defined as effectiveness (ϵ) (right ordinate). The effectiveness de-

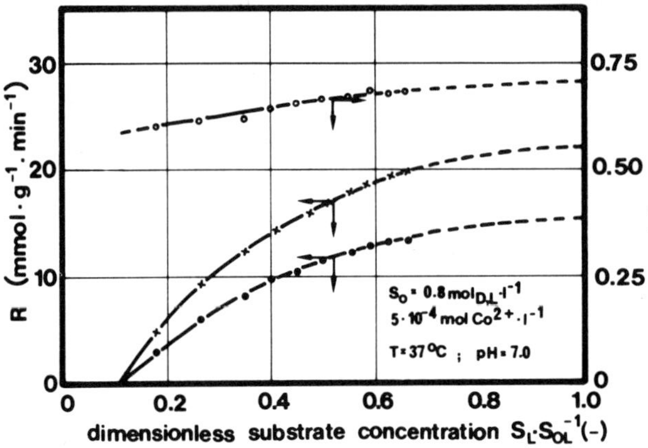

Fig. 34. Reaction rate of free enzyme (x-x-x) and native enzyme in membrane reactors (•-•-•) and catalyst efficiency in membrane reactors (○-○-○) as functions of dimensionless substrate concentration

creases slightly with increasing conversion. The value for initial rate conditions can be extrapolated and was estimated to be about 70%. The value for the effectiveness of the catalyst in continuous operation is needed for the economic analysis. The operational stability of the catalyst was determined in a long-term experiment. For this purpose the apparatus shown in Fig. 33 was operated continuously for 15 days. The experimental conditions are summarized in Table 11. The deactivation constant, directly calculated from the recorded flux of additional catalyst solution, is shown as a function of operation time in Fig. 35. The gradually decreasing value is due to the improved catalyst retention when the protein coat on the membrane is formed. The effect is not only due to better retention, but also to the fact that the enzyme coat on the membrane is more favourable for new still active enzyme than the pure membrane itself, thus new catalyst is protected from deactivation caused by the membrane. At steady state a deactivation constant of 0.0344 d^{-1} results in comparison with a value of 0.020 d^{-1} for "free" soluble acylase. It should be stated that practically one month would be needed for the protein concentration in the membrane reactor to double.

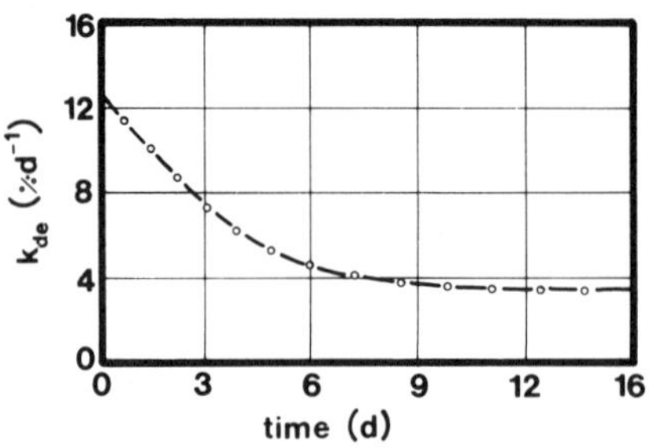

Fig. 35. Deactivation constant as a function of operation time in a membrane reactor (for operation condition see Table 11)

Table 11. Operation conditions of the extended time test to measure the deactivation constant

Racemate concentration	(mmol · l^{-1})	600
Temperature	(°C)	37
pH	(–)	7.0
Effector, Co^{2+}	(mmol · l^{-1})	0.5
Reactor volume	(l)	1.07
Membrane area	(m^2)	1
Membrane cut-off	(MG)	70,000
Flow-rate	(ml)	9.6
Residence time	(min)	111
Initial amount of enzyme	(g)	0.835
Flushed amount of enzyme	(g)	~0.350
Residual amount of enzyme	(g)	0.485
Enzyme concentration	(g · l^{-1})	0.453
Conversion	(%)	80.2
Measured reaction rate	(mmol · l^{-1} · min^{-1})	2.16
Calculated reaction rate	(mmol · l^{-1} · min^{-1})	3.75
Effectivity	(%)	57.6
Deactivation constant k_{de}	(d^{-1})	0.0344

One argument against this type of membrane reactor might be that for this type of kinetics, a plug flow reactor is more favourable. Plug flow can be approximated using a cascade. Practically it would be expensive to install a separation unit in each stage of a cascade. Therefore the reactor system shown in Fig. 36 was developed[60].

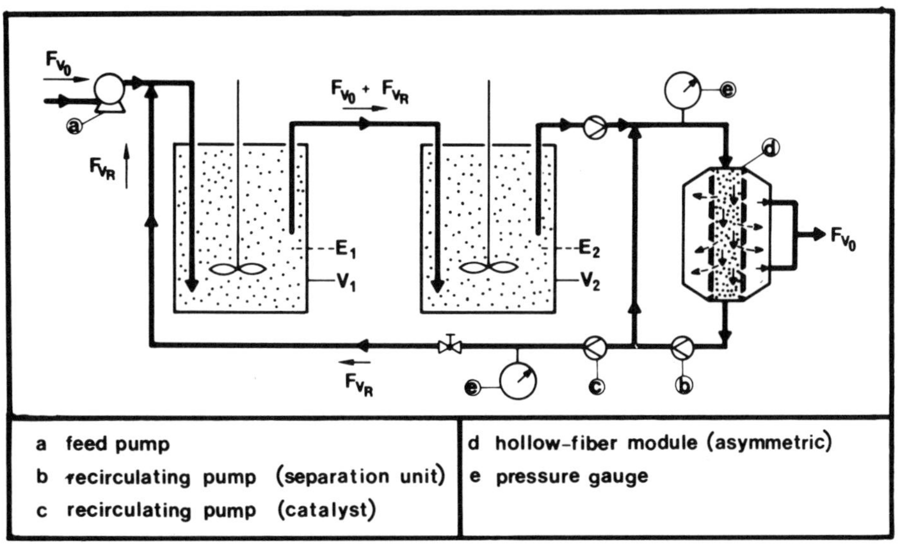

a feed pump	d hollow-fiber module (asymmetric)
b recirculating pump (separation unit)	e pressure gauge
c recirculating pump (catalyst)	

Fig. 36. Flow chart of a membrane reactor cascade with catalyst recirculation using only one separation unit

Fresh substrate solution with a volumetric flow F_{vo}, together with a recycle flow F_{vR} from the catalyst pump enters the first tank of volume V_1 with an enzyme concentration E_1. The second tank has the same volume ($V_2 = V_1$) and enzyme concentration ($E_2 = E_1$). The third stage has only 1/10 of the Volume ($V_3 = V_2/10$). The enzyme concentration E_3 differs from that in the first two tanks. A product stream F_{vo} leaves the stage across a bunch of hollow fibers while the catalyst is retained and recycled to the first tank with a flux F_{vR}.

If this flux F_{vR} is zero, the whole of the catalyst will accumulate in the last stage and no conversion will be achieved in the first two stages. If, on the other hand the catalyst pump is operated at very high rates, the catalyst is distributed uniformly along the whole cascade. In this state the whole system behaves as a single CSTR. Both extremes will lead to the same reactor performance. In the first described alternative there is a CSTR of small volume but high enzyme concentration, in the second case there is a large well-mixed volume with low enzyme concentration. Between those two extremes there must obviously be an optimum.

As separation unit a bunch of 44 hollow fibers with 1.5 mm inner diameter and a cut-off of 10,000 daltons was used[a].

The hollow-fiber module was operated as indicated in Fig. 36. Owing to the comparatively large inner diameter of the hollow fibers and the high linear velocity of the fluid in the inner core, no obstruction of fibers occurred. The cascade was operated under the conditions summarized in Table 12.

Table 12. Operation conditions of cascade

V_1	(l)	0.287	S_0	($mmol_{D,L} \cdot l^{-1}$)	500
V_2	(l)	0.287	T	(°C)	37
V_3	(l)	0.0287	pH	(–)	7.0
F_{vo}	($ml \cdot h^{-1}$)	200	Co^{2+}	($mmol \cdot l^{-1}$)	0.5
τ	(h)	3.0			

Since the third stage has only 1/10 of the volume of the earlier stages, the performance of an ideal three-stage cascade will result when the catalyst concentration in the third stage is ten times higher than that in the first two stages. In this case the product of residence time and enzyme concentration is equal in all three stages. This case is achieved when Eq. (29) holds:

$$(F_{vo} + F_{vR})/F_{vR} = 10 \,. \tag{29}$$

For the total amount of enzyme (E_m) the following expression can be used:

$$E_m = V_1 \cdot E_1 + V_2 \cdot E_2 + V_3 \cdot E_3 \,. \tag{30}$$

[a] Berghof, Tübingen, FRG, special design

Setting $V_1 = V_2$ and $E_1 = E_2$ Eq. (31) follows:

$$E_m = 2 \cdot V_2 E_2 + V_3 E_3 \, . \tag{31}$$

At steady-state the catalyst concentrations are constant.

For the third stage Eq. (32) holds:

$$0 = (F_{vo} + F_{vR}) \cdot E_2 - F_{vR} \cdot E_3 \, . \tag{32}$$

Equation (32) may be solved for E_2 or E_3:

$$E_2 = \frac{E_m}{2\,V_2 + V_3 \cdot \dfrac{F_{vo} + F_{vR}}{F_{vR}}} \, , \tag{33}$$

$$E_3 = \frac{E_m}{2\,V_2 \cdot \dfrac{F_{vR}}{F_{vo} + F_{vR}} + V_3} \, . \tag{34}$$

Experiments were carried out using a total amount of 100 mg acylase and a constant substrate flow rate of 200 $cm^3 \cdot h^{-1}$. Conversion was measured after each stage polarimetrically.

The actual enzyme concentration in each stage was calculated from kinetic measurements by means of samples taken from each stage. At steady-state a sample of 1 cm^3 was taken and diluted with 4 cm^3 fresh substrate solution. The conversion in this diluted sample was followed with another polarimeter. The rate was estimated by extrapolation to conversion of zero. Using the kinetic model, the amount of enzyme present in the sample was calculated. All data are summarized in Table 13. Moreover, the results are shown graphically in Fig. 37. The continuous lines are calculated using an average effectiveness of the catalyst of 42% as found in the experiments. As can be seen from the figure, at low rates of catalyst-recycling most of the conversion is achieved in the third stage. On the other hand, at high rates of catalyst-recycling there is only a small increase in conversion in the second and third stages. The deviation of the experimental results from the calculated values for the lowest rate of catalyst-recycling can be explained by the fact that, in this case, the catalyst concentration in the last stage is comparatively high, so that a greater amount is adsorbed at the membrane and not available for effective homogeneous catalysis. At higher recycling rates the catalyst-recycling can be ex- in the last stage will become smaller and smaller so that this effect disappears gradually. At a rate $F_{vR}/F_{vo} = 0.111$ an identical product of catalyst concentration and residence time for each stage will theoretically be reached. This would be the optimal catalyst distribution for a first order reaction. In this case the maximal conversion is achieved at a rate 0.125, indicating that higher recycling rates of catalyst are more favourable, owing to the fact that the reaction order only increases slowly with increasing conversion from zero to 1.

Table 13. Results of cascade-operation (initial amount of enzyme = 100 mg)

F_{vR}	$(ml \cdot h^{-1})$	10	20	60
U_0	(%)	3.6	7.3	18.4
U_1	(%)	23.2	36.9	52.7
E_{m1}	(mg)	7.20	12.17	19.00
$(E_{m1}/E_m)_{meas}$	(%)	18.4	28.1	43.2
$(E_{m1}/E_m)_{calc}$	(%)	24.4	32.2	41.1
U_2	(%)	44.1	61.3	73.9
E_{m2}	(mg)	8.62	12.92	18.17
$E_{m2}/E_m)_{meas}$	(%)	22.0	29.9	41.3
$E_{m2}/E_m)_{calc}$	(%)	24.4	32.2	41.1
U_3	(%)	75.6	80.0	79.9
E_{m3}	(mg)	23.32	18.17	6.84
$E_{m3}/E_m)_{meas}$	(%)	59.6	42.0	15.5
$E_{m3}/E_m)_{calc}$	(%)	51.2	35.6	17.8
$E_{m\ eff}$	(mg)	39.14	43.26	44.01

2.5.2 Reactors for Carrier Fixed Enzymes

The kinetic investigations have shown that a reactor with plug flow should be used. Acylase fixed to aluminium silicate can be used in tubular reactors at reasonable particle Reynolds numbers in contrast with catalysts fixed on polymer carriers. This is why this fixed bed has a better permeability than layers of polymer particles which are highly compressible.

The flow pattern of a tubular reactor filled with carrier fixed acylase was investigated at a particle Reynolds number of 0.7 (where film diffusion effects can still be neglected), showing that a layer 16 cm deep would be enough to give a flow pattern like that in a 10-stage cascade. A layer of 16 cm would, of course, not be enough to achieve reasonable conversion with the maximal possible volumetric enzyme concentration in a tubular reactor.

The reactor length necessary can be estimated as follows:

Equation (22) is used once more:

$$\frac{\tau \cdot E}{S_0} = \int_o^U \frac{1}{R}\, dU\,. \tag{22}$$

An initial substrate concentration of 400 $mmol_{D,L} \cdot l^{-1}$ is used as an example. In Fig. 23 the reciprocal value of the reaction rate is plotted as a function of conversion. By means of numerical integration[61] the value for the left side of Eq. (22) can be calculated using a conversion of 90% as example

$$\frac{\tau \cdot E}{200} = \int_o^{0.9} \frac{1}{R}\, dU = 11.26 \text{ min} \cdot g \cdot mmol^{-1}\,.$$

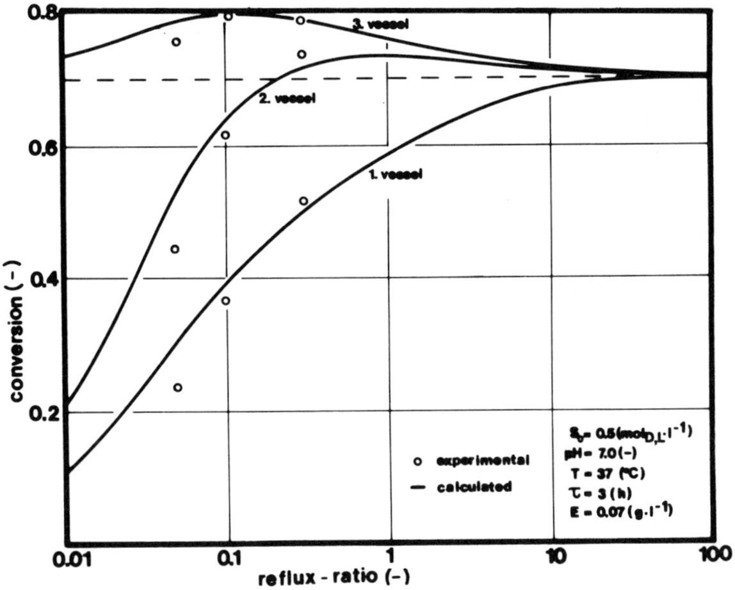

Fig. 37. Conversion as a function of the reflux-ratio

With the initial substrate concentration mentioned, a product

$$\tau \cdot E = 2252 \text{ min} \cdot g \cdot l^{-1}$$

would be necessary. With an achievable volumetric enzyme concentration of about 500 $g \cdot l^{-1}$ this would require a residence time of about 4.5 min. The minimal linear velocity can now be calculated with the condition that the particle Reynolds number should not drop below 0.7:

$$Re_K = \frac{u_l\,(cm \cdot s^{-1}) \times 0.0284\,(cm)}{0.908 \times 10^{-2}\,(cm^2 \cdot s^{-1})} = 0.7$$

with an average particle diameter $\bar{d}_K = 0.0284$ (cm)
and a kinematic viscosity $\nu = 0.908 \times 10^{-1}$ ($cm^2 \cdot s^{-1}$).

The linear velocity for the empty tube is calculated as:

$$u_l = 0.224 \text{ cm} \cdot s^{-1} \,.$$

Taking into account a void volume of about 80%, an effective linear velocity of u_{eff} = 0.28 cm $\cdot\, s^{-1}$ follows. The reactor length L can be calculated using Eq. (35):

$$L = \tau \cdot u_{eff} \tag{35}$$

In this case a reactor length of about 75 cm is calculated. A tubular reactor of length 100 cm was designed. The tube was filled with 36 g (250–315 μ) of carrier fixed acylase. The height of the fixed bed finally became 90.5 cm due to compression during initial operation.

The tubular reactor is shown schematically in Fig. 38. A dosing pump (a) conveys the substrate across a sterile filter (b) through a thermostat (c) to the top of the tubular reactor. A layer of glass beads (d) is applied to achieve a flat flow pattern even at the top of the catalyst layer. Samples could be taken at distances of 15.5, 53, and 90.5 cm respectively as shown in the figure. A thermostatting jacket is used to achieve a uniform temperature throughout the reactor. Pressure drop is measured by means of two manometers.

The catalyst concentration was 638 g $\cdot$ l^{-1} (void volume). The total void volume of the fixed bed was 56.45 cm^3. For a minimal particle Reynolds number of 0.7, residence times up to 5.16 min were possible. Experiments were carried out using initial substrate

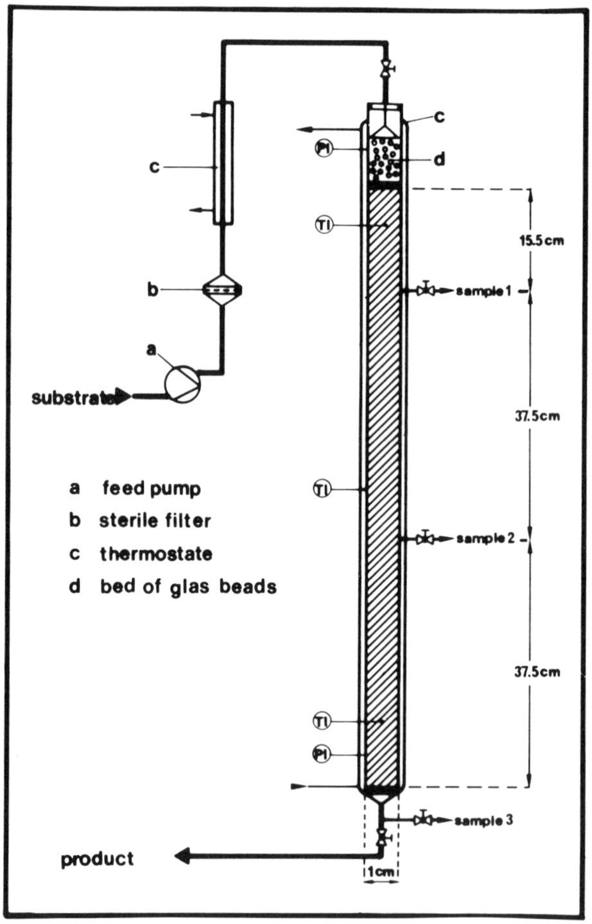

Fig. 38. Plug flow reactor for carrier fixed enzyme

concentrations of 200, 400, 600, and 800 mmol $_{D,L}$ · l^{-1}. Samples were taken as illustrated in Fig. 38. The results are shown in Fig. 39. Three sets of points can be seen according to the three different places where samples were taken. The continuous lines were calculated using the kinetic measurements described in Part 2.4.2 and Eq. (22). In Fig. 39 the conversion is shown as function of residence time, the product of residence time and volumetric activity, respectively. Especially the latter expression is useful since it is independent of the particular catalyst sample. The product can be considered as a modified Damköhler number.

Only measurements in a loop reactor are necessary to calculate the conversion as a function of residence time when the effective catalyst concentration in the tubular reactor in comparison to the loop reactor is taken into account. New catalyst samples can easily be used without knowledge of kinetic parameters. Using the numerical integration method, only the condition has to be proved that no film diffusion effects occur.

Pressure drop was described using Blake's law[62)]

$$\psi = \frac{\epsilon^3}{1-\epsilon} \cdot \frac{\Delta p}{\rho \cdot u_l^2} \cdot \frac{d_K}{L} \tag{36}$$

where ϵ (=) void volume
Δp ($N \cdot m^{-2}$) pressure drop
ρ ($kg \cdot m^{-3}$) density
u_l ($m \cdot s^{-1}$) linear velocity referred to the empty tube
d_K (m) particle diameter
L (m) reactor length .

For a laminar flow pattern Eq. (37) is commonly in use to describe the pressure resistance number:

$$\psi = \frac{\text{const}}{\text{Re}} . \tag{37}$$

The Reynolds number used here is defined by Eq. (38):

$$\text{Re} = \frac{\text{Re}_K}{1-\epsilon} . \tag{38}$$

Using Eqs. (36) and (37), an expression for a "specific" pressure drop may be developed:

$$\frac{\Delta p}{L \cdot \nu} = \text{const} \cdot \frac{(1-\epsilon)^2}{\epsilon^3} \cdot \frac{\rho \cdot u_l}{d_K^2} . \tag{39}$$

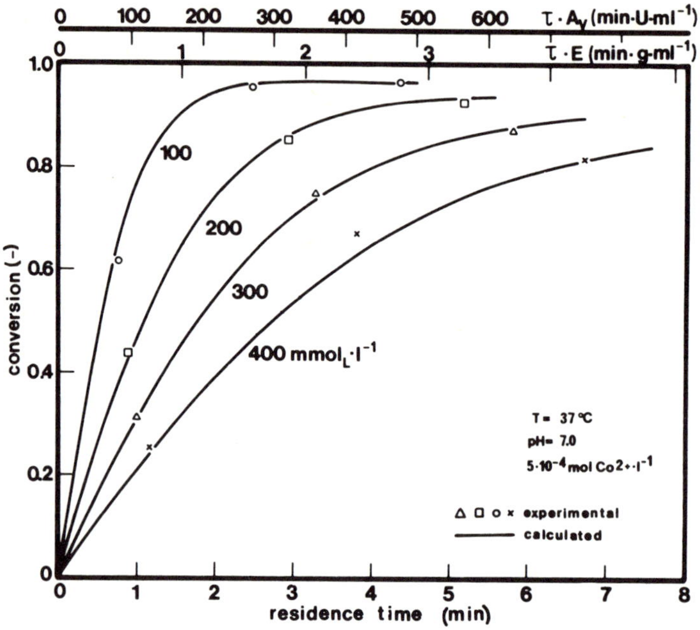

Fig. 39. Conversion as a function of residence time in a plug flow reactor (parameter: substrate concentration)

This expression is given as a function of the linear velocity (referred to the empty tube) in Fig. 40. A straight line results as predicted when the reactor length as well as the kinematic viscosity of the reaction mixture are known. For further details refer to[39, 59].

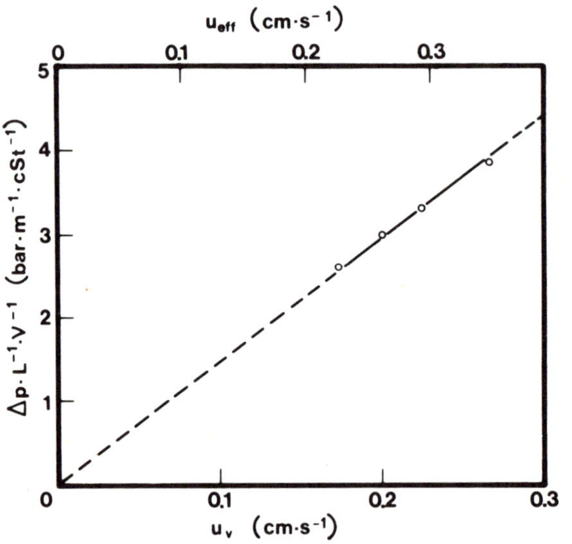

Fig. 40. Specific pressure drop as a function of space velocity (effective – u_{eff}, related to void tube – u_V)

3 Economic Analysis

The results of the experimental investigations are the basis for the economic analysis. First the kinetic results are summarized with special respect to the significance of the model parameters. The cost structure will be investigated afterwards, finally followed by a reaction engineering and economic optimization.

3.1 Modelling

For practical application a kinetic model should contain as few parameters as possible with a physico-chemical significance as high as possible. In terms of a mechanistic model experimental results can be more easily understood, but it should be kept in mind that it is hardly ever possible to determine the "true" kinetics.

The individual steps of the kinetic development are summarized in Table 14. In the first step an equation was chosen following Michaelis and Menten. This expression is useful to describe initial rates up to a concentration level of 10 mmol $\cdot$ l^{-1}. For integral runs the reduction of the driving force due to the reverse reaction has to be taken into account as indicated by the second step. The non-stereospecific substrate inhibition is formulated in step 3. This is done in terms of a non-competitive inhibition. If a racemic substrate mixture is used, the inhibition effect of the D-enantiomer remains constant, whilst the inhibition effect of the L-enantiomer diminishes with increasing conversion. The product inhibition is taken into account in the last step.

An analogous model could be applied for carrier fixed acylase, too. The corresponding model parameters have been shown in Table 10. The apparent k_m value is enlarged in comparison to the value for the soluble enzyme due to the reduced substrate saturation, whilst the apparent inhibition constant for the competitive product inhibition is enlarged due to the enhanced product saturation within the catalyst particle.

Table 14. Steps of model-development

Model-development	
Step 1	$R = \dfrac{V'_{max} \cdot S'_L}{k'_m + S_L}$
Step 2 with	$S'_L = S_L - \dfrac{(S_{OL} - S_L)^2}{K}$
Step 3 with	$V'_{max} = \dfrac{V_{max}}{\left(1 + \dfrac{S_L + S_{OD}}{k_{in}}\right)}$
Step 4 with	$k'_m = k_m \left(1 + \dfrac{S_{OL} - S_L}{k_{ic}}\right)$

In the case of soluble acylase the model can be applied for concentration levels of substrate from 50 to 400 $mmol_L \cdot l^{-1}$ in the corresponding racemate. In the case of carrier fixed acylase the model is valid in a concentration range from 200 to 400 $mmol_L \cdot l^{-1}$ in the corresponding racemate. In this case more precise predictions of reactor performance can be achieved using methods of numerical integration as shown in Fig.39.

Because non-linear regression analysis was used for parameter estimation and sufficient data were available for statistical analysis, a variance-covariance analysis was possible. Non-linear least-square fit was carried out using Eq. (40):

$$QS = \sum_{i=1}^{m} (R_i - \hat{R}_i(\underline{Y}, \underline{\theta}))^2 \overset{!}{=} \text{Min} \tag{40}$$

where QS = sum of squares
m = number of data sets
R = rate (measured)
$\hat{R}$ = rate (calculated)
$\underline{Y}$ = vector of experimental conditions (conversion and concentration level)
$\underline{\theta}$ = vector of parameters .

In Table 15 the results of the variance-covariance analysis are summarized for soluble acylase. The parameters are given with their confidence intervals (for a probability of 68% corresponding to ± one standard deviation). Moreover, the correlation coefficient matrix $\underline{C}$ is formulated. For further, especially mathematical details refer to[63].

From Table 15 it can be seen that there is a very high correlation between the k_m value and the product inhibition constant.

Table 15. Results of variance-covariance analysis (4-parameter model, native acylase)

Parameter	V_{max} $mmol \cdot g^{-1} \cdot min^{-1}$	k_m	k_{in}	k_{ic}
		$mmol \cdot l^{-1}$		
Mean value	34.0	0.77	1450	2.8
Confidence interval	± 0.53	±0.53	± 79	±1.1
Matrix of correlation coefficients	1.000	0.872	0.942	-0.860
	0.872	1.000	0.846	-0.999
	0.942	0.846	1.000	-0.841
	- 0.860	-0.999	-0.841	1.000

This can easily be understood from Eq. (28):

$$R = E \cdot \frac{V_{max}}{\left(1 + \frac{S_L + S_D}{k_{in}}\right)} \cdot \frac{S - \frac{(S_{oL} - S_L)^2}{K}}{k_m \left(1 + \frac{S_{oL} - S_L}{k_{ic}}\right) + S_L} \; . \tag{28}$$

Especially for high product concentration the inequality (41) holds:

$$1 \ll (S_{oL} - S_L) . \tag{41}$$

Therefore, it is sensible to use the ratio of k_m and k_{ic} instead of two different kinetic parameters, as formulated in Eq. (42):

$$R = E \cdot \frac{V_{max}}{\left(1 + \frac{S_L + S_D}{k_{in}}\right.} \cdot \frac{S_L - \frac{(S_{oL} - S_L)^2}{K}}{k \cdot (S_{oL} - S_L) + S_L} . \tag{42}$$

The results of a variance-covariance-analysis using the 3-parameter model are given in Table 16.

Table 16. Results of variance-covariance analysis (3-parameter model, native acylase)

Parameter	V_{max} $mmol \cdot g^{-1} \cdot min^{-1}$	k –	k_{in} $mmol \cdot l^{-1}$
Mean value	32.5	0.280	1770
Confidence interval	± 0.23	±0.0058	± 56
Matrix of	1.000	–0.783	0.707
correlation	– 0.783	1.000	0.289
coefficients	0.707	0.289	1.000

As could be predicted, the sum of squares only increases slightly (owing to the high correlation between two kinetic parameters in the 4-parameter model), but the elements of the correlation coefficient matrix are much more favourable. The confidence intervals of the parameters justify the use of the model for further calculations. A similar variance-covariance analysis has been carried out with the results of kinetic measurements using carrier fixed acylase[39].

The influence of inadequately developed kinetic models on the design of a CSTR is shown in Fig. 41. The rate is given as function of fractional conversion according to the various models represented by incorporation of the four steps in Table 14. The dashed straight line represents the convective term in the mass balance of the CSTR. The point of intersection represents the operating point of the reactor. The first line is calculated using Michaelis-Menten kinetics. In the second line additionally the euqilibrium is taken into account. In the third step, the substrate inhibition measured from initial rate experiments is also considered. Finally the product inhibition from integral runs is used to complete the model so that a sufficient fit of data is obtained. The intersections of the dashed vertical lines with the abscissa represent the achievable conversion predicted by the different models, showing that insufficiently developed models lead to erro-

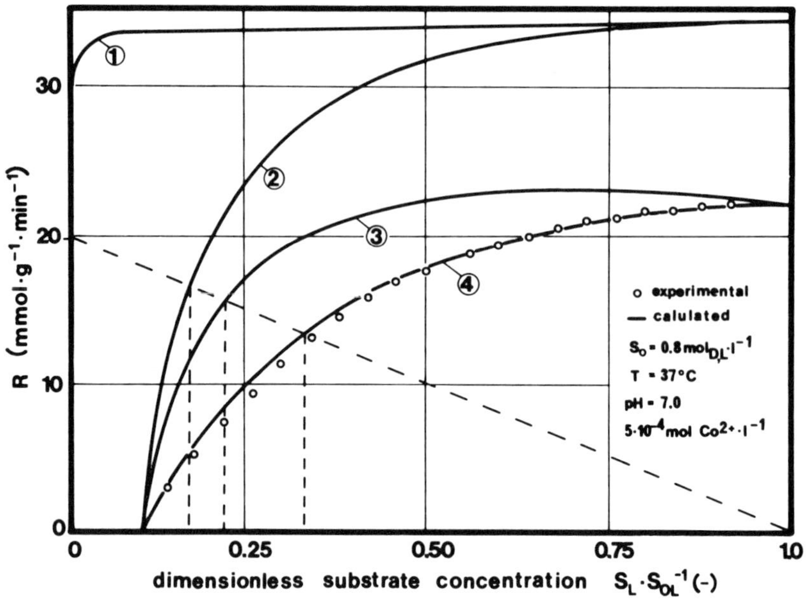

Fig. 41. Operating points of a CSTR depending on model development

neous predictions. At steady-state for a CSTR Eq. (43) holds:

$$\frac{S_0 \cdot U}{\tau} = -R(S_0, U) \ . \tag{43}$$

The left side of the equation represents the convective term, a straight line as function of conversion. The right side of the equation – the kinetic expression – is a non-linear function of conversion and initial substrate concentration level. Equation (43) is a cubic function with respect to conversion U when the completely developed model is used. The function

$$U = f(S_0, \tau)$$

is solved iteratively. Thus the performance of a cascade can be predicted, if it is considered that the conversion after vessel n can be used as initial condition for vessel (n + 1).

For plug flow conditions Eq. (28) has to be integrated. For further details see the Appendix.

Figure 42 shows computed results for conversion as a function of residence time, when a substrate concentration of 1000 $mmol_{D,L} \cdot l^{-1}$ and a catalyst effectiveness of 50% is taken into account. From the graph – valid for soluble acylase – the influence of the flow pattern on the reactor performance can be seen. A three-stage cascade is

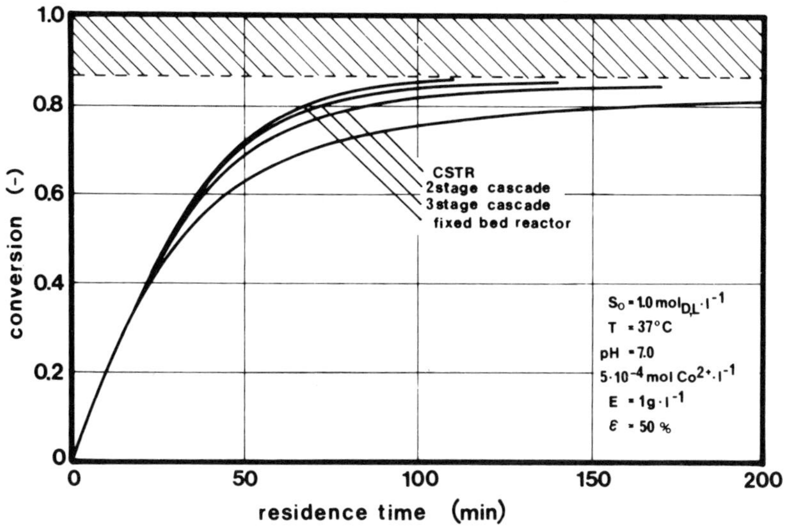

Fig. 42. Conversion as a function of residence time and type of reactor

obviously already a very good approximation of plug flow conditions. The main improvement is achieved if a two-stage cascade is applied instead of a single CSTR, as is normally found[64b)]. For these two reactor types Fig. 42 is extended to a three-dimensional graph taking into account the concentration level additionally as shown in Fig. 43. Figure 42 can be regarded as the projected area on the right-hand side (according to a CSTR and a two-stage cascade) of Fig. 43. The influence of the flow pattern can be shown in another way as illustrated in Fig. 44. The activity needed in relation to the activity necessary in a tubular reactor is given as a function of the fraction for a single CSTR and a two-, resp. three-stage cascade. The hatched part of the graph indicates the region not obtainable due to the state of equilibrium. As for Fig. 43 the greatest improvement is found for the transition from a single CSTR to a two-stage cascade. In Fig. 45 the actual productivity referred to the productivity at initial rate conditions is given as a function of the fraction (S_L/S_{oL}). As parameter the type of catalyst is used (continuous lines for soluble acylase, dashed lines for carrier fixed acylase). For each type of catalyst two different flow patterns are taken into account (CSTR conditions and plug flow conditions). It can be seen that the main factor is the flow pattern. However using soluble acylase, it is not quite so important to approximate plug flow conditions owing to the absence of mass transfer limitations. This is also the reason for a slower decrease of the observed productivity when soluble acylase is used instead of the carrier fixed type.

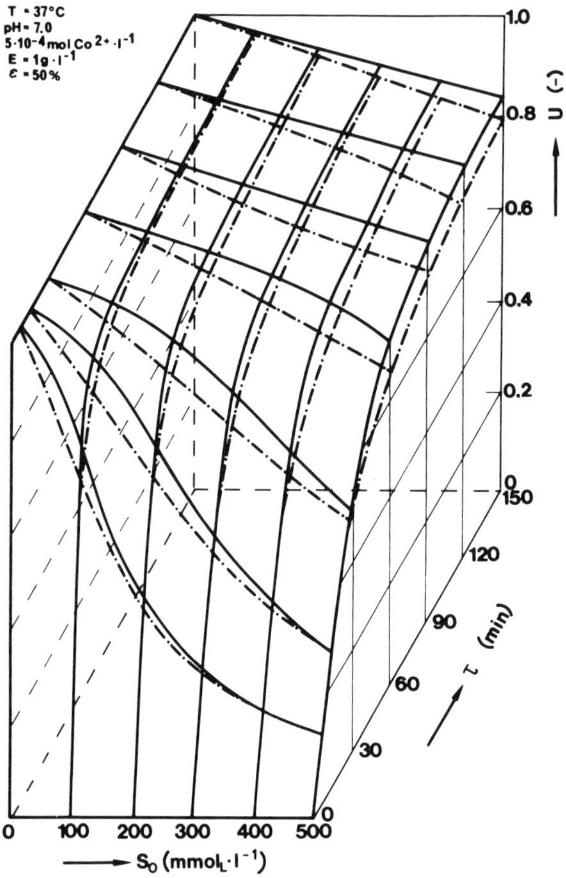

Fig. 43. Dependence of conversion on residence time and substrate concentration in membrane reactors —·—· CSTR, —— 2-stage cascade

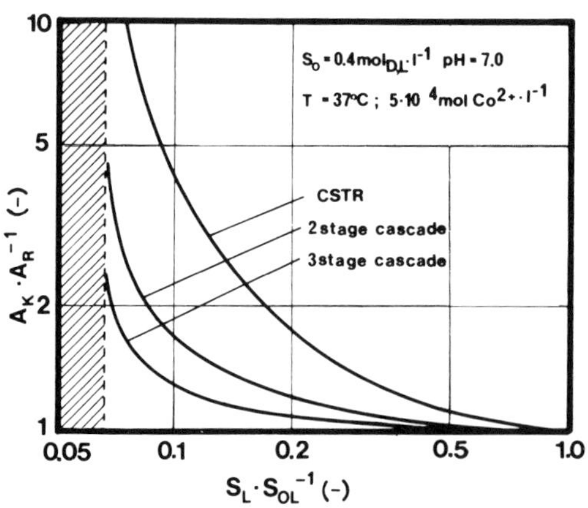

Fig. 44. Catalyst-activity requirement of different types of reactors compared with a plug flow reactor

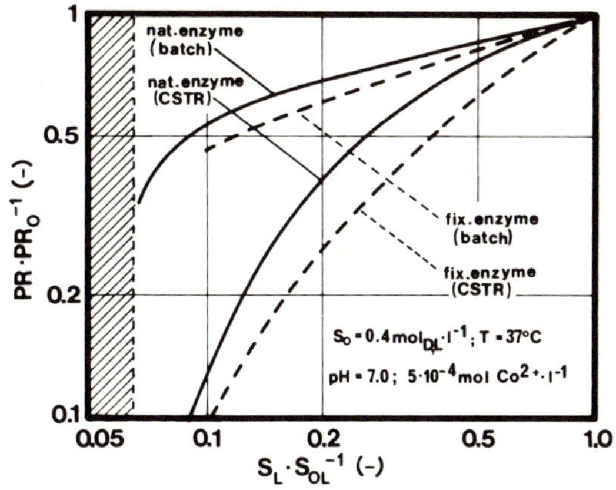

Fig. 45. Comparison of native and carrier fixed acylase in different modes of operation

3.2 Cost Structure

In this part the influence of the most important cost parameters on the total proportional cost for the production of L-methionine by means of acylase from N-acetyl-D,L-methionine will be investigated. Cost for raw material, catalyst and separation (caused by the energy needed to concentrate the product solution) are taken into account. The following cost factors were used:

price of raw material	(KFR)	=	$10\ \mathrm{DM}\cdot \mathrm{kg}^{-1}$
price of enzyme	(KFE)	=	$10\text{–}30\ \mathrm{DM}\cdot \mathrm{g}^{-1}$
price for separation	(KFT)	=	$0\text{–}0.3\ \mathrm{DM}\cdot \mathrm{l}^{-1}$
product price	(KFP)	=	$78\ \mathrm{DM}\cdot \mathrm{kg}^{-1}$.

Raw material price and product price are valid for ton-scale. The enzyme price is valid for a catalyst of the activity described here. The price differs very much with the amount needed. Separation cost is based on the energy cost necessary for evaporation of the solvent water. Calculation is carried out in term of specific costs per unit weight of product. The total proportional costs per unit-weight of product are obtained, when the sum of the total expenditures is divided by the capacity:

$$KG = \frac{1}{L}\left(\frac{d\,KR}{dt} + \frac{d\,KE}{dt} + \frac{d\,KT}{dt}\right) \tag{44}$$

where KG	= total proportional cost	$\mathrm{DM}\cdot \mathrm{kg}_p^{-1}$
L	= capacity	$\mathrm{kg}_p \cdot \mathrm{time}^{-1}$
KR	= cost caused by raw material	DM
KE	= cost caused by the catalyst	DM
KT	= cost caused by separation	DM .

For the cost due to raw material per unit time Eq. (45) holds:

$$\frac{d\,KR}{dt} = F_v \cdot 2 \cdot S_o \cdot MS \cdot KFR \tag{45}$$

$$DM \cdot min^{-1} = l \cdot min^{-1} \cdot kg \cdot mol^{-1} \cdot DM \cdot kg^{-1}$$

where F_v = volumetric flow ($l \cdot min^{-1}$)
MS = molecular weight of substrate ($kg \cdot mol^{-1}$) .

The factor 2 arises from the stoichiometry since this calculation is done on the basis of the L-enantiomer not taking into account the possibility of reracemization of the D-enantiomer.

For the cost due to the catalyst per unit time Eq. (46) holds:

$$\frac{d\,KE}{dt} = F_v \cdot (\tau \cdot E) \cdot FDE \cdot KFE \tag{46}$$

$$DM \cdot min^{-1} = l \cdot min^{-1} \cdot min \cdot g \cdot l^{-1} \cdot min^{-1} \cdot DM \cdot g^{-1}$$

where: FDE = deactivation constant (min^{-1}) .

This cost increases with increasing deactivation constant, enzyme concentration and reactor volume (or, which is the same, product of volumetric flow and residence time). Equation (46) holds exactly when the total enzyme activity is kept constant by means of additional supply of fresh catalyst according to the deactivation in the reactor – as is possible in case of the membrane reactor.

The separation cost was calculated as the energy cost for the case when the product stream is concentrated until saturation with respect to L-methionine at 100 °C. The product is obtained by crystallization at 0 °C. Since the solubility of L-methionine at 100 °C is 0.72 mol $\cdot l^{-1}$ and at 0 °C 0.18 mol $\cdot l^{-1}$ the separation yield is 75%. For the cost due to separation per unit time Eq. (47) holds:

$$\frac{d\,KT}{dt} = (F_v - F_c) \cdot KFT \tag{47}$$

$$DM \cdot min^{-1} = l \cdot min^{-1} \cdot DM \cdot l^{-1}$$

where: F_c = volumetric flow of the concentrated product solution ($l \cdot min^{-1}$) .

This cost results as product of the cost factor and the volumetric flow of the overhead product of the evaporator.

For the volumetric flow of the concentrated product solution Eq. (48) is used:

$$F_c = F_v \cdot \frac{P}{P_c} = F_v \cdot \frac{S_o \cdot U}{P_c} \tag{48}$$

where P = product concentration (before evaporator) $(\mathrm{mol} \cdot \mathrm{l}^{-1})$
P_c = product concentration (after evaporator) $(\mathrm{mol} \cdot \mathrm{l}^{-1})$.

If Eq. (48) is used for the term F_c in Eq. (47), Eq. (49) results:

$$\frac{d\,KT}{dt} = \left(F_v - F_v \cdot \frac{S_o\,U}{P_c}\right) \cdot KFT \tag{49}$$

where $P_c = 0.72\ (\mathrm{mol} \cdot \mathrm{l}^{-1})$

$$\frac{d\,KT}{dt} = F_v \left(1 - \frac{S_o\,U}{0.72}\right) \cdot KFT\,.$$

The capacity is defined by Eq. (50):

$$L = F_v \cdot (S_o \cdot U) \cdot MP \cdot 0.75 \tag{50}$$

$$\mathrm{kg} \cdot \mathrm{min}^{-1} = \mathrm{l} \cdot \mathrm{min}^{-1} \cdot \mathrm{mol} \cdot \mathrm{l}^{-1} \cdot \mathrm{kg} \cdot \mathrm{mol}^{-1}\,.$$

The factor 0.75 is due to the separation yield.

If Eqs. (45), (46), (49), and (50) are used for the corresponding quantities in Eq. (44), Eq. (51) results:

$$KG = \frac{1}{S_o \cdot U \cdot MP \cdot 0.75}\left[2 \cdot S_o \cdot MS \cdot KFR + (\tau \cdot E) \cdot FDE \cdot KFE + \left(1 - \frac{S_o\,U}{0.75}\right) \cdot KFT\right]. \tag{51}$$

Specific cost per unit weight of product according to raw material cost, catalyst cost and separation cost can be derived from Eq. (51):

$$ROH = \frac{2 \cdot MS \cdot KFR}{U \cdot MP \cdot 0.75} \qquad (\mathrm{DM} \cdot \mathrm{kg}^{-1})\,, \tag{52}$$

$$ENZ = \frac{(\tau \cdot E) \cdot FDE \cdot KFE}{S_o\,U \cdot MP \cdot 0.75} \qquad (\mathrm{DM} \cdot \mathrm{kg}^{-1})\,, \tag{53}$$

$$TRE = \frac{KFT}{S_o\,U \cdot MP \cdot 0.75} - \frac{KFT}{MP \cdot 0.72 \times 0.75} \qquad (\mathrm{DM} \cdot \mathrm{kg}^{-1})\,. \tag{54}$$

One realizes that the specific raw material cost only depends on conversion. In the case of the separation cost the concentration level has also to be considered. Only the enzyme cost depends on residence time and thus on the reactor type. Mathematically an effectiveness of the catalyst less than 100% is taken into account as increased residence time. The different costs are illustrated by means of the following graph (Fig. 46), which shows the specific raw material cost as a function of conversion. According to the state of equilibrium the line is dashed for high conversions.

In Fig. 47 the specific separation cost is given as a function of conversion with the initial substrate concentration level as parameter. Separation cost will decrease with increasing concentration level since less solvent has to be evaporated.

The specific enzyme costs for a membrane reactor with the behaviour of a CSTR are given as function of conversion in Fig. 48. These values were obtained using an enzyme price of 15 DM · g^{-1}, an effectivity of 50% and a deactivation constant of 0.02 d^{-1}. The

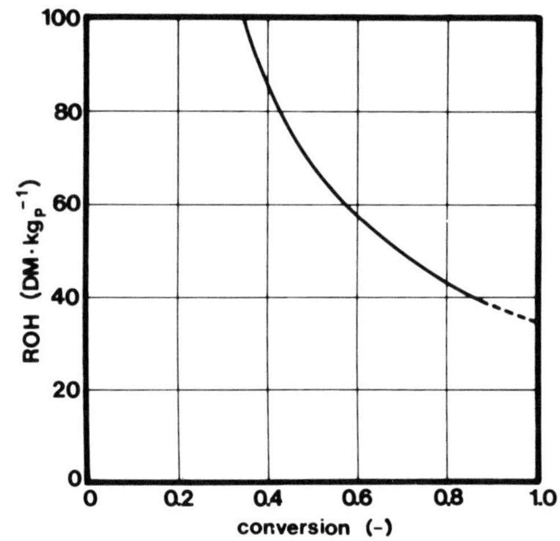

Fig. 46. Cost of raw material per unit weight of product as a function of conversion

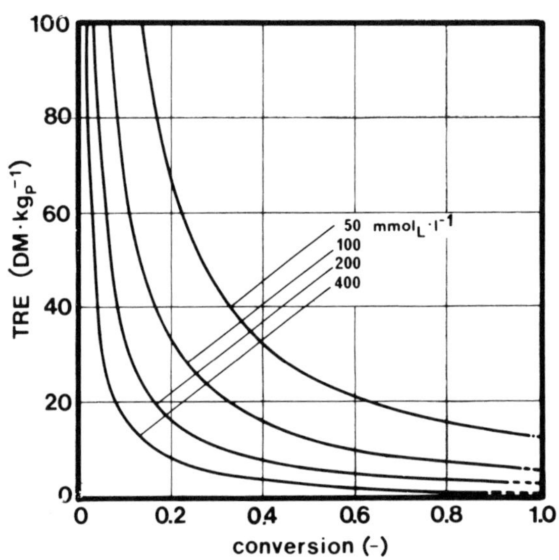

Fig. 47. Cost of separation per unit weight of product as a function of conversion at different starting substrate concentrations

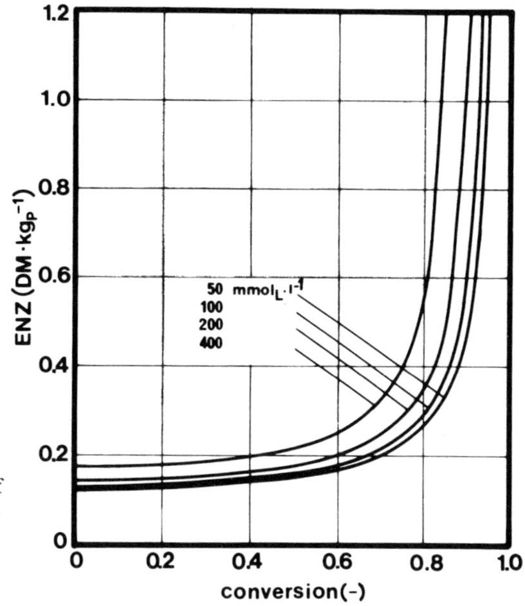

Fig. 48. Cost of catalyst per unit weight of product as a function of conversion at different starting substrate concentrations

costs increase with increasing concentration level owing to substrate inhibition. At a higher concentration level the costs rise earlier with increasing conversion due to the state of equilibrium.

In Fig. 49 a superposition of specific raw material cost and enzyme cost together with separation cost, is given. This curve is only valid for a concentration level of

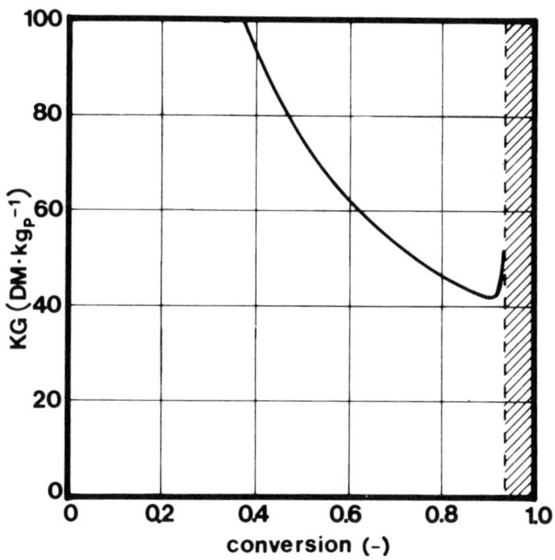

Fig. 49. Total product cost as a function of conversion for S_0 = 0.2 mol · l^{-1} (native enzyme, membrane reactor)

200 mmol$_L$ · l^{-1} (in the corresponding racemate). It can be seen that a cost minimum occurs near the equilibrium conversion. This is caused by the relatively high stability of the catalyst so that it is worthwhile to aim for a high utilization of the raw material. Taking into account the separation cost, the cost minimum shifts to a slightly lower conversion.

Total proportional cost as a function of conversion and concentration level is given in Fig. 50. A 2-dimensional minimum occurs. This is because, kinetically, low concentrations would be favourable, but on the other hand low concentrations cause high separation costs. In Fig. 50 a "wall" indicating the maximal achievable conversion depending on the concentration level can be seen.

In Fig. 51 the specific enzyme cost is plotted as a function of conversion. The continuous lines are valid for different membrane reactors operating with soluble acylase,

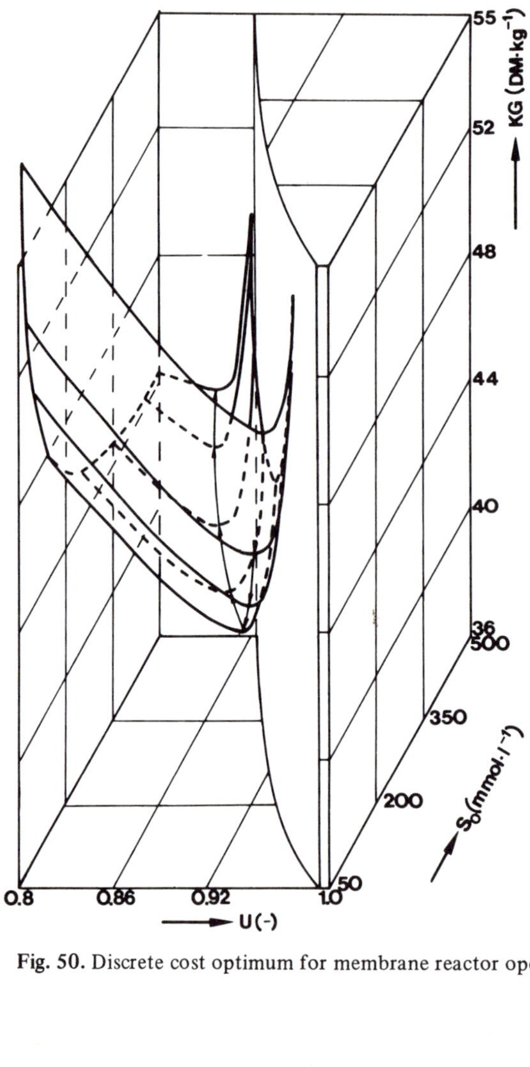

Fig. 50. Discrete cost optimum for membrane reactor operation (for cost parameter see text)

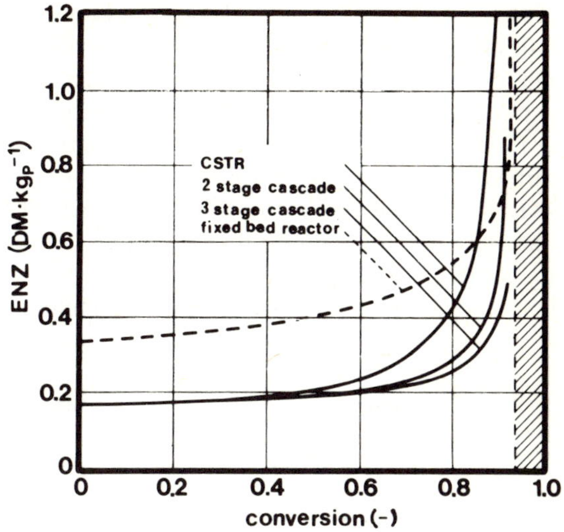

Fig. 51. Cost of catalyst per unit weight of product as function of conversion (S_0 = 0.2 mol · l^{-1})

the dashed line is valid for a tube reactor with carrier fixed acylase. The given data for a concentration level of 200 $mmol_L\ l^{-1}$ (in the corresponding racemate) are based on experimental values for catalyst stability and effectiveness:

soluble acylase:	effectivity = 70%
	stability = 0.0344 d^{-1}
carrier-fixed acylase:	effectivity = 5.77%
	stability = 0.0056 d^{-1} .

Only the cost of the enzyme protein is taken into account for the carrier-fixed type of catalyst (15 DM · g^{-1}). It can be seen that at initial rate conditions the specific cost for the catalyst using the carrier fixed type of catalyst is about twice as high as when soluble acylase is used in a membrane reactor. This is because the stability of the carrier fixed acylase is about six times higher, but the effectiveness is about 12 times lower. At high conversion the specific catalyst cost is very sensitive to the flow pattern. Up to a conversion of 85% the single-stage membrane reactor is more favourable than the plug flow reactor with this carrier fixed acylase. When a two-stage membrane reactor is used, the intersection of the corresponding line with the curve for the carrier fixed catalyst is found to be at 91% conversion. For a three-stage membrane cascade the point of intersection cannot be shown graphically.

3.3 Optimization

The optimization task is to minimize the function:

$$KG = f(S_o, U)_{KFR,\ KFE,\ KFT,\ FDE,\ ETA} \overset{!}{=} \text{Min} . \qquad (55)$$

The experimentally determined values for catalyst effectiveness and stability were taken into account. The influence of the reactor type is considered by means of the residence time needed for a certain conversion. In order to obtain a cost sensitivity analysis, the minimal total proportional costs and the corresponding optimal operation conditions were determined as functions of the catalyst price and the price for separation. This was carried out at a constant raw material price of 10 DM · kg^{-1}. The first comparison is given for a membrane reactor which behaves as a CSTR and a tube reactor with carrier fixed acylase in Fig. 52. For this comparison only protein cost is taken into account for the carrier fixed type of catalyst. In Fig. 52 two curved areas can be recognized. Each point on these areas represents a single optimum with different optimal reaction condi-

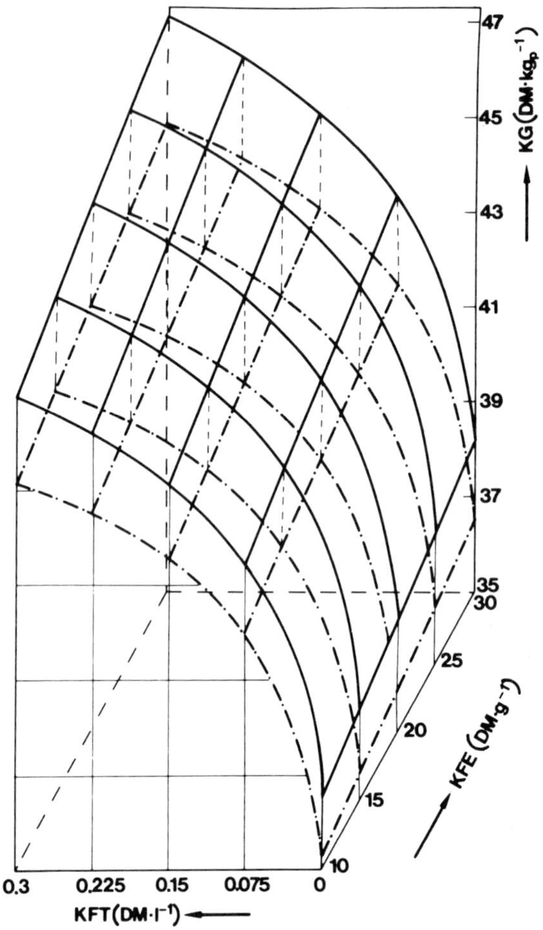

Fig. 52. Comparison of cost sensitivity of a single membrane reactor (soluble acylase) with a plug flow reactor (carrier fixed acylase) (no price-increase during fixation) —— membrane reactor (soluble acylase) —·—· plug flow reactor (carrier fixed acylase)

tions. Furthermore, it can be seen that the influence of a changing catalyst-price would be comparatively small in comparison with a change in the price for separation. Obviously a membrane reactor which behaves like a CSTR is less favourable than a tube reactor with this type of carrier fixed acylase. This is due to the fact that no costs for carrier and preparation of the carrier fixed enzyme have been taken into account.

Even on this basis the membrane concept will become more favourable than a tube reactor if a three-stage membrane reactor cascade is used. This is illustrated in Fig. 53. It can be seen that the cost area for the membrane reactor is located below the corresponding area for the tubular reactor.

The last comparison of economic efficiencies is carried out with a very special type of carrier fixed enzyme. The demanded hydrodynamic stability of the catalyst led to a

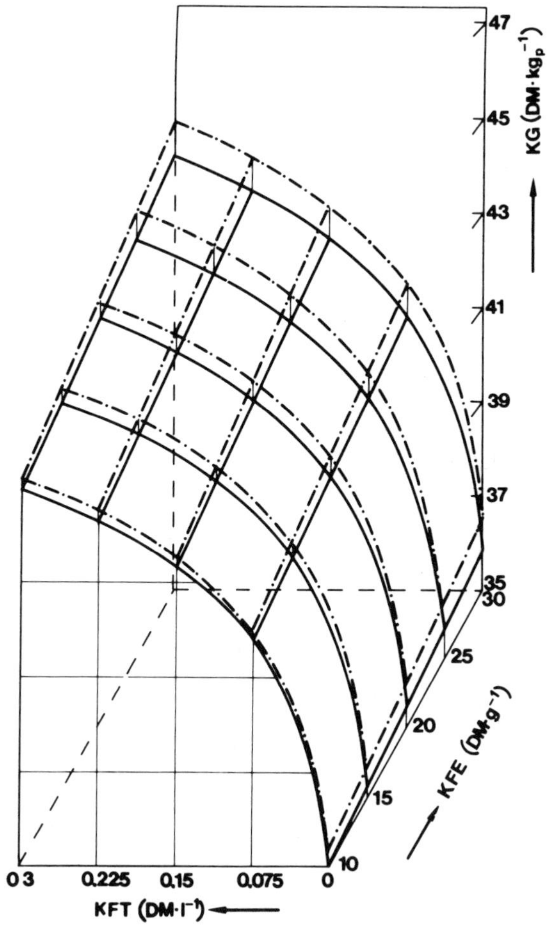

Fig. 53. Comparison of cost sensitivity of a 3-stage membrane reactor cascade (native acylase) with a plug flow reactor (carrier fixed acylase) (no price-increase during immobilization) —— 3-stage membrane reactor (soluble enzyme), —·—· plug flow reactor (carrier fixed acylase)

low activity yield during fixation. Therefore the price of the carrier fixed enzyme per unit activity may be actually too high even on the basis of protein-content. In order to avoid speculation on actual prices for carrier fixed acylase, a more general comparison will be given.

Starting from the native acylase described here, the carrier fixed acylase may yield an activity which can be expressed by the enzyme effectiveness ETA and a stability which can be expressed by the deactivation constant FDE. ETA and FDE may be combined as a so-called "immobilization efficiency" IE, which is defined by Eq. (56):

$$IE = \frac{ETA}{FDE} \tag{56}$$

This constant is also valid for other kinds of immobilization and may be interpreted as a modified kind of endurance.

Assuming that the kinetic constants, experimentally established for the carrier fixed acylase investigated, are still valid and taking the same prices for substrate (10 DM $\cdot$ kg^{-1}), separation (0.075 DM $\cdot$ l^{-1}) and enzyme protein (15 DM $\cdot$ l^{-1}), the catalysts can be compared over the whole range of immobilization efficiency without a particular assumption concerning the prices of carrier fixed enzymes.

The immobilization efficiency of both the catalysts studied is in case of the native acylase (ETA = 0.7, FDE = 0.0344 d^{-1}) IE = 20.34 d and for the carrier fixed acylase (ETA = 0.0577, FDE = 0.0056) IE = 10.3 d.

Knowing the minimal total proportional cost KG for different modes of operation with native acylase, a factor can be calculated, how much more expensive a carrier fixed acylase operating in a fixed bed can be than the native acylase in different reactors – based on the protein-price. This is shown in Fig. 54 comparing the carrier fixed catalyst with a single membrane reactor and a 3-stage cascade of membrane reactors operating with soluble acylase.

If the immobilization efficiency of the carrier fixed acylase used during this study is taken into consideration (marked by the vertical line at IE = 10.3 d in Fig. 54) it can be seen, that this type of catalyst can be 2.9 times more expensive than the native type operating in a single membrane reactor behaving like a CSTR. This factor changes to 0.9 if compared with the 3-stage cascade of membrane reactors as may be deduced by comparison with Fig. 53. If the immobilization efficiency were the same for both kinds of catalyst (indicated by the vertical line IE = 20.34 in Fig. 54) the carrier fixed type could be 1.7 or 5.75 times more expensive than the soluble type in a 3-stage cascade or a single membrane reactor respectively. The normal immobilization efficiency of carrier fixed enzymes should be in the range of IE = 70 d. If this efficiency could be reached with acylase, this type could be approximately 5 times more expensive than the native acylase in a 3-stage cascade to produce L-methionine at equal cost.

In Fig. 55 another comparison is shown. The total proportional cost per unit weight of product is plotted versus the immobilization efficiency for various modes of continuous operation. Besides a comparison of actual costs in the case of those catalysts studied, it is to be seen that the total cost depends very much on the flow pattern. If

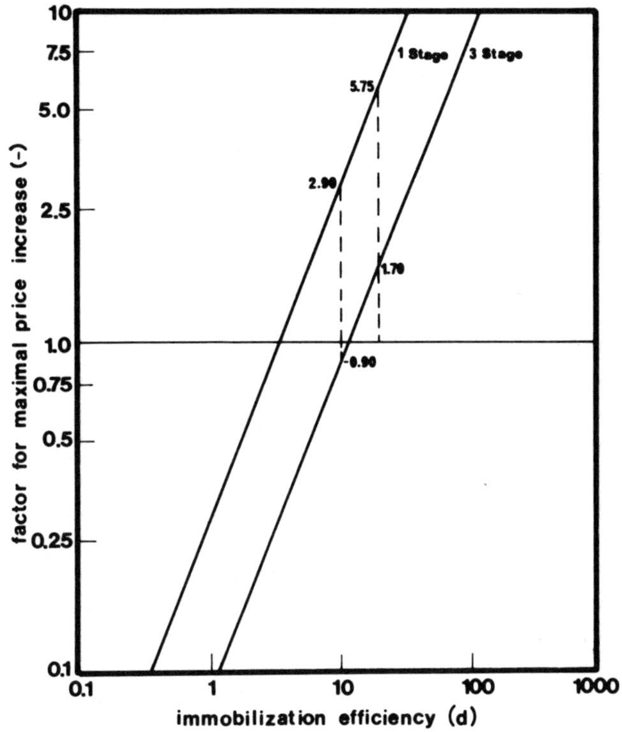

Fig. 54. Maximal price-ratio of carrier fixed acylase in plug flow reactors to native acylase in single and 3-stage membrane reactors as function of immobilization efficiency (KFR = 10 DM $\cdot$ kg^{-1}, KFE = 15 DM $\cdot$ g^{-1}, KFT = 0.075 DM $\cdot$ l^{-1})

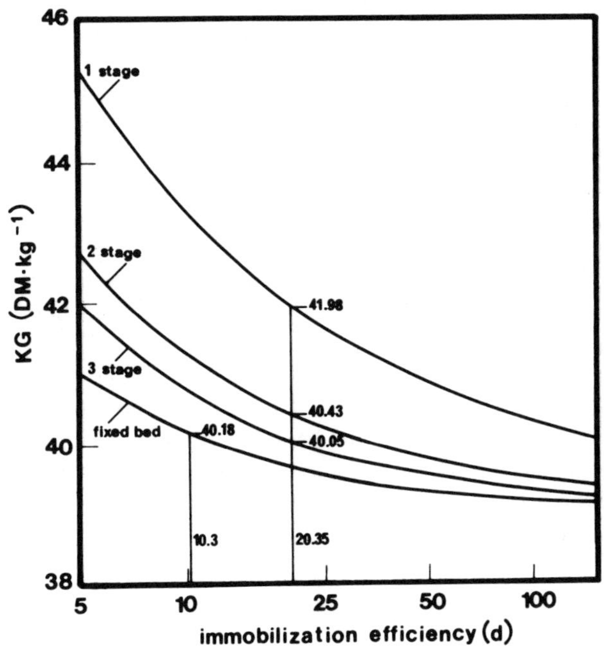

Fig. 55. Dependence of total proportional cost on immobilization efficiency for different kinds of acylase reactors

native enzymes are very stable – or immobilization cannot be carried out with high immobilization efficiency – the 3-stage membrane reactor cascade will be as efficient as the fixed bed reactor, not taking into account expenditures for catalyst development, carrier and labour.

Concluding Remarks

Biocatalysts are powerful aids to carry out many chemical reactions. Owing to their high activity and excellent selectivity technical applications also seem to be very promising in spite of the temperature sensitivity, pH-sensitivity, and the relatively high prices.

For process development purposes, combined studies on physical properties and thermodynamic properties as well as kinetic investigations are necessary. In principle well-known chemical reaction engineering methods can be applied. To get reliable data for further design tasks, the relevant operating conditions must be chosen. Kinetic results obtained at physiological conditions seem to be seldom transferable to technical applications. For parameter estimation non-linear regression analysis combined with a statistical analysis of the data is recommended. The graphical methods commonly in use should be restricted to preliminary model testing. For an economic analysis precise data for catalyst effectiveness and stability are needed. It is also possible to obtain meaningful data for catalyst stability in cases where no suitable kinetic model is known. This can be achieved by numerical integration methods using experimental rate measurements, but may lead to quite erroneous predictions, if measurements of decreasing conversion are directly used to calculate catalyst stability. This is especially dangerous at high initial degrees of conversion.

In spite of numerous methods for enzyme immobilization, only a few technical applications are known at present. This might be due to the fact that immobilization know-how is still rather expensive and that many samples do not possess properties which are desired for technical applications, such as high physical strength, suitable hydrodynamic properties, high volume specific activity and reasonable activity yield. In these circumstances the membrane concept seems to be a very promising alternative. Thus a homogeneous catalysis can be carried out continuously. It is even possible to approximate plug flow pattern in a multistage cascade with one single separation unit using catalyst recycling. The membrane concept becomes even more advantageous when co-enzyme dependent systems or multi-enzyme systems are considered. Owing to the homogeneous character of the catalysis no transport limitations occur. Since co-enzymes are in principle transport metabolites between different enzyme systems membrane technology may open the way to continuous use of co-enzymes as soon as soluble co-enzymes with artificially increased molecular weight are available.

Acknowledgements

The authors are indebted to Prof. Dr. K. Schügerl, Institut für Technische Chemie der TU - Hannover for his generous support. We wish to thank Boehringer - Mannheim Co., West-Germany, for having kindly placed the carrier fixed acylase to our disposal. This work has been financially supported by the „Bundesministerium für Forschung und Technologie" of the FRG government.

Appendix

Mass Balance of CSTR

The kinetic expression Eq. (28) is used. The mass balance at steady-state can be formulated as:

$$\frac{S_o \cdot U}{\tau} = -R \tag{A-1}$$

or

$$\frac{S_o \cdot U}{\tau} = ETA \cdot E \cdot \frac{V_{max}}{\left(1 + \frac{S_o(2-U)}{k_{in}}\right)} \cdot \frac{S_o(1-U) - \frac{S_o^2 \cdot U^2}{K}}{k_m\left(1 + \frac{S_o \cdot U}{k_{ic}}\right) + S_o(1-U)} . \tag{A-2}$$

Equation (A-2) can be solved for $\tau \cdot E$:

$$\tau \cdot E = \frac{S_o \cdot U}{ETA} \cdot \frac{\left(1 + \frac{S_o(2-U)}{k_{in}}\right)}{V_{max}} \cdot \frac{k_m\left(1 + \frac{S_o \cdot U}{k_{ic}}\right) + S_o(1-U)}{S_o(1-U) - \frac{S_o^2 \cdot U^2}{K}} . \tag{A-3}$$

Equation (A-2) is cubic with respect to the conversion U. The equation is solved for U by means of iterative methods.

Mass Balance for a Plug Flow Reactor

As kinetic expression Eq. (28) is used. The mass balance can be formulated as:

$$S_o \cdot \frac{dU}{dt} = R \tag{A-4}$$

or

$$S_0 \cdot \frac{dU}{dt} = ETA \cdot E \cdot \frac{V_{max}}{\left(1 + \frac{S_0(2-U)}{k_{in}}\right)} \cdot \frac{S_0(1-U) - \frac{S_0^2 \cdot U^2}{K}}{k_m\left(1 + \frac{S_0 U}{k_{ic}}\right) + S_0(1-U)} . \tag{A-5}$$

In order to solve Eq. (A-5) for $\tau \cdot E$, the expression has to be integrated. Equation (A-5) can be simplified by combining several coefficients in the following form:

$$E \cdot dt = \frac{d + e \cdot U + U^2}{a + b \cdot U + c \cdot U^2} \cdot dU \tag{A-6}$$

where a - e constants .

The total integral can be separated in three parts:

$$E \int_0^\tau dt = d \cdot \int_0^U \frac{1}{a + b \cdot U + c \cdot U^2} dU + e \cdot \int_0^U \frac{U}{a + b \cdot U + c \cdot U^2} dU$$

$$+ \int_0^U \frac{U^2}{a + b \cdot U + c \cdot U^2} . \tag{A-7}$$

Using

$$X = a + b \cdot U + c \cdot U^2 , \tag{A-8}$$

$$Q = 4ac - b^2 , \tag{A-9}$$

it follows for the condition $Q < 0$

$$d \cdot \int_0^U \frac{1}{X} dU = \left| d \cdot \frac{1}{\sqrt{-Q}} \log \left(\frac{2c \cdot U + b - \sqrt{-Q}}{2c \cdot U + b + \sqrt{-Q}}\right) \right|_0^U , \tag{A-10}$$

$$e \cdot \int_0^U \frac{U}{X} dU = \left| e \cdot \frac{1}{2c} \cdot \log X \right|_0^U - \frac{b}{2c} \cdot \int_0^U \frac{1}{X} dU , \tag{A-11}$$

$$\int_0^U \frac{U^2}{X} dU = \left| \frac{U}{c} \right|_0^U - \left| \frac{b}{2c^2} \cdot \log X \right|_0^U + \frac{b^2 - 2ac}{2c^2} \int_0^U \frac{1}{X} dU . \tag{A-12}$$

For Eqs. (A-10)-(A-12) the inverse functions cannot be formulated. To solve the system for

$$U = f(S_o, E \cdot \tau)$$

iterative methods have to be applied.

Nomenclature

Symbols

A	activity	$mmol \cdot min^{-1}$
A_m	activity (mass specific)	$mmol \cdot g^{-1} \cdot min^{-1}$
A_v	activity (volume specific)	$mmol \cdot l^{-1} \cdot min^{-1}$
C	concentration	$mol \cdot l^{-1}$
$\underline{C}$	matrix of correlation coefficients	–
d	length of light path	cm
d_k	particle diameter	μ
E	enzyme concentration	$g \cdot l^{-1}$
E_λ	extinction	–
E_m	amount of enzyme	g
ENZ	catalyst cost per unit weight of product	$DM \cdot kg^{-1}$
ETA	catalyst effectiveness	–
F_c	flow rate of concentrated product stream	$l \cdot min^{-1}$
F_{H^+}	titratable fraction of hydrogen ions	–
F_v	flow rate	$l \cdot min^{-1}$
F_{vo}	flow rate (new feed)	$ml \cdot min^{-1}$
F_{vR}	flow rate (recirculation)	$ml \cdot min^{-1}$
F_{vs}	specific flow rate	$ml \cdot min^{-1} \cdot m^{-2}$
FDE	economic deactivation constant	$d^{-1}, \% \cdot d^{-1}$
ΔH	molar enthalpy	$kJ \cdot mol^{-1}$
I	concentration of inhibitor	$mmol \cdot l^{-1}$
IE	immobilization efficiency	d^{-1}
k	kinetic constant	–
k_{de}	deactivation constant (kinetic)	$d^{-1}, \% \cdot d^{-1}$
k_m	Michaelis-Menten constant	$mmol \cdot l^{-1}$
k_{ic}	constant of competitive inhibition	$mmol \cdot l^{-1}$
k_{in}	constant of noncompetitive inhibition	$mmol \cdot l^{-1}$
K	equilibrium constant	$mmol \cdot l^{-1}$
KE	cost of enzyme	DM
KFE	price of enzyme	$DM \cdot g^{-1}$
KFR	price of substrate (raw material)	$DM \cdot kg^{-1}$
KFT	price of evaporating 1 l solute (thermal energy)	$DM \cdot l^{-1}$
KG	total proportional cost	$DM \cdot kg^{-1}$
KR	cost of raw material	DM
KT	separation cost	DM
L	length	cm, m
L	capacity	$kg \cdot d^{-1}$
MP	molecular weight of product	$kg \cdot mol^{-1}$
MS	molecular weight of substrate	$kg \cdot mol^{-1}$
$[M]_\lambda^T$	molecular rotation	$° \cdot l \cdot mol^{-1} \cdot m^{-1}$
$[M_u]_\lambda^T$	$[M]_\lambda^T$ at U = 100%	$° \cdot l \cdot mol^{-1} \cdot m^{-1}$
P	product concentration	$mmol \cdot l^{-1}$
Δp	pressure drop	bar
PR	integral productivity	$kg \cdot g^{-1} \cdot d^{-1}$
R	reaction rate	$mmol \cdot g^{-1} \cdot min^{-1}$
Re_K	Reynolds number of particle	
ROH	product specific cost of raw material	$DM \cdot kg^{-1}$

S	substrate concentration	$mmol \cdot l^{-1}$
t	time	s, min, h, d, a
$t_{1/2}$	half-life	d
T	temperature	°C, K
TRE	product specific cost of separation	$DM \cdot kg^{-1}$
u	velocity	$cm \cdot s^{-1}$
U	conversion	–, %
V_f	void volume	ml, l
V_{max}	maximal reaction rate	$mmol \cdot g^{-1} \cdot min^{-1}$
V_R	reactor volume	l
β	buffer capacity	$mol \cdot l^{-1}$
ϵ	catalyst effectiveness	–
ϵ_M	molar extinction coefficient	$cm^2 \cdot mol^{-1}$
η	dynamic viscosity	cP, $g \cdot s^{-1} \cdot cm^{-1}$
θ	dimensionless time (t/τ)	–
λ	wave length	nm
ν	kinematic viscosity	cSt, $cm^2 \cdot s^{-1}$
ρ	density	$g \cdot cm^{-3}$
τ	residence time	min
ψ	drag coefficient	–
ω	angular velocity	$rad \cdot s^{-1}$

Indices

c	concentrated
D	– of D-enantiomer
D, L	– of racemate
e	final
eff	effective
L	– of L-enantiomer
max	maximal
o	initial
ot	open-tube
P	product
R	tube
S	substrate
V	related to void volume

References

1. Zaborsky, O.R.: Immobilized enzymes. Cleveland, O.: CRC Press 1973
2. Pitcher, W.H.: In: Immobilized enzymes for industrial reactors. Messing, R.A. (ed.). New York: Academic Press 1975
3. Helmrich, H.: Dissertation, TU-Hannover, FRG 1974
4. Halwachs, W.: Dissertation, TU-Hannover, FRG 1976
5. Wandrey, C., Halwachs, W., Weiss, R., Schügerl, K.: Vth Int. Ferm. Symp., Berlin 1976
6. Kezdy, F.J., Bender, U.L.: Biochem. *1*, 6 (1962)
7. Schonbaum, G.R., Zerner, B., Bender, U.L.: J. Biol. Chem. *236*, 11 (1961)
8. Bender, U.L., et al.: J. Amer. Chem. Soc. *88*, 24 (1966)
9. Ford, J.R., Chambers, R.P., Cohen, W.: Biochim. Biophys. Acta *309*, 175 (1973)
10. Commission Biochemical Nomenclature: Enzyme Nomenclature, Elsevier, Amsterdam, Netherland 1973

11. Lilly, U.D., Dunhill, P.: Process Biochem. *6*, 29 (1971)
12. Vieth, W.R., Venkatasubramanian, K.: Chem. Tech. *677*, 1 (1973)
13. Vieth, W.R., Venkatasubramanian, K., Constantinides, A., Davidson, B.: In: Immobilized enzyme principles, Vol. 1. New York: Academic Press 1976
14. Wingard, L.B.: Biotechnol. Bioeng. Symp. *3*, 3 (1972)
15. Lilly, U.D., Regan, D.L., Dunhill, P.: In: Enzyme engineering. Pye, E.K., Wingard, L.B. (ed.), Vol. 2. New York: Plenum Press 1974
16. Havewala, N.B., Pitcher, W.H.: In: Enzyme engineering. Pye, E.K., Wingard, L.B. (ed.), Vol. 2. New York: Plenum Press 1974
17. Lim, H.C., Emigholz, K.F.: In: Enzyme engineering. Pye, E.K., Wingard, L.B. (ed.), Vol. 3. New York: Plenum Press 1977
18. Lilly, M.D., Smith, S.W., Dunhill, P.: Vth Int. Ferm. Symp., Berlin 1976
19. Chibata, I.: Proc. I.S.F.M., 75 (1972)
20. Mitz, M.A., Schlüter, R.J.: Biochim. Biophys. Acta *27*, 168 (1958)
21. Potapow, V.M.: Wiss. Z. Techn. Hochsch. Chem. Leuna-Merseburg *11*, 1 (1967)
22. Lowry, W.T., Vercellotti, J.R., Carrell, A.S.: Carbohydrate Res. *28*, 93 (1973)
23. Wandrey, C.: Chem. Ing. Tech. *48*, 537 (1976)
24. Möller, K.G., Wandrey, C.: Eur. J. Appl. Microbiol. *3*, 81 (1976)
25. Blass, D.A.: Anal. Biochem. *71*, 405 (1976)
26. Wandrey, C., Hönig, W., Kula, M.R.: Eur. J. Appl. Microbiol. *3*, 257 (1977)
27. In: Handbook of chemistry and physics. Weast, R.C. (ed.), 56th ed. Cleveland, O.: CRC Press 1975
28. Seel, F.: Grundlagen der analytischen Chemie, 4th ed. Weinheim: Verlag Chemie 1968
29. Tosa, T., Mori, T., Fuse, N., Chibata, I.: Enzymologia *31*, 214 (1966)
30. Tosa, T., Mori, T., Fuse, N., Chibata, I.: Enzymologia *31*, 225 (1966)
31. Tosa, T., Mori, T., Fuse, N., Chibata, I.: Enzymologia *32*, 153 (1967)
32. Tosa, T., Mori, T., Fuse, N., Chibata, I.: Biotechnol. Bioeng. *9*, 603 (1967)
33. Tosa, T., Mori, T., Fuse, N., Chibata, I.: Agr. Biol. Chem. *33*, 1047 (1969)
34. Tosa, T., Mori, T., Chibata, I.: Agr. Biol. Chem. *33*, 1053 (1969)
35. Tosa, F., Mori, T., Chibata, I.: Enzymologia *40*, 50 (1970)
36. Tosa, T., Mori, T., Chibata, I.: J. Ferment. Technol. *49*, 522 (1971)
37. Sato, T., Mori, T., Tosa, T., Chibata, I.: Arch. Biochem. Biophys. *147*, 788 (1971)
38. Mori, T., Sato, T., Tosa, T., Chibata, I.: Enzymologia *43*, 213 (1972)
39. Wandrey, C.: Habilitationsschrift, TU-Hannover, FRG 1977
40. Bruns, F.H., Schulze, C.: Biochem. Z. *336*, 162 (1962)
41. Cheng-Wu Chi, Orekhovich, V.N.: Biochimija *23*, 772 (1958)
42. Svedberg, T., Pedersen, K.O.: The ultracentrifuge. Oxford: Calderon Press 1959
43. Lehninger, A.L. (ed.): Biochemie, p. 124. Weinheim: Verlag Chemie 1975
44. Kördel, W., Schneider, F.: Biochim. Biophys. Acta *445*, 446 (1976)
45. Jaworek, D.: In: Enzyme engineering. Pye, E.K., Wingard, L.B. (ed.), Vol. 2. New York: Plenum Press 1974
46. Jokote, Y., Fujita, M., Schimura, G., Noguchi, S., Kimura, K., Samejima, H.: Agr. Biol. Chem. *39* (8), 1545 (1975)
47. Jaworek, D.: pers. commun.
48. Michaelis, L., Menten, M.L.: Biochem. Z. *49*, 333 (1913)
49. Briggs, G.E., Haldane, J.B.S.: Biochem. J. *19*, 338 (1925)
50. Hougen, O.A., Watson, K.M.: In: Chemical process principles, Vol. 3. New York: Wiley and Sons 1947
51. Langmuir, I.: J. Amer. Chem. Soc. *30*, 1742 (1908)
52. Hooke, R., Jeeves, T.A.: J. Assoc. Comp. Mach. *8* (2), 212 (1961)
53. Nelder, J.A., Mead, R.: Computer J. *7*, 441 (1965)
54. Marquardt, D.W.: Soc. Ind. Appl. Math. *11*, 431 (1963)
55. Lineweaver, H., Burk, D.: J. Amer. Chem. Soc. *56*, 658 (1934)

56. Eddie, G.S.: J. Biol. Chem. *146*, 85 (1942)
57. Hofstee, B.H.J.: Science *116*, 329 (1952)
58. Mounter, W.A., Diem, L.T.H., Bell, F.E.: J. Biol. Chem. *233*, 403 (1958)
59. Flaschel, E.: Dissertation, TU-Hannover, FRG, 1976
60. Borchert, A.: Diplomwork, TU-Hannover, FRG 1976
61. Zurmühl, R.: In: Praktische Mathematik für Ingenieure und Physiker, 5th ed., p. 229. Berlin, Heidelberg, New York: Springer 1965
62. Blake, F.C.: Trans. AIChE *14*, 415 (1922)
63. Draper, N., Smith, H.: Applied regression analysis. New York: Wiley and Sons 1966
64. Levenspiel, O.: Chemical reaction engineering, 2nd ed., a) p. 110; b) pp. 136, 137. New York: Wiley and Sons 1972
65. Chambers, R.P., Cohen, W., Baricos, W.H.: Methods in enzymology *44*, 291 (1976)
66. Pasek, A., Skachova, H., Hanus, J., Kucera, J.: Chemicke listy/svazek *71*, 1053 (1977)
67. Wandrey, C., Flaschel, E.: Chem.-Ing.-Techn. *49* (3), 257 (1977)

The Rational Design of Affinity Chromatography Separation Processes

David J. Graves
Department of Chemical and Biochemical Engineering
The University of Pennsylvania, Philadelphia, PA 19104, U.S.A.

Yun-Tai Wu
Chemicals and Plastics Division, Union Carbide Corporation
Bound Brook, NJ 08805, U.S.A.

A number of simple models of affinity chromatography are presented which can be used individually or in combination to predict such results as 1) when a column will become saturated with enzyme, 2) when a peak will emerge from a column and how sharp it will be, 3) how much purification one can expect under a given set of conditions, and 4) what concentration of a competitive inhibitor is needed to remove enzyme from a column. Kinetic problems can be important in affinity separations, and a predictive equation which is given allows one to estimate when such problems are likely to be seen.

1 Introduction

Affinity chromatography is one of the most powerful separation methods that the biochemical engineer has at his disposal. Enzymes, receptors, antigens, antibodies, hormones, binding proteins, whole cells and other substances which possess a specific reversible binding site are amenable to highly selective purification. In contrast to classical techniques such as gel filtration or fractional precipitation, which rely on small differences in molecular size, isoelectric point, etc., affinity procedures utilize specific chemical reactivity which very few if any of the contaminants will have in common with the desired product. Hundreds of such separations have been devised and described in the literature within the past five years. However, since the enzyme purification is normally only an intermediate step in an investigation, when one finds a method for a particular enzyme, it is normally not an optimal scheme which one would employ on a large scale. In industrial scale preparations, minimal loss, maximum speed, the smallest amount of (usually) expensive column packing, etc. are important practical considerations.

For those who wish some introductory material on affinity chromatography, there is a book by Lowe and Dean[1)], numerous papers and reviews by Cuatrecasas (e.g.[2, 3)]) a prolific pioneer in the field, a volume in the *Methods in Enzymology* series[4)] and a collection of papers from an ACS symposium[5)]. The purifications carried out during 1972 have been summarized in one work[6)]. In this review we will not further discuss descriptive aspects of the process such as technique, etc. Instead, we take the starting assumption here that a basic affinity procedure has been found either experimentally or through a literature search and we try to answer some questions an engineer might face in evolving an efficient and economical process. This usually involves both mathematical modeling and the experimental measurement of several parameters where these are not known or cannot be readily estimated. We treat each of these areas separately and in turn.

2 Theoretical Description of Affinity Chromatography

Despite the large number of papers dealing with the practical application of affinity chromatography, theoretical analyses are quite rare. Two models which have been formulated in some detail are a paper by Wankat[7)] and a previous treatment by us[8)]. The former was based on a countercurrent distribution model similar to one originated by Martin and Synge[11)] and the latter was based on what amounts to multiple extraction of a single batch of adsorption gel. The techniques used in these two papers are different, but the philosophy involved in deriving them was similar. Each describes somewhat different aspects of affinity chromatography as a process.

We base our presentation here on several different types of models. First, we show how a simple statistical model of affinity chromatography can give rough estimates for most of the desired quantities. Next, we summarize and expand the results of our previous work on single-stage processes to indicate how they could be used in a multi-stage model of an affinity column such as Wankat's. There are two reasons for doing this. First, Wankat's model was based on surface adsorption of enzyme, which is not too appropriate for most separations in gels. Second, and perhaps more important, is the fact that in very strong adsorption processes such as those that might be typified by antigen-antibody and hormone-receptor systems, the dissociation rate constant can be very low even after the equilibrium dissociation constant K_i has been raised to a reasonably high level or a strong competitive inhibitor has been added[9)]. In such cases, stirring a batch of affinity beads with eluant is a more efficient way of recovering product than passing solution through a column (although Lowe and Dean[10]) have suggested stopping the flow through a column for some time to allow dissociation of the complex). Our batch model is directly applicable to such experiments. Finally, we discuss kinetic factors which one may wish to incorporate into a more rigorous model of affinity chromatography.

Our nomenclature will follow the convention that "enzyme" refers to the substance which is to be purified, regardless of what it actually is. We will use "ligand" to refer to the bound affinity molecule and "inhibitor" to refer to a soluble species which can compete with the ligand for enzyme binding. Conventional chromatography has given us the terms "mobile phase" which we will occasionally refer to as "eluant" when its composition is such that if effects removal of the enzyme from the ligand and "stationary phase" which, of course, in our case is the affinity gel.

2.1 Equilibrium Aspects of Affinity Chromatography

Whatever model of affinity chromatography one develops, he must take the equilibrium relationships between enzyme and affinity gel into consideration. Here we present a number of such relationships, both rigorous and simplified, which are utilized in Sect. 2.2 and 2.3 in particular chromatography models. Although kinetic aspects of adsorption also can be important, they are considerably more difficult to incorporate into a model directly because of their mathematical complexity. Consequently, kinetics are treated separately in Sect. 2.4 and kinetic effects are approximated in the statistical model.

Assuming that in a thin slice of an affinity column one has a volume v within the gel accessible to penetration by the enzyme and a void volume V of solution phase, and that a simple reaction is taking place

$$E + L \underset{k-1}{\overset{k_1}{\rightleftharpoons}} EL \, . \tag{1}$$

One can show that

$$\frac{EL}{L_0} = \frac{1}{2} [B] [1 - \sqrt{1-A}] \, , \tag{2}$$

where

$$B = \frac{K_i (V-v)}{L_0 \, v} + \frac{E_0 \, V}{L_0 \, v} + 1 \, , \tag{3}$$

and

$$A = \frac{4 \, E_0 \, V}{B^2 L_0 \, v} \, . \tag{4}$$

In these equations, E, L, and EL are the equilibrium concentrations of enzyme, ligand, and enzyme-ligand complex respectively. It is assumed that all of the enzyme was initially present in volume V at a concentration E_0 and that the ligand was in v at concentration L_0. The dissociation constant K_i is defined as

$$K_i = \frac{k_1}{k_{-1}} = \frac{(E)(L)}{(EL)} . \tag{5}$$

For most practical purposes, the quantity A is much less than 1 and Eq. (2) can be rearranged and simplified to give

$$\frac{\text{bound enzyme}}{\text{total enzyme}} \cong \frac{(EL)v}{(E_0)V} = \frac{L_0\, v}{C + L_0\, v} , \tag{6}$$

where

$$C = K_i\,(V + v) + E_0\, V . \tag{7}$$

This form shows that a hyperbolic adsorption of enzyme similar to a Langmuir adsorption isotherm or the saturation kinetics of a Michaelis-Menten enzymatic reaction is to be expected for affinity adsorption of the simplest type. By rearrangement of Eqs. (6) and (7) it is equally easy to show similar behavior for the saturation of the gel phase.

$$\begin{matrix}\text{fractional gel}\\ \text{saturation}\end{matrix} = \frac{(EL)}{(L_0)} \cong \frac{E_0\, V}{D + E_0\, V} , \tag{8}$$

where

$$D = K_i\,(V + v) + L_0\, v . \tag{9}$$

Since typical maximum ligand concentrations are 1–10 mM, and void volume V is roughly equal to the gel volume v, it is easy to show from (6) that for effective enzyme adsorption, the value of K_i must be 10^{-3} M or less. This simple theoretical result has been borne out in practice by many different investigations. Conversely, Eq. (8) tells us that a very large volume of enzyme solution (with a typical $E_0 = 10^{-6}$M) would be needed to effectively saturate a gel with moderate capacity. Other relationships can be derived from the identities

$$\text{total moles of enzyme} \equiv (E_0)V , \tag{10}$$

$$\text{moles ligand bound enzyme} \equiv (EL)v \equiv V(E_0 - E) - v(E) , \tag{11}$$

$$\text{moles free enzyme in gel} \equiv (E)v ,$$

$$\text{moles free enzyme in external solution} \equiv (E)V . \tag{12}$$

In the special case where the gel volume is negligibly small and the number of enzyme moles is much less than the number of ligand moles, a very simple analogy of Eq. (6) can be derived

$$\frac{\text{bound enzyme}}{\text{total enzyme}} = \frac{L_0/K_i}{1 + L_0/K_i} . \tag{13}$$

The quantity L_0/K_i is seen as a key parameter, and this will be evident in some of our further derivations. Figure 1 graphically depicts this behavior. As can be seen, a value for $L_0/K_i = 1$ represents 50% binding, with higher values giving more and lower values less interaction with the ligand.

2.2 A Statistical Model of the Chromatography Process

A wide variety of models for column chromatography have been proposed, ranging from very simple to very complex. Since virtually no application of such models to affinity processes has been made in the past though, the simplest models appear most appropriate. Then contain the smallest number of parameters, and their predictions are easiest to compare with real experimental data. As some deficiencies in their formulation become apparent, they can then be modified in the appropriate ways.

Among the available formulations, the simple model originally used by Martin and Synge[11)] and later elaborated somewhat by Giddings[12)] appears most useful. It has the advantage that a single set of calculations can be used to describe both the behavior of a column subjected to a pulse input of enzyme (case 1) and the behavior during constant infusion of enzyme at a fixed concentration (case 2). The relationship between the results of these two types of experiments is shown in Fig. 2. The application of a limited

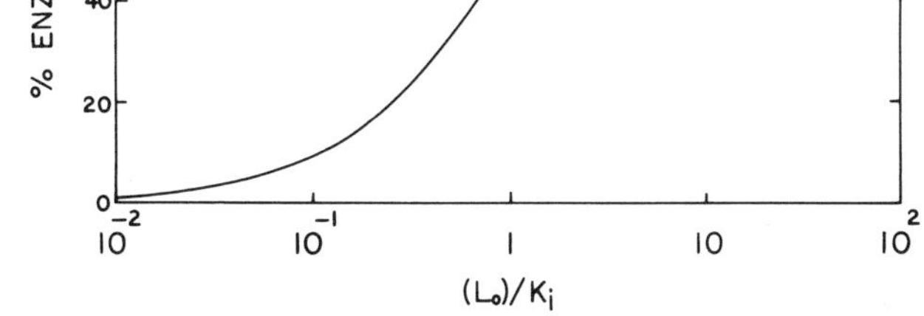

Fig. 1. When the total moles of enzyme are much less than those of ligand, the percentage of bound enzyme is a simple function of the parameter group L_0/K_i

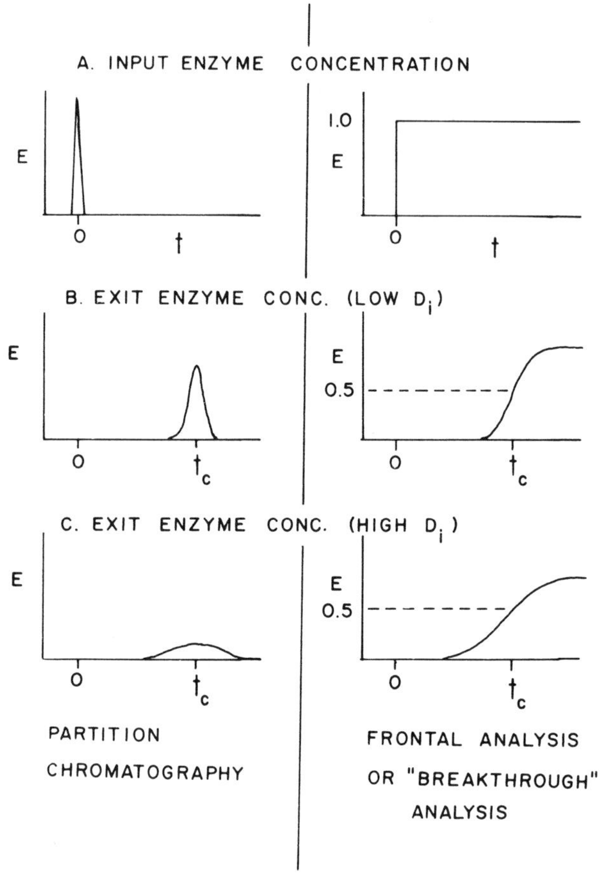

Fig. 2. The response of an affinity column to either a pulse or a step input of enzyme can be described by two parameters, t_c/t_0 and D_i, according to the simplest statistical model of chromatography

amount of enzyme to a good affinity column (high L_0/K_i) will result in a thin sharp zone of enzyme near the top of the column. The emergence of a peak at a later much lower L_0/K_i then can be expected to resemble the case where a small pulse was added to the column initially. If such a narrow pulse of enzyme or a continuous infusion of enzyme is applied to a column starting at $t = 0$, the peak will emerge (case 1) or the column exit concentration will have rise to 50% of its final value (case 2) at time t_c. The width of the peak and the degree of tailing in the step function response are each a function of the single dispersion parameter D_i. The greatest deficiency in the model is that it predicts symmetricallyshaped peaks, while those in real affinity experiments sometimes have a fairly sharp leading edge and a disperse tail. However, we will return to this problem in Sect. 2.3 and 2.4 to show how the model could be improved in this aspect.

The simple formulation of Giddings shows that this peak emergence time t_c is simply the time t_0 for the void volume to appear divided by the fractional time R a species remains in the mobile phase, i.e.

$$t_c = t_0/R\ . \tag{14}$$

The fractional time R can be assumed to equal the fractional distribution in the mobile phase

$$R = \frac{\text{enzyme in mobile phase}}{\text{total enzyme}}\ . \tag{15}$$

From Eqs. (6) and (7) and from the definition of K_i, one can show that

$$R = \frac{E}{E_0} \cong \frac{1}{1 + \frac{v}{V}\left(1 + \frac{L_0}{K_i}\right)}\ . \tag{16}$$

This derivation assumes that the total moles of enzyme which has been in contact with a given number of ligand moles is small. That is, the uptake of enzyme by affinity material is assumed linear in concentration. If it is recognized that the quantity $1 + \frac{L_0}{K_i}$ is the distribution coefficient between gel and mobile phase, Eq. (16) becomes identical to the classical equation of Martin and Synge[11] for partition chromatography. Thus, the relative peak emergence time or breakthrough time for 50% inlet concentration (similarly the elution volume V_c relative to the void volume V) is simply

$$\frac{t_c}{t_0} = \frac{V_c}{V} \cong 1 + \frac{v}{V}\left(1 + \frac{L_0}{K_i}\right)\ . \tag{17}$$

It should be noted that an expression equivalent to Eq. (17) was derived by Dunn and Chaiken[13]. A second of their equations predicting the influence of a soluble competitive inhibitor, as we will see later, is also consistent with our formulations for low enzyme levels.

In the special case that the gel has zero volume (surface adsorption of enzyme), and again applying the restriction that the number of enzyme moles is much smaller than the number of ligand moles, it may be shown in a similar fashion from Eq. (13) that

$$R = \frac{E}{E_0} = \frac{1}{1 + L_0/K_i} \tag{18}$$

which is analogous to Eq. (16) for this special case, and similarly in analogy to Eq. (17)

$$\frac{t_c}{t_0} = \frac{V_c}{V} = 1 + \frac{L_0}{K_i}\ . \tag{19}$$

Lowe and Dean[10] have given such an expression for the retention time in an affinity column. It is, as we have said, only valid for a zero volume adsorbent such as the case of an affinity sorbent coated on non-porous glass beads. Since $\frac{v}{V}$ is about 1.4 for beds such as Sepharose 4B, from a comparison of the Eqs. (17) and (19), it is clear that significantly better results could be obtained in difficult separations with non-porous systems. For instance if $\frac{L_0}{K_i} = 1$, a coated column will have a peak emergence time twice the impurity emergence time whereas in an ordinary gel column it will be only 1.6 times the impurity peak time. With even moderate width peaks, the increased possibility of enzyme and contaminant overlap is obvious. For high ratios of enzyme moles to ligand moles, an alternate expression for R can be used

$$R = \frac{E}{E_0} \cong \frac{E_0 V - L_0 v}{E_0 (V + v)} \,. \tag{20}$$

This expression assumes that the ligand is totally saturated with enzyme, and as is obvious from the numerator, must not be used when there is less enzyme than ligand. For rough approximations, Eq. (16) can be used when E_0V is less than or equal to L_0v and (20) when it is greater. Equation (16) becomes very accurate at low E_0V/L_0v ratios and (20) is accurate at high ratios. Both equations overestimate the uptake of enzyme by ligand in intermediate ranges and thus give R values which are too low. In intermediate cases ($0.1\ L_0v < E_0v < 5.0\ L_0v$), use of the exact solution from equation (2) is, of course, preferable. EL/L_0 from Eq. (2) can be substituted in Eq. (21) to give R

$$R = \frac{V}{V + v} - \left(\frac{EL}{L_0}\right)\left(\frac{L_0}{E_0}\right)\left(\frac{v}{V + v}\right) . \tag{21}$$

Of these solutions, Eq. (16) will usually be found to be the most generally useful.

We can now roughly estimate the time of breakthrough for an affinity chromatography column using Eq. (16) or (21) and classical results from other similar types of partition chromatography. If a constant concentration of enzyme is continuously applied to a column, then the outlet concentration will theoretically have an error function profile[14] given by

$$\frac{E}{E_0} = \frac{1}{2}\left[1 + \operatorname{erf}\left(\frac{\theta - t_c}{\tau_c\sqrt{2}}\right)\right], \tag{22}$$

where E is the concentration at time θ relative to the inlet concentration E_0; t_c is the time when the concentration is exactly half of its final value. As mentioned previously, this time corresponds to the time when a peak would occur if a pulse rather than a step input in concentration had been made. The quantity τ_c is the standard deviation of the peak width which can be related to operating characteristics of the column.

In order to use Eq. (22), we need an expression for this standard deviation, which depends on dispersion processes occurring within the column. Giddings[12] considers a multitude of processes including diffusion within the mobile and stationary phases, adsorption rates, etc. These processes affect the height of a theoretical equilibrium stage H in the column. However, if we assume that the primary process giving rise to temporal or volumetric dispersion of the enzyme is diffusion within the stationary phase (our affinity gel), we may use his equation 4.6-43:

$$H = \frac{R\,(1-R)\,d_p^2}{30\,(1-\phi)\,D_s}\,v_1\,, \tag{23}$$

where d_p is the gel particle diameter, ϕ is the fraction of enzyme molecules within the gel which would be bound at equilibrium, D_s is the enzyme diffusivity in the gel and v_1 is the linear velocity of mobile phase (cm s^{-1}). It may be shown easily that the fraction of bound molecules is given by

$$\phi \cong \frac{\dfrac{L_0}{K_i}}{1 + \dfrac{L_0}{K_i}}\,. \tag{24}$$

Finally, the quantity τ_c is given by

$$\tau_c = t_c\sqrt{H/L_c}\,, \tag{25}$$

where L_c is the column length. Making the appropriate substitutions and using Eq. (16) for R, we have

$$D_i = \sqrt{2}\,\frac{\tau_c}{t_c} \cong \sqrt{\frac{d_p^2}{15\,D_s}\,\frac{v}{V}\,\frac{1}{t_0}}\left\{\frac{1 + \dfrac{L_0}{K_i}}{1 + \dfrac{v}{V}\left(1 + \dfrac{L_0}{K_i}\right)}\right\}. \tag{26}$$

It must be remembered that Eq. (26) is valid only for low ratios $\frac{E_0 V}{L_0 v}$. Inserting typical values (d_p = 0.01 cm, $D_s = 10^{-7}$ cm^2 s^{-1}, v/V = 1.41) we find that the bracketed term ranges from 0.67 to 0.71 as $\frac{L_0}{K_i}$ ranges from 10 to 1000. Approximating this term as 0.67, we can then write

$$D_i \cong 6.8\,(t_0)^{-1/2}\,. \tag{27}$$

For high E_0V/L_0v ratios (> 1), similar substitutions lead to

$$D_i \cong \left\{\frac{d_p^2}{15\,D_s}\,\frac{1}{t_0}\left[\frac{V}{V+v} - \frac{L_0}{E_0}\left(\frac{v}{V+v}\right)\right]\left(1 + \frac{L_0}{E_0}\right)\left(\frac{v}{V+v}\right)\left(1 + \frac{L_0}{K_i}\right)\right\}^{1/2}. \tag{28}$$

Unfortunately the value of this last expression can vary over a wide range even at constant d_p, D_s, t_0, v, and V, so a simplified version such as (27) is not available for an excess of enzyme. Equation (22) can be written in two forms with either E/E_0 or t/t_c as the independent variable:

$$\frac{E}{E_0} \cong \frac{1}{2}\left[1 + \operatorname{erf}\left(\frac{t/t_c - 1}{D_i}\right)\right] \tag{29}$$

and

$$\frac{t}{t_c} \cong 1 + D_i \operatorname{erf}^{-1}\left(\frac{2E}{E_0} - 1\right). \tag{30}$$

Although tables of the error function are tabulated in a number of places, for computational simplicity one may wish to make use of a simple approximation[15] such as

$$\operatorname{erf}(T) \cong 1 - 2\left[\exp\left[\frac{-15}{T-7}\right] + 1\right]^{-1}. \tag{31}$$

The entire concentration-time curve can be estimated from (29) or (30) after substituting the appropriate value of D_i. Also, since the breakthrough curve shape is directly related to that of a peak (given a narrow input pulse), the shape of an emerging peak in normal chromatography can be estimated as well

$$\frac{E}{E_0} \cong \frac{1}{D_i\, t_c \sqrt{\pi}} \exp\left[-\frac{(t/t_c - 1)^2}{D_i^2}\right]. \tag{32}$$

As an example of an application, we can calculate the time at which the exit concentration for a column is 10% of the inlet concentration with a constant inlet E_0. We will do this for three flow rates to demonstrate the effect of flow. If $L_0/K_i = 100$, Eq. (17) gives $t_c/t_0 = 143.4$. Now for void volume elution times t_0 of 100, 1000, and 10^4 s, Eq. (30) using Eq. (27) for D_i gives t/t_c values of 0.39, 0.81, and 0.94 respectively. The 10% breakthrough concentration ($E/E_0 = 0.1$) thus would be achieved in 1.55 h, 32.27 h, and 15.6 days for these three cases. The t/t_c values give an indication of how disperse the breakthrough curve is, since they give the ratio of the time when E is 10% of E_0 to the time when it is 50% of E_0.

In the first case the curve is very disperse and in the last it is quite sharp. Faster flow rates lead to increasingly poorer column utilization. One note of caution we should add is that the time to traverse tubing and detector beyond the column must be subtracted from experimental data to find the true column exit time t_c.

Perhaps more interesting is the comparison of predictions from this crude theoretical model with real data. Kasai and Ishii[16] have studied the affinity chromatography of bovine trypsin on glycylglycyl L arginine Sepharose and have given enough data that comparisons can be attempted. Table 1 shows their data and computed values for t_c/t_0, the peak emergence or $E/E_0 = 0.5$ breakthrough time, and gives values for t/t_c at $E/E_0 = 0.1$. The predicted peak emergence time agrees within a factor of 1.45 to 1.6 with ex-

Table 1. Comparison of theory and data of Kasai and Ishii[16)]

	Case 1	Case 2[a]
K_i	2.3×10^{-4} M	3.4×10^{-3} M
L_0	2.2×10^{-3} M	2.2×10^{-3} M
E_0	5×10^{-7} M	5×10^{-7} M
V + v	5.1 ml	5.1 ml
v[b]	2.98 ml	2.98 ml
V[b]	2.12 ml	2.12 ml
V + ext. vol.	7.0 ml	7.0 ml
Enzyme	β trypsin	β trypsin
$E_0 V/L_0 v$	1.62×10^{-4}	1.62×10^{-4}
Flow rate	3 ml/h	3 ml/h
t_0	2544 s	2544 s
Elution vol.[c]	59.4 ml	15.0 ml
Corrected elut. vol.[d]	54.5 ml	10.1 ml
Volume at $E/E_0 = 0.10$	52.8 ml	–[f]
Corrected vol. at $E/E_0 = 0.10$[d]	47.9 ml	–[f]
t_c/t_0 experimental	25.7	4.8
t_c/t_0 predicted[e]	15.9	3.3
t/t_c at $E/E_0 = 0.1$ experimental	0.88	–[f]
t/t_c at $E/E_0 = 0.1$ predicted	0.84	–[f]

[a] Chromatography in presence of benzyloxycarbonyl arginine (4×10^{-3} M) gave this effective increased K_i value
[b] Void fraction assumed 0.415
[c] Time at which $E/E_0 = 0.5$
[d] External tubing volume subtracted: V + Ext. Vol. – V = 5.88 ml
[e] Since $E_0 V/L_0 v << 1$, Eqs. (17) and (27) were used
[f] Data too difficult to read accurately from graph

perimental data, and the predicted peak width (from t/t_c) is remarkably close to the observed result.

There are a number of uncertainties in the experimental results including a secondary interaction between the enzyme and gel admitted by the authors. Lower K_i values resulting from such an interaction would bring our predictions closer to the observed elution times. Our predictions are also, of course, based on equations which are only approximate at best. Nevertheless, the agreement is satisfactory enough that the equations given should be considered if one wishes to predict how a particular affinity column will function. Unfortunately, most authors do not give enough data to permit comparisons such as this one.

A number of other types of calculations can be made using the equations given in this section. For impurities, the quantity L_0/K_i (hopefully) will be equal to or only slightly greater than zero. Substituting this value in the preceding equations, e.g. (17) and (30), should give the impurity peak emergence time and peak shape. One can calculate the impurity peak emergence profile and thus see approximately how long one should wash a column before attempting to elute the enzyme of interest. One caution which should be observed is that the gel volume v which is accessible to impurity may differ from that for enzyme if the two species differ greatly in molecular weight. With the large pores of Sepharose 4 B, such an eventuality is unlikely, but for other cases it may be important.

When eluting with salt gradients, pH gradients, temperature changes, or other techniques which can be assumed to result only in a change in K_i, one simply uses the modified K_i value in Eqs. (17) and (32) to calculate the peak emergence time and peak shape. When eluting with a soluble competitive inhibitor, one can employ a similar tactic provided that he defines a new apparent dissociation constant

$$K_{da} = \frac{(E + EI)\,(L)}{(EL)}, \tag{33}$$

where (I) is the soluble inhibitor concentration and (EI) soluble complexed enzyme. If the soluble inhibitor has a dissociation constant K_s

$$K_s = \frac{(E)\,(I)}{(EI)}. \tag{34}$$

Then it may be shown quite easily that

$$K_{da} = K_i \left[1 + \frac{I_0}{K_s}\right] \tag{35}$$

provided that the inhibitor is so concentrated that its initial concentration (I_0) is approximately equal to the final equilibrium concentration (I). Now one simply uses this value K_{da} in place of K_i in equations such as (17) and (32) as before. This relationship makes evident the fact that one can use a relatively poor soluble inhibitor (high K_s) for elution provided that he makes it concentrated enough. It is also interesting to note that I_0/K_s, a ratio exactly analogous to L_0/K_i, defines the effect of a competitive inhibitor on the system.

The best overall confirmation of the effectiveness of our simple model is experimental work by Dunn and Chaiken[13)] and Chaiken and Taylor[17)]. As mentioned previously, their equations for elution volume are equivalent to our expression (17) either by itself or in combination with (35). They showed that a rearrangement of the expression predicting a linear relationship between the elution volume and the inhibitor concentration was very nicely obeyed by ribonuclease on an affinity column. We should caution, however, that not all enzymes are apt to be as ideal in their behavior as this particular one was.

In summary, one can use an appropriate model based on simple statistical concepts to calculate breakthrough times, peak emergence times under eluting and non-eluting conditions, peak shapes and breakthrough curve shapes. Purification factors can be calculated from the overlap of enzyme and impurity peaks. Although such a model is relatively crude, it provides quick answers to most of the important questions one is faced with in designing an efficient affinity procedure. Once some real experimental data become available, the model can be modified to provide still better estimates for refinement of the procedures. An example of such an advanced affinity chromatography model, which, however, is probably not yet appropriate for application because of its complexity, is that of Denizot and Delaage[18].

2.3 Batch or Single Stage Calculations of Affinity Procedures

As we have stated, batch equilibrium relationships are needed if one wishes to build up a multi-stage model of a column procedure or to determine how a real batch process would compare with a column procedure. Although batch processes have not been commonly used in the past, a simple calculation shows why they may be in the future. A number of important potential affinity systems based on such interactions as those in antibody-antigen complexes have very low dissociation rate constants.

For example, Smith and Skubitz[19] have measured the dissociation kinetics of antibody-hapten complexes and have found rate constants k_{-1} of 1.9 to 170 x 10^{-4} s^{-1}. The reaction is first order, and as is well known, one can for such a case define a half time $t_{1/2}$ when the concentration of complex is half of its original value

$$t_{1/2} = \frac{0.693}{k_{-1}} \, . \tag{36}$$

For the lower of the two rates given, the half time is 3647 s or ~1 h. More than 4 h would be required for 95% dissociation, assuming that the released substances were prevented from recombining (for instance with a soluble complexing agent) and that there were no diffusional process slowing the rate of equilibration. In fact, the diffusional problem cannot be entirely eliminated. In Sect. 2.4 we will return to this example to show how the true equilibrium time easily can be 40 h or even 400. In a column washing process, a peak removed under these conditions would be exceedingly dilute and therefore difficult to handle. Equilibrating a batch of affinity adsorbant with an eluting solution might be a preferable way to recover material in similar situations.

Our model of such a batch process is built up in the following way. A volume v of gel is equilibrated with volume V of enzyme solution, resulting in a saturation given by Eq. (2) or less exactly by (8), (13) etc. We presume that next it is contacted sequentially with one or more washes, each of volume w. The dissociation constant is unchanged during this process, and the purpose of washing is simply to remove impurity. Next, the gel is contacted one or more times with eluting solution, each time of volume V'. Here we assume that two possibilities exist. The first is that the eluting solution changes the enzyme-ligand dissociation constant to K_i', the second is that the solution contains a soluble competing inhibitor of concentration I_0 and having a dissociation constant K_s. We have given the results for this type of calculation previously[8, 20] and here we simply summarize the important results and present a few additional figures which should prove useful in conducting a similar analysis. For details of the derivation, one may refer to the original work.

2.3.1 Elution by a Change in K_i

For a single batch elution at a modified K_i equal to K_i' and following a single adsorption step (with no intermediate washes), one can show that

$$f_R = \frac{\text{recovered enzyme in } V'}{\text{initial enzyme in } V} = \left(\frac{V'}{V' + v}\right)$$

$$\left[1 - \frac{L_0 v\,(K_i v + L_0 v)}{K_i\,(V + v) + L_0 v\;K_i'\,(V + v) + L_0 v}\right.$$

$$\left. - \left(\frac{V}{V + v}\right)\left(1 - \frac{L_0 v}{K_i\,(V + v) + L_0 v}\right)\right]. \tag{37}$$

The recovery is improved by low K_i, low V/v, high K_i', and high V'/v values. An S-shaped plot of enzyme recovery versus K_i' is obtained which is similar to Fig. 1. The upper asymptote is set by the $V'/(V' + v)$ ratio and one can determine rather easily that successful operation requires $\frac{L_0}{K_i} > 100$ (for good binding to the gel) and $\frac{L_0}{K_i'} < 1$ (for good removal from the gel). For a material, such as contaminant, which undergoes no specific binding, the appropriate expression is simply

$$f_{CR} = \left(\frac{v}{V + v}\right)\left(\frac{V'}{V' + v}\right). \tag{38}$$

When n washing steps are included in the sequence, the relationship for enzyme recovery becomes

$$f_R \cong \left(\frac{V'}{V' + v}\right)\left[1 - f_{BE} - \sum_{i=0}^{n} f_i\right], \tag{39}$$

where f_{BE} is given by

$$f_{BE} \cong \frac{L_0 v\,(K_i v + L_0 v)^{n+1}}{[K_i\,(V + v) + L_0 v]\,[K_i'\,(V' + v) + L_0 v]\,[K_i\,(w + v) + L_0 v]^n} \tag{40}$$

and where the quantity f_i is given by

$$f_i = \left(\frac{w}{w + v}\right)\left\{1 - f_{B(n)} + (1 - f_0)\sum_{j=1}^{n-1}\frac{(n-1)}{(n-1-j)!j!}\left(\frac{-w}{w + v}\right)^j\right.$$

$$\left. - \sum_{j=1}^{n-1} f_{(B)j}\left[\sum_{k=0}^{n-1-j}\frac{(n-1-j)!}{(n-1-j-k)!k!}\left(\frac{-w}{w + v}\right)^{k+1}\right]\right\} \tag{41}$$

with

$$f_0 = \left(\frac{V}{V+v}\right)\left[1 - \frac{L_0 v}{K_i (V+v) + L_0 v}\right] \tag{42}$$

and

$$f_{B(n)} = \frac{L_0 v (K_i v + L_0 v)^n}{[K_i (V+v) + L_0 v]\,[K_i (w+v) + L_0 v]^n}\,. \tag{43}$$

The quantity $f_{B(j)}$ is defined exactly the same as $f_{B(n)}$ with the index j replacing n. The quantity $\sum_{i=0}^{n} f_i$ in Eq. (39) represents the total enzyme lost during all washing, and f_{BE} is the enzyme still bound following the elution. This summation is an interesting quantity to see visually, and we have shown it in Fig. 3 for a specific case, namely $V/v = w/v = 1$. It is obvious that for $L_0/K_i > 1000$, there is practically no enzyme loss even after extensive washing, but at values of 10 or less, the loss quickly becomes intolerable. The analogy to Eq. (38) for contaminant recovery after n washes is simply

$$f_{CR} = \left(\frac{v}{V+v}\right)\left(\frac{V'}{V'-v}\right)\left(\frac{v}{w+v}\right)^n. \tag{44}$$

The exponential appearance of n makes clear how effective a number of washing steps can be. There is a great increase in complexity as we go from Eq. (37) (no washing steps) to Eq. (39) (multiple washing steps). One might expect, then, that progression to multiple elutions with multiple washes would become unbearably complex. Fortunately, such is not the case. If we denote the fractional recovery in elution j by $f_{R(j)}$, then

$$f_{R(1)} = f_R\,, \tag{45}$$

$$f_{R(2)} = \left(\frac{V'}{V'+v}\right)\left[1 - f_{BE(2)} - f_{R(1)} - \sum_{i=0}^{n} f_i\right], \tag{46}$$

$$f_{R(m)} = \left(\frac{V'}{V'+v}\right)\left[1 - f_{BE(m)} - \sum_{j=0}^{m-1} f_{R(j)} - \sum_{i=0}^{n} f_i\right], \tag{47}$$

where m represents the mth elution. The new quantity $f_{BE(j)}$ is that portion remaining bound following the jth elution, and is given by

$$f_{BE(j)} = f_{BT(n)} \frac{L_0 v\,[K_i' v + L_0 v]^{j-1}}{[K_i' (V'+v) + L_0 v]^j} \tag{48}$$

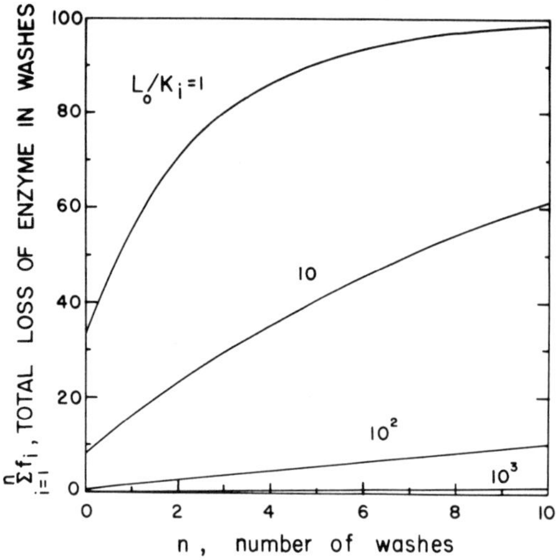

Fig. 3. When L_o/K_i is small, considerable enzyme loss can occur in washing (batch case: $v/V = v/w = 1$)

with

$$f_{BT(n)} = \frac{[K_i v + L_0 v]^{n+1}}{[K_i (V + v) + L_0 v]\,[K_i (w + v) + L_0 v]^n} \,. \tag{49}$$

These last two equations differ slightly from those originally given because they (more correctly) consider both bound and entrapped enzyme. The actual numerical results from Eq. (47) differ insignificantly from those found with our earlier definition of $f_{BE(j)}$. As will be obvious from the repeated-fraction nature of Eq. (44), the appropriate contaminant recovery equation for the jth elution with n washes is simply

$$f_{CR(j)} = \left(\frac{v}{V + v}\right)\left(\frac{V'}{V' + v}\right)\left(\frac{v}{w + v}\right)^n \left(\frac{v}{V' + v}\right)^{j-1} \,. \tag{50}$$

In order to make these rather abstract results more understandable, we present in Figs. 4 and 5 calculated values for the total combined fractions of enzyme recovered (dashed curves) and the purification factor (solid). The latter is defined as the enzyme to contaminant ratio following the multi-step process divided by its initial ratio. Two cases are shown: $L_0/K_i = 100$ (Fig. 4) and 1000 (Fig. 5). Higher ratios are essentially identical to Fig. 5 and lower ratios have not been considered since they result in such excessively large losses of enzyme during washing.

2.3.2 Elution with a Competitive Inhibitor

To calculate the elution which can be achieved by the addition of a competitive inhibitor, one can simply use the definition of an effective dissociation constant K_{da} given in Eq. (35) and systematically substitute this quantity for K_i' in the equations which we have just given for elution by a change in K_i. In our previous work we also derived an asymptotic expression for the enzyme which is complexed by soluble inhibitor, and

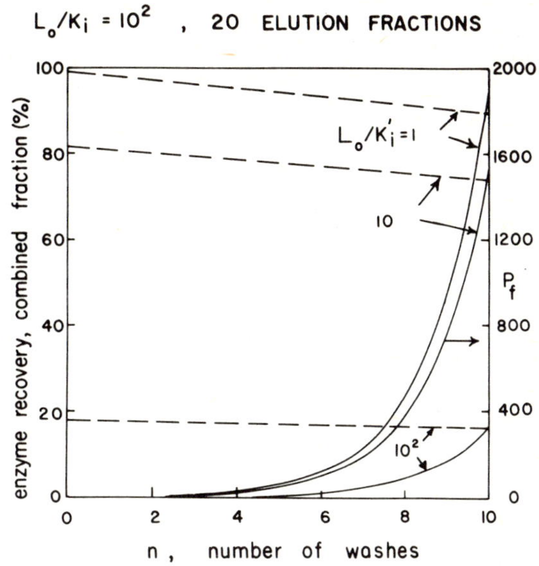

Fig. 4. Following n washes a batch of affinity beads is eluted 20 consecutive times to give the total recovery (dashed) and purification factor (solid, right ordinate) shown here ($v/V = v/w = 1$)

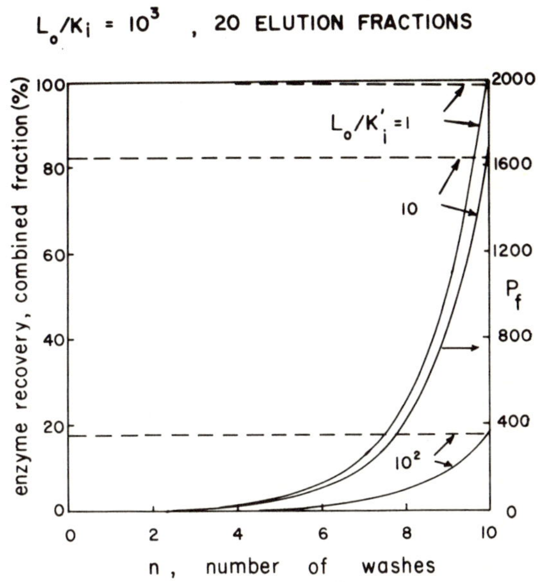

Fig. 5. Results similar to those in Fig. 4 but for a case of stronger initial binding. Note that the loss on washing is significantly reduced

equations similar to those in Sect. 2.3.1 can be derived from these results as an alternative approach.

The simplified equation for complexed enzyme (EI) which will be removed in the elution volume V′ as a fraction of the initial amount in the gel following any washing steps, v(EI), is

$$f_E = \frac{V'\,(EI)}{v\,(E_i)} = \left(\frac{p}{p+1}\right)\left[\frac{p\,I_0/K_s}{p\,I_0/K_s + 1 + p + L_0/K_i}\right], \tag{51}$$

where p is the ratio V'/v. The concentration (E_i) is really a pseudo-concentration representing both bound and gel-entrapped enzyme. This concentration may be related to the amount of enzyme originally exposed to the affinity gel through

$$\frac{v\,(E_i)}{V\,(E_0)} = f_{BT(n)} \tag{52}$$

for n washes, where $f_{BT(n)}$ has been given previously [Eq. (49)]. Note that the ubiquitous quantity L_0/K_i and a similar quantity I_0/K_s appear in these equations. The initial factor $p/(p+1)$ determines the asymptotic recovery on a given wash. Figure 6 from Eq. (51) shows how these factors influence recovery for the case p = 1. As one can see, the stronger the binding between enzyme and gel (higher L_0/K_i) the higher I_0/K_s must be to obtain good recovery. As a general rule, I_0/K_s should be 10 ot 100 times as large as L_0/K_i for efficient enzyme removal when p = 1, or 1 to 10 times L_0/K_i if p = 10.

The recovery on the first elution with inhibitor is then

$$f_{R(1)} = f_{BT(n)} f_E , \tag{53}$$

and for subsequent elutions, m

$$f_{R(m)} = f_{R(m-1)} \left\{ \frac{\left(\frac{p}{p+1}\right) \frac{I_0}{K_s} + 1 + p + \frac{L_0}{K_1}}{p\,\frac{I_0}{K_0} + 1 + p + \frac{L_0}{K_i}} \right\} . \tag{54}$$

The overall purification factor can be found as it was previously, namely through the relationship

$$p_f = \frac{\sum_{j=1}^{m} f_{R(j)}}{\sum_{j=1}^{m} f_{CR(j)}} . \tag{55}$$

In order to see the influence of washing and elution, we have presented Figs. 7–9 for L_0/K_i values of 10, 100, and 1000 respectively. Following the initial binding step (B),

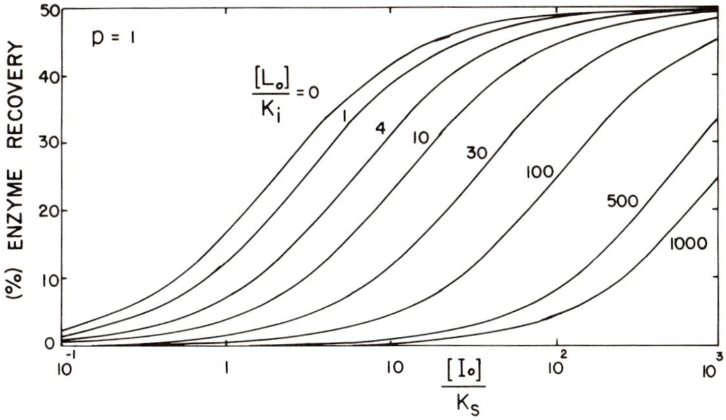

Fig. 6. Enzyme removal with a competitive inhibitor may be characterized by the group I_0/K_S. The stronger the initial binding (L_0/K_i), the higher this group must be to achieve good elution

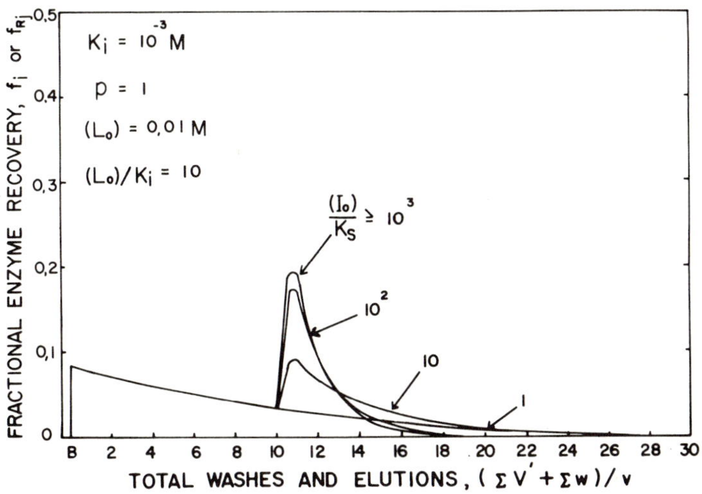

Fig. 7. Simulation of column chromatography by multiple batch washing (10 times) followed by multiple batch elution with a competitive inhibitor (20 times). Higher values of I_0/K_S give sharper enzyme peaks

10 washes (w = v) are sequentially carried out, followed by 20 elutions (p = 1), containing inhibitor at various levels (I_0/K_s from 1 to 10^5). The influence of inhibitor effectiveness on the ease with which enzyme can be recovered and the loss of enzyme during the initial washing steps can be readily appreciated from these figures. Sharp discontinuities have been eliminated in the results by drawing smooth curves through the data points. Although multiple batch extractions are certainly a rather poor model for a column chromatography process, these curves show a strong resemblance to chromatographic

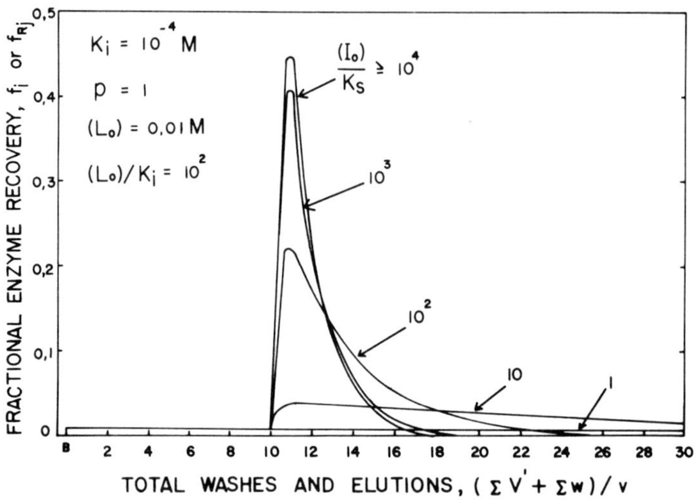

Fig. 8. Simulation similar to Fig. 7 but with stronger initial binding. Note the reduced loss of enzyme during washing

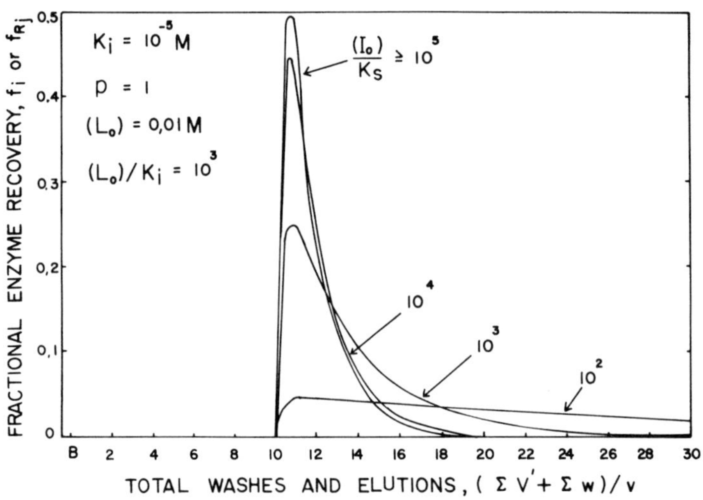

Fig. 9. Simulation similar to Fig. 7 but with still stronger initial binding. There is negligible enzyme loss during washing

peaks. They may be useful for those studying affinity columns in the following sense: the "tailing" seen in each case given here, since it represents the results of a batch process, represents the *minimum* tailing which one would observe under the same experimental conditions from a real affinity column. In this sense these results are complementary to those in Sect. 2.2, because the statistical peaks presented in that section were ideally symmetrical, while these have a sharp leading edge and a diffuse tail. Similar "chromatography simulations" have been given previously[8)] for the case of a change in the K_i value. The results given here could be extended rather easily to multiple-batch-in-series models, and such models should be quite effective in simulating a real column process.

2.4 Kinetic Problems Arising from Diffusion-Reaction Processes in Single Beads

In Sect. 2.2 we provided a very simple model of chromatography. It was assumed then [Eq. (23)] that kinetic processes resulting in dispersion could be represented simply as a term $(1 - \phi)$ reducing the diffusion coefficient. Such an assumption is not really rigorous, but it was expedient. Here we will show that a slightly more realistic model of simultaneous diffusion and reaction within a single bead predicts that the adsorption and desorption rates from a single bead can differ and that the term $(1 - \phi)$ can range from its calculated value up to 1. This in itself is important and aids in the understanding of why a chromatography peak can have a sharp leading edge and a diffuse trailing edge. Furthermore, our simple model will allow one to use the forward and reverse rate constants for association and dissociation between the ligand and enzyme (if they are known or can be estimated) to determine whether diffusion effects in a single bead can lead to experimental difficulties in a given system.

As one example of why such information is needed, we can cite the work of Lowe, Harvey, and Dean[21)] who found that column geometry, ligand concentration, flow rate etc. seemed to have an influence on the strength of binding to an affinity column. They also found[22)] that the apparent binding constant measured under column conditions differed from that found in a batch study. Although it is possible that the particular multi-subunit enzyme they used had unusual equilibrium binding properties, we believe that many of the effects they saw can be understood if one takes into account some of the kinetic aspects of enzyme ligand interaction.

Although it would be desirable to base our model on Eq. (1), for simplicity in solving the resulting partial differential equations of diffusion and reaction, we have chosen to use the simpler reaction

$$E \underset{k_{-1}}{\overset{k_1'}{\rightleftharpoons}} EL \,. \tag{56}$$

If one defines the pseudo-first-order rate constant by

$$k_1' = k_1 \, L_0 \tag{57}$$

then this reaction is equivalent to the second-order case when L_0 is much larger than E and thus is effectively constant. As is evident, the model becomes less correct as the bead approaches saturation. We have previously examined the saturation rate of such a bead when it is suddenly placed in a large bath of enzyme at concentration E_0 both for the case of no surface resistance[23] and a finite surface resistance to diffusion[24]. The latter might represent a stagnant boundary layer of fluid surrounding the bead. Although we have presented the solutions previously, the results were given in a rather abstract form which made them difficult to use except for one who is familiar with such mathematical treatments. Here we simplify the results to a few simple equations and graphs which should be relatively easy for anyone with even a modest mathematical background and skills to use. If we assume temporarily that there is no surface resistance to transport into a bead, we can show that the initial rate of enzyme uptake is given by

$$\text{Rate} = \frac{dm}{dt} = 4\left[\pi \frac{1 + K_e}{t}\right]^{1/2} - 4\pi , \tag{58}$$

$$\frac{t}{1 + K_e} \leqslant 1$$

where the time t is a dimensionless time defined in real time units θ by

$$t = \frac{D_s}{r^2}\theta \tag{59}$$

for a diffusivity D_s and bead radius r. The dimensionless equilibrium constant K_e has been defined for convenience as the now-familiar ratio

$$K_e = \frac{L_0}{K_i} . \tag{60}$$

Equation (58) and other equations containing K_e are, of course, applicable to plain beads containing no ligand simply by setting $K_e = 0$. If we first assume that the reaction is instantaneously rapid, then at a time $\frac{t}{1 + K_e} > 0.1$ Eq. (58) becomes less correct but may be replaced by a truncated infinite series with good accuracy

$$\text{Rate} = \frac{dm}{dt} = 8\pi\left[e^{-\frac{\pi^2 t}{1+K_e}} + e^{-\frac{4\pi^2 t}{1+K_e}} + e^{-\frac{9\pi^2 t}{1+K_e}}\right] . \tag{61}$$

$$\frac{t}{1 + K_e} > 0.1$$

Thus Eqs. (58) and (61) together describe the total course of material uptake for a non-reactive bead or one where the rate constant k_1 is very high. Similarly the integrals of these equations give the total amount of enzyme taken up by the bead (both bound and free).

$$\frac{m}{m_{eq}} = \frac{6}{\pi^{1/2}} \left[\frac{t}{1 + K_e}\right]^{1/2} - 3 \frac{t}{1 + K_e} \tag{62}$$

$$\frac{t}{1 + K_e} \leqslant 0.1$$

and

$$\frac{m}{m_{eq}} = 1 - \frac{6}{\pi^2} \left[e^{-\frac{\pi^2 t}{1+K_e}} + \frac{1}{4} e^{-\frac{4\pi^2 t}{1+K_e}} + \frac{1}{9} e^{-\frac{9\pi^2 t}{1+K_e}} \right] . \tag{63}$$

$$\frac{t}{1 + K_e} > 0.1$$

Equations (62) and (63) have been divided by the equilibrium amount which eventually is taken up by the bead for convenience to give the fractional absorption.

$$m_{eq} = \frac{4}{3} \pi (1 + K_e) . \tag{64}$$

Also, the quantities m in Eqs. (58), (60), (62), and (63) are dimensionless concentrations of both free and complexed enzyme and are based on the relationship

$$m = \frac{4}{3} \pi \frac{E}{E_0} (1 + K_e) . \tag{65}$$

With Eq. (65), real concentrations of free enzyme in the bead E relative to the real E_0 in the fluid around the bead can be found if desired. In some cases one may wish to use equations for m rather than m/m_{eq}, and these are

$$m = 8 \pi^{1/2} (1 + K_e)^{1/2} t^{1/2} - 4 \pi t \tag{66}$$

$$\frac{t}{1 + K_e} > 0.1$$

and

$$m = \frac{4}{3} \pi (1 + K_e) - \frac{8}{\pi} (1 + K_e) \left[e^{-\frac{\pi^2 t}{1+K_e}} + \frac{1}{4} e^{-\frac{4\pi^2 t}{1+K_e}} + \frac{1}{9} e^{-\frac{9\pi^2 t}{1+K_e}} \right] . \tag{67}$$

$$\frac{t}{1 + K_e} > 0.1$$

In order to make the significance of these equations clear, we have prepared Tables 2 and 3 which present the fractional saturation at reduced times of 4×10^{-3} to 114, representing real times of 1 s to 10 h if $D_s = 10^{-7} cm^2 s^{-1}$ and $r = 50 \mu$. The first table represents an infinitely fast reaction and the second represents one which is infinitely slow but still has the capacity $4 \pi (1 + K_e)/3$. That is, Table 3 is the ratio of Eqs. (66) or (67) with $K_e = 0$ to Eq. (64) with K_e set at its desired value (the one given in the table). The results show that one can have moderately slow uptake even with an infinitely fast reaction provided that L_0/K_i (that is, K_e) is high enough. For an infinitely slow reaction, of course, even a modest K_e value means that the amount taken up is only a negligible fraction of its total capacity. The bead becomes saturated with enzyme after only a minute or so, but the reactive ligand has taken up none of the enzyme.

Table 2. Fractional enzyme uptake in a bead with infinitely fast reaction

K_e \ t	m/m_{eq}			
	4×10^{-3}	0.240	14.4	144
0	0.202	0.943	1.00	1.00
10	0.064	0.435	1.00	1.00
10^2	0.021	0.158	0.851	1.00
10^3	0.007	0.056	0.363	0.852
10^4	0.002	0.017	0.124	0.363

Table 3. Fractional enzyme uptake in a bead with infinitely slow reaction

K_e \ t	m/m_{eq}			
	4×10^{-3}	0.240	14.4	144
0	2.0×10^{-1}	9.4×10^{-1}	1.00	1.00
10	1.8×10^{-2}	8.6×10^{-2}	9.1×10^{-2}	9.1×10^{-2}
10^2	2.0×10^{-3}	9.3×10^{-3}	9.9×10^{-3}	9.9×10^{-3}
10^3	2.0×10^{-4}	9.4×10^{-4}	1.0×10^{-3}	1.0×10^{-3}
10^4	2.0×10^{-5}	9.4×10^{-5}	1.0×10^{-4}	1.0×10^{-4}

This second table of results would appear to be purely fictitious since an infinitely slow reaction with a finite equilibrium constant is almost a contradiction. However, this table is still useful, because our prior work showed that in real cases with a finite rate constant, one begins at a rate equivalent to the infinitely slow reaction rate equation and at some point in time crosses over to the infinitely fast reaction rate equation.

During this "crossover period" the rate of enzyme uptake by the bead is roughly constant and is given by

$$\frac{dm}{dt} = R_c = 4.2 \left[\frac{1}{1 + 0.1\,K_1}\right]^2 K_1 + 12 \left[\frac{0.2\,K_1}{1 + 0.2\,K_1}\right]^2 K_1^{1/2} , \tag{68}$$

where

$$K_1 = \frac{r^2}{D_s} k_1 L_0 . \tag{69}$$

A set of results which has been numerically computed for the cases $K_1 = 1000$ and $K_1 = 1$ is given in Fig. 10. One can now approximate the entire adsorption curve by a portion of the lower curve, a horizontal line representing the constant "crossover" rate, and a portion of the upper curve. The entire time course of a bead's saturation thus can be approximately predicted. To do this it is necessary to find the intersections of the upper and lower curves with the line defined by Eq. (68). The calculations of the intersections have been done and they are presented in Fig. 11. One simply draws a vertical line at the proper K_1 value and finds the starting time for crossover t_{cs} from the intersection of this vertical with the $K_e = 0$ curve and the finishing time t_{cf} from the intersection with the curve for the appropriate K_e.

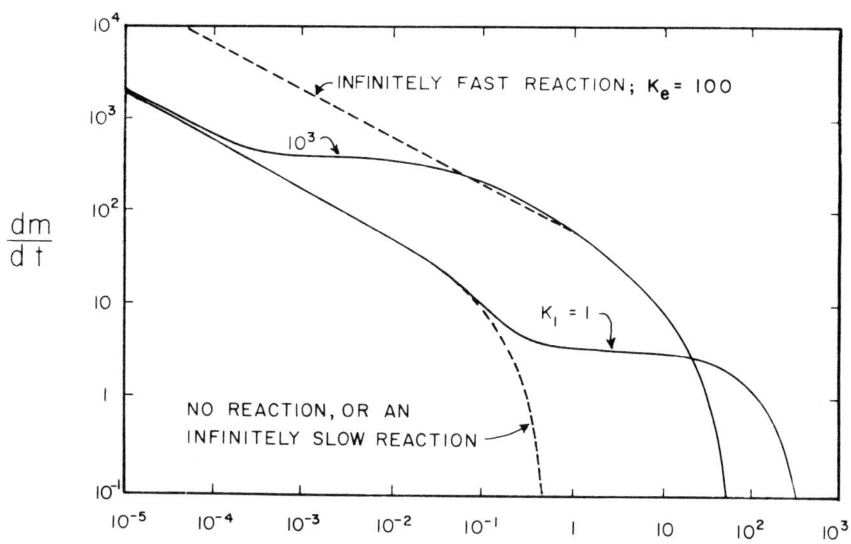

Fig. 10. The rate of enzyme uptake or loss from reactive and unreactive beads for the cases $K_e = 100$ and $K_e = 0$. Finite reaction rates result in a roughly constant "crossover" rate between the two limiting cases

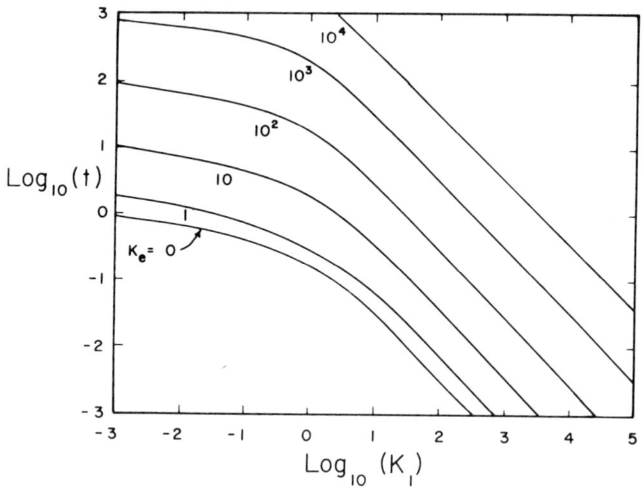

Fig. 11. If the dimensionless rate constant K_1 is known, the beginning and ending times for "crossover" can be found by drawing a vertical line on this graph

The method in which one does these calculations may seem a bit confusing to the reader unless he studies this section carefully, so we will present a simple flow chart leading through them. First, however, the final complication which one may add if desired is that of an external film resistance due to fluid surrounding a bead or any other similar resistance. Fortunately, all that is necessary is to add an upper horizontal line repressing this film-limiting rate

$$\frac{dm}{dt} = R_{fl} = \frac{4\pi r}{D_s} k_L = \frac{4\pi r}{l}, \tag{70}$$

where the coefficient k_L in cm s^{-1} may be found from standard correlations[25)] or if one has an approximate estimate of the boundary layer thickness 1, he can use the last equality. A 10 μ thick film on a 100 μ diameter bead would give an upper limiting rate of about 63, for example. For the constant rates given by R_c and R_{fl}, the total mass uptake if one started at time zero would be

$$m_c = R_c\, t, \tag{71}$$

$$m_{fl} = R_{fl}\, t. \tag{72}$$

And of course if one followed these rates or those given by (58) and (61) between times t_1 and t_2, then the mass uptake would have to be calculated twice, once for t_1 and once for t_2, with the lower limit being subtracted from the upper to properly represent the integral. The flow chart (Fig. 12) leads one easily through the correct calculations. Other than the integral evaluation at two times, another reason why so many calculations are

INITIAL CALCULATIONS		
CALCULATED PARAMETER	EQUATION	SPECIAL CONDITIONS
R_{fl}	(70)	
$t = t_{ff}$	(58) or (61)	$dm/dt = R_{fl}$
$t = t_{fs}$	(58) or (61)	$dm/dt = R_{fl}$; $K_e = 0$
$t = t_{cs}$	FIGURE 11	$K_e = 0$
$t = t_{cf}$	FIGURE 11	
$m_1 = m_{fl}$	(72)	$t = t_{fs}$
$m_2 = m$	(66) or (67)	$t = t_{cs}$
$m_3 = m_c$	(71)	$t = t_{cf}$
$m_4 = m_{fl}$	(72)	$t = t_{ff}$
$m_{L1} = m$	(66) or (67)	$t = t_{fs}$; $K_e = 0$
$m_{L2} = m_c$	(71)	$t = t_{cs}$
$m_{L3} = m$	(66) or (67)	$t = t_{cf}$
$m_{L4} = m$	(66) or (67)	$t = t_{ff}$

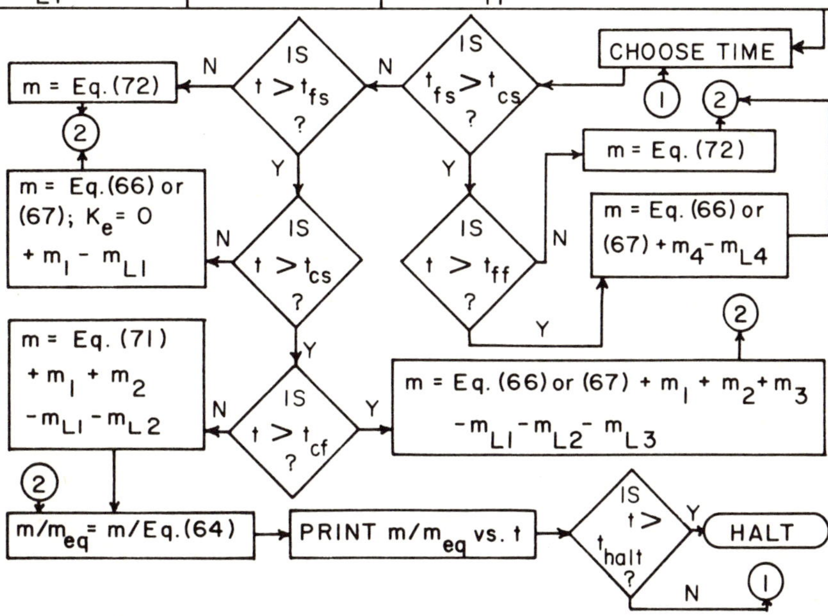

Fig. 12. Using this flowchart and the appropriate equations, the entire time course of adsorption and desorption in a bead can be determined

necessary is that different cases result depending on whether R_c is greater or less than R_{fl}.

So far we have considered only adsorption of enzyme by a bead. However, it may be shown that the desorption case is mathematically identical to the adsorption case provided that K_e is constant and that one makes the following substitutions:

$$-K_{-1} \rightarrow K_1 , \tag{73}$$

$$-K_1 \rightarrow K_{-1} , \tag{74}$$

$$(1 - E) \rightarrow E , \tag{75}$$

$$(K_e - EL) \rightarrow EL , \tag{76}$$

where the arrow in each of these expressions signifies "replaces". In other words, these substitutions are correct for a pulse moving through a column under constant conditions (Case 1, Sect. 2.2). If K_i is altered to K_i' or to K_{da} by some change in conditions, of course, the value of the modified reverse rate constant should be employed. The solutions in m_d and $d\,m_d/d\,t$ are given by

$$m_d = (1 + K_e) - m , \tag{77}$$

$$\frac{d\,m_d}{d\,t} = -\frac{d\,m}{d\,t} . \tag{78}$$

In these latter two expressions, m_d is the amount of enzyme (free and combined) in the bead during the desorption process. The importance of these substitutions is that since K_1 and K_{-1} are interchanged, the times at which one crosses from the lower to the upper rate curve in Fig. 10 can be drastically different for adsorption and desorption. Since the rates on the upper and lower curves are equal at a time ratio of $1 + K_e$ and since K_e can be 10, 100, or even higher, this difference can be large. Examining equations such as (62) and (63), one can see that the quantity $1 + K_e$ serves to modify the time scale inversely. In the "crossover" region, the rate behaves as if the time-scale factor $1 + K_e$ changes from unity to its calculated value. By returning to the definition of $1 - \phi$, we see that it is identically the reciprocal of $1 + K_e$. The reason for our introductory statement to Sect. 2.4 that the quantity $1 - \phi$ can vary from unity to its calculated value now should be clear.

The application of this kinetic analysis to a real affinity chromatography model would be quite difficult. What the equations in this section allow one to do, however, is to estimate the time required for adsorption and for desorption in a single bead. If these times are longer than the time scale of an experiment, then kinetic problems can be anticipated. In examining the literature, one finds statements such as[22] "Under column conditions, the affinity ... is increased about 10^4 times and it appears that ... the behavior of the enzyme system deviates from that of the freely reversible equilibrium found under batchwise conditions." When one realizes that a batch equilibrium involves only adsorption and chromatography requires desorption as well, the possibility of a slow desorption step preventing one from observing the true equilibrium constant of the system during a column experiment is easy to understand. Part of the reason why most enzymologists have not appreciated these problems is that they are not familiar with simultaneous reaction-diffusion processes, but perhaps equally important is the fact that these problems are almost unique to affinity chromatography. Enzymes have low diffusivities, are diluted relative to the immobilized ligand concentration, and bind very strongly to the ligand. One or more of these qualities are absent in ion exchange, other types of chromatography, etc. and kinetic problems normally are not encountered. We would urge any one trying to learn something from a column affinity experiment at least to find the time from Fig. 11 at which "crossover" begins (t_{cs}) and ends (t_{cf}) for the forward and reverse rate constants of his system. Then one can use Eqs. (62) and (63) to find roughly the fractional bead saturation at given times, setting the quantity K_e to zero at $t < t_{cs}$ and to its correct value after t_{cf}. If the constants are not known and cannot be estimated, then experimental batch adsorption *and* desorption studies are a feasible alternative. If either the calculations or experiments suggest a potential kinetic problem, appropriate precautions or preventive measures then can be applied.

In closing this section we return to the reference we made at the beginning of section c to the fact that desorptive equilibration could take 40 or even 400 h for the specific example given. If $k_{-1} = 1.9 \times 10^{-4}$, $K_{-1} = 4.8 \times 10^{-2}$ (for $r = 50\ \mu$ and $D_s = 10^{-7}\ cm^2\ s^{-1}$). Now if $L_0 = 10^{-3}$ M and $K_i = 10^{-6}$ M, $K_e = 10^3$. Figure 11 then indicates that the crossover region will begin at a reduced time of less than 1 and continue until a reduced time of more than 500, or real times of 4 min to 35 h. Obviously, for a higher ligand concentration or lower K_i value, this time could be lengthened considerably.

3 Experimental Determination of Important Parameters

Fortunately, the experimental aspects of affinity chromatography are much better developed than the theoretical aspects. Although there are a number of traps which one must avoid in obtaining data, most of the necessary techniques and pitfalls are similar to those which are familiar to enzymologists. These methods are covered in a number of standard biochemistry texts. Consequently, we will limit ourselves to a brief review of methods which are particularly relevant to measuring the physical parameters which appear in our equations.

The free solution diffusivity or diffusion coefficient is not measured in biochemical work as frequently as it should be, although it is an important parameter in any process involving mass transfer. It most frequently appears as a byproduct of results obtained with the ultracentrifuge. The easiest way to obtain a value for D_s, the diffusivity within the gel phase, is to assume that it equals this diffusivity and then use data which is compiled in a number of references (e.g.[26)]) for free-solution diffusivity. Normally, the gel pores of Sepharose 4B are much larger than the proteins which penetrate them, so this approximation is not a bad one. With very large diffusing molecules or a material such as Sephadex with much smaller pores, one should be much more cautious, and experimental measurement is advisable. A multitude of experimental techniques for measuring this parameter in polymers is given by Crank and Park[27)] and a number of their methods might be applied to affinity gel systems.

One technique which we have successfully used is perhaps worth mentioning since it is relatively simple and makes use of standard biochemical instrumentation. A special flow-through spectrophotometer cuvette was constructed which contained a Sepharose slab 1.8 mm thick, about 21 mm high, and 6.5 mm wide. A flow channel in contact with one of the large faces of the gel was 1.3 mm thick. The cell was placed in a spectrophotometer so that the light beam passed through a quartz window, the flow channel, the gel, and finally a second quartz window. If buffer solution is pumped continuously through the flow channel of the cell and at some time t_1 the circulated solution is suddenly changed to the protein of interest, the optical density at 280 nm suddenly jumps to a higher value and then gradually increases over the time course of about 1 h to still higher values. The sudden jump, of course, is due to protein in a cell with effective thickness 1.3 mm, and the gradual increase is due to protein penetrating the gel. Eventually, the cell path length is effectively 1.3 + 1.8 mm, representing both flow channel and gel thicknesses. The technique and theory are presented in more detail elsewhere[20)], but in summary, between about 2 and 30 min, an excellent straight line is obtained from a plot of A_{280} versus $\sqrt{\theta}$. The diffusivity is then found from

$$D_s = \left(\frac{\text{slope}}{2\,A_s}\right)^2, \tag{79}$$

where A_s if the absorbance of the protein solution contacting the gel when it is measured in a 1 cm path length cell.

An example of data from such an experiment is given in Fig. 13. The calculated diffusivity in this case was 4.06×10^{-6} cm^2 s^{-1}. The enzyme was carbonic anhydrase, and

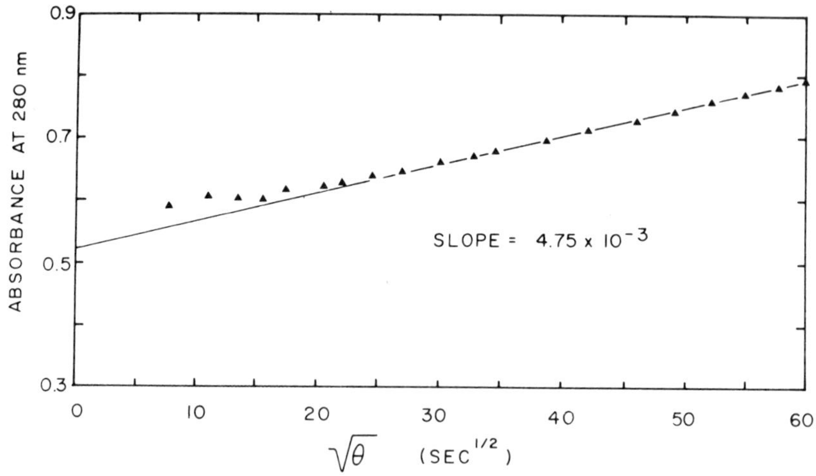

Fig. 13. A simple technique for measuring diffusivity in gels can be implemented with a spectrophotometer. Sample data for a reactive bead are given here

the gel contained the bound ligand p-aminomethylbenzene-sulfonamide. For a non-reactive gel, the measured diffusivity was 1.39×10^{-6} cm^2 s^{-1}. The fact that the diffusion coefficient was increased while we have previously spoken of a decreased rate of equilibration may seem contradictory, but it is not. The *capacity* of the bead to absorb enzyme is increased much more then the *rate* of uptake. Consequently, the entire process of equilibration is slowed down, and we have approximated this effect in Eq. (23) as a decrease in the *effective* diffusion coefficient.

Ackers[28)] has used similar procedures to scan intact columns containing gels in bead form, so this alternative may be tried by one wishing to work with unmodified Sepharose 4B. His technique offers the advantage that the bead configuration is more representative of the real column, but the disadvantage that considerably more light scattering results from the beads than from a flat gel slab. In either case some ligands also absorb ultraviolet light strongly. Our experience with flat gel slabs has resulted in background absorbance values of one to two optical density units. An absorbing solution or partial mask in the reference cell position of the spectrophotometer can be used to compensate for this background, provided that it is not too severe.

The internal gel volume plus void volume v + V can be found under column conditions by applying a non-interacting protein pulse of the same molecular weight as the desired enzyme to the column and measuring the peak emergence time. The chief worry here is how one can be sure that his "inert" protein truly does not interact through hydrophobic and/or ionic attraction with the affinity gel. Operating at various pH values with and without salt, etc. would give one more confidence in his results. The void volume V is obtained in a similar way using a very high molecular weight tracer such as Blue Dextran 2000. Although our theory does not explicitly consider the effect of affinity plus (e.g.) ionic interactions, we would also advise testing the enzyme of interest on the column under non-binding conditions while these other experiments are being done. A large excess of inhibitor, added salt, altered pH, etc. would serve to create these "non-binding" conditions. Such experiments would give one more confidence in the true specific nature of the affinity binding, or they might uncover a secondary non-specific interaction which could be eliminated by slight changes in operating conditions to give a "pure" affinity system. Such studies are quite important because a number of recent reports[17, 29)] have indicated hitherto unrecognized hydrophobic or ionic interactions superimposed on the biospecific adsorption phenomenon. Obviously the true effective (combined) K_i would be needed to apply any of our equations properly.

The measurement of enzyme-inhibitor binding when both are in solution is a classical problem which may be handled by determining the altered kinetics of the enzyme, by equilibrium dialysis, by ultrafiltration, sedimentation, or gel filtration. Bush and Alvin[30)] have cataloged various methods. The gel filtration technique, originated by Hummel and Dreyer[31)] has been recommended recently[32)] as being less prone to difficulties in systems where the protein can dissociate into subunits.

For measuring enzyme-ligand binding, a batch equilibrium experiment could be carried out or one could do a column experiment with such low enzyme concentrations that he could be sure equations such as (17) were applicable. The batch experiment is preferable as a means of assuring that equilibrium has been attained and of avoiding the redundancy of using our theory to check itself. Data from such batch studies may be subjected to a double reciprocal plot[33] or a Scatchard plot[34] to find L_0 and K_i.

4 Present Status and Future Trends

Affinity chromatography is beginning to emerge from its status as a child prodigy in the biochemical separations field. As it matures into what may well become one of the most reliable and useful tools of the biochemical engineer, it should increasingly provide manufacturers with better and lower cost products "off-the-shelf". During this period we forsee that analytical and predictive aspects of the process will begin to catch up with the spectacular laboratory successes it has achieved to date. Industrial-scale suppliers of enzymes, who need mathematical approaches to maximize their production economies, should soon begin to develop models far more accurate and sophisticated than those which have been offered here. If our suggestions stimulate interest in predicting the outcome of affinity separations, they will have served an important and useful function. Indeed, the rapid demise of our simple models is even to be hoped for, since that will mean that better experimental data and models have become available. Such a development could not but be beneficial to all who work with enzymes and other substances contained in that complex entity we call the cell.

5 Acknowledgment

D. J. G. gratefully acknowledges the support of the Alexander von Humboldt Foundation and the Fulbright Commission during the 1976–77 academic year, which enabled him to work on this and several other projects. Research support at the Max Planck Institute for Biophysical Chemistry (Abteilung Membranen) and by Professor Klaus Mosbach at Lund University are also much appreciated. Discussions with a number of colleagues, particularly Dr. W. Vaz, were most useful. Mrs. G. Dauda typed the manuscript.

6 Nomenclature

English

A	A parameter defined by Eq. (4)
A_s	Absorbance of a protein solution in a 1 cm path length cell
B	A parameter defined by Eq. (3)
C	A parameter defined by Eq. (7)

d_p	The diameter of a gel bead
D	A parameter defined by Eq. (9)
D_i	The dispersion parameter in chromatography, Eq. (26)
D_s	The protein diffusivity within the gel
E	The equilibrium concentration of enzyme
E_i	The pseudo-concentration of bound plus entrapped enzyme
EI	The equilibrium concentration of enzyme-inhibitor complex
E_0	The initial enzyme concentration before equilibration
EL	The equilibrium concentration of enzyme-ligand complex
f_{BE}	The fraction of initial enzyme remaining bound to ligand after n washes and one elution
$f_{BE(j)}$	Defined as for f_{BE} but after j elution steps, Eq. (48)
$f_{B(n)}$	The fraction of initial enzyme remaining bound to ligand after n washing steps, Eq. (43)
$f_{BT(n)}$	The fraction of enzyme bond and entrapped within the gel after n washing steps, Eq. (49)
f_{CR}	The fraction of contaminant recovered, Eqs. (38) and (44)
$f_{CR(j)}$	The fraction of contaminant recovered in elution j
f_E	The enzyme recovered during an elution as a fraction of the enzyme in the gel following all washing steps
f_i	The fraction of enzyme lost in washing step (2.3.1)
f_0	The fraction of enzyme failing to bind during the binding step
f_R	The fraction of initial enzyme recovered in V′ after n washes and one elution
$f_{R(m)}$	The fraction of initial enzyme recovered in V′ after with elution m after n washes
H	The height of a theoretical equilibrium stage
I	The soluble inhibitor concentration at equilibrium
I_0	The initial concentration of soluble inhibitor
k_1	The second-order forward rate constant, Eq. (1)
k_1'	The first-order forward rate constant, Eq. (56)
k_{-1}	The first-order reverse rate constant, Eqs. (1) and (56)
K_{da}	The apparent dissociation constant, Eq. (35)
K_e	The ratio of L_0 to K_i
K_i	The enzyme-ligand dissociation constant, Eq. (5)
K_i'	The altered value of K_i during elution
K_s	The enzyme-inhibitor dissociation constant, Eq. (34)
K_1	A dimensionless rate constant, Eq. (69)
K_{-1}	A dimensionless rate constant with k_{-1} replacing k_1, Eq. (69)
L	The equilibrium free ligand concentration
L_c	The length of the chromatography column
L_0	The initial ligand concentration
m	The amount of bound plus free enzyme in a bead, Eq. (65)
m_c	The integral amount of m during crossover conditions, Eq. (71)
m_d	The quantity of m during a desorption process, Eq. (77)
m_{eq}	The amount of bound plus free enzyme at equilibrium, Eq. (64)
m_{fl}	The integral amount of m during film-limiting conditions, Eq. (72)
P	The ratio of V′ to v
P_f	The purification factor, Eq. (55)
r	The radius of a bead
R	The fractional time a molecule remains in the mobile phase during chromatography, Eq. (15)
R_c	The roughly constant rate of enzyme gain or loss during "crossover" conditions, Eq. (68)
R_{fl}	The external mass transfer film-limiting rate, Eq. (70)
t	A dimensionless time, Eq. (59)
t_c	The time required to elute an enzyme peak from a column or the time during column saturation when the exit concentration rises to 50% of the inlet concentration
t_{cs}	The time at which the "crossover" period starts
t_{cf}	The time at which the "crossover" period ends

t_0	The time required to elute the void volume of a column
$t_{1/2}$	The half-time for a first-order reaction, Eq. (36)
T	An arbitrary variable, Eq. (31)
v	The inter-gel volume penetrable by enzyme
v_1	The linear velocity of mobile phase
V	The void volume in a column surrounding the beads or the volume of solution in contact with the beads during a batch experiment
V'	The volume of eluting solution during a batch experiment
V_c	The volume of solution needed to elute a peak from a column
w	The volume of a washing solution during a batch experiment

Greek

ϕ	The fraction of enzyme molecules bound to ligand at equilibrium, Eq. (24)
θ	Real time
τ_c	The standard deviation of peak width, Eq. (25)

7 Literature Citations

1. Lowe, C.R., Dean, P.D.G.: Affinity chromatography. New York: Wiley Interscience 1974
2. Parikh, I., Cuatrecasas, P.: In: Methods of protein separation. Catsimpoolas, N. (ed.), Vol. 1, p. 225. New York: Plenum Press 1975
3. Cuatrecasas, P.: Advances Enzymol. *36*, 29 (1972)
4. Jacoby, W.B., Wilchek, M. (ed.): Methods enzymol., Vol. 34. New York: Academic Press 1974
5. Dunlap, R.B. (ed.): Immobilized biochemicals and affinity chromatography. New York: Plenum Press 1974
6. May, S.W., Zaborsky, O.R.: Sep. and Purif. Methods *3*, 1 (1975)
7. Wankat, P.: Analyt. Chem. *46*, 1400 (1974)
8. Graves, D.J., Wu, Yun-Tai: In: Methods enzymol. Jacoby, W.B., Wilchek, M. (eds.), Vol. 34, p. 140. New York: Academic Press 1974
9. Dunn, B.M., Chaiken, I.M.: Biochemistry *14*, 2343 (1975)
10. Lowe, C.R., Dean, P.D.G.: Biochemistry *14*, 66 (1975)
11. Martin, J.P., Synge, R.L.M.: Biochem. J. *35*, 1358 (1941)
12. Giddings, J.C.: Dynamics of chromatography: principles and theory, Part 1. New York: Marcel Dekker 1965
13. Dunn, B.M., Chaiken, I.M.: Proc. Nat. Acad. Sci. USA *72*, 4840 (1975)
14. Giddings, J.C.: Proc. Nat. Acad. Sci. USA *72*, 34 (1975)
15. Algie, S.H.: Analyt. Chem. *49*, 186 (1977)
16. Kasai, K., Ishii, S.: J. Biochem. *77*, 261 (1975)
17. Chaiken, I.M., Taylor, H.C.: J. Biol. Chem. *251*, 2044 (1976)
18. Denizot, F.C., Delaage, M.A.: Proc. Nat. Acad. Sci. USA *72*, 4840 (1975)
19. Smith, T.W., Skubitz, K.M.: Biochem. *14*, 1496 (1975)
20. Wu, Yun-Tai: A rate and equilibrium analysis of affinity separation processes. PhD Thesis, University of Pennsylvania 1975
21. Harvey, M.J., Lowe, C.R., Craven, D.B., Dean, P.D.G. (part 2, p. 335), Lowe, C.R., Harvey, M.J., Dean, P.D.G. (parts 3, p. 341 and 4, p. 347), Harvey, M.J., Lowe, C.R., Dean, P.D.G. (Part 5, p. 353): Eur. J. Biochem. *41* (1974)
22. Lowe, C.R., Harvey, M.J., Dean, P.D.G.: Eur. J. Biochem. *42*, 1 (1974)
23. Wu, Yun-Tai, Zakian, V., Graves, D.J.: Chem. Eng. Sci. *31*, 153 (1976)
24. Wu, Yun-Tai, Graves, D.J.: Submitted for publication
25. Thomas, J.M., Thomas, W.J.: Introduction to the principles of heterogeneous catalysis, p. 469. London: Academic Press 1967

26. Sober, H.A. (ed): Handbook of biochem, second ed., p. C-10. Cleveland: Chemical Rubber Co. 1970
27. Crank, J., Park, G.S.: Diffusion in polymers. New York: Academic Press 1968
28. Ackers, G.K.: Methods protein sep. *2*, 1 (1976)
29. O Carra, P., Barry, S., Griffen, T.: Biochem. Soc. Trans. *1*, 289 (1973)
30. Bush, M.T., Alvin, J.D.: Ann. N.Y. Acad. Sci. *226*, 36 (1973)
31. Hummel, J.P., Dreyer, W.J.: Biochim. Biophys. Acta *63*, 530 (1962)
32. Cann, J.R., Hinman, N.D.: Biochem. *15*, 4614 (1976)
33. Gawronski, T.H., Wold, F.: Biochem. *11*, 442 (1972)
34. Scatchard, G.: Ann. N.Y. Acad. Sci. *51*, 660 (1949)

Note Added in Proof

An important point which should have been made is that some types of linkages between the affinity ligand and the support gel are subject to spontaneous breakage over long periods of time (see for example T.C.J. Tesser and G.I. Tesser: Experientia *30*, 1228 (1974)). In critical applications, one would be well advised to remeasure ligand concentrations at the end of a long series of experiments. An additional excellent review of some of the practical applications of affinity chromatography which should have been mentioned in the introduction is that by M. Wilchek and C. Hexter: Methods in Biochem. Anal. *23*, 347 (1974). Since our manuscript was submitted, a book on affinity chromatography which is almost encyclopedic in its coverage has been published. (Jaroslava Turková: Affinity Chromatography, Elsevier Scientific Publ. Comp., Amsterdam 1978.) This text would be the logical starting place for any literature search on specific published techniques or applications in affinity chromatography. Another recent review has been written by K. Mosbach: Adv. in Enzymol. *46*, 205 (1978).

Molecular Biology Biochemistry and Biophysics

Editors: A. Kleinzeller, G. F. Springer, H. G. Wittmann

Volume 1
J. H. van't Hoff: **Imagination Science.** Translated into English, with Notes and a General Introduction by G. F. Springer. 1967. 1 portrait. VI, 18 pages.
ISBN 3-540-03933-3

Volume 2
K. Freudenberg, A. C. Neish: **Constitution and Biosynthesis of Lignin.** 1968. 10 figures. IX, 129 pages.
ISBN 3-540-04274-1

Volume 3
T. Robinson: **The Biochemistry of Alkaloids.** 1968. 37 figures. X, 149 pages. ISBN 3-540-04275-X

Volume 4
A. S. Spirin, L. P. Gavrilova: **The Ribosome.** 1969. Out of print

Volume 5
B. Jirgensons: **Optical Activity of Proteins and Other Macromolecules.** 2nd revised and enlarged edition 1973. 71 figures, IX, 199 pages. ISBN 3-540-06340-4 (The title of the first edition was: Optical Rotatory Dispersition of Proteins and Other Macromolecules)

Volume 6
F. Egami, K. Nakamura: **Microbial Ribonucleases.** 1969. 5 figures. IX, 90 pages. ISBN 3-540-04657-7

Volume 8
Protein Sequence Determination. A Sourcebook of Methods and Techniques. Editor: S. B. Needleman. 2nd revised and enlarged edition 1975. 80 figures. XVIII, 393 pages. ISBN 3-540-07256-X

Volume 9
R. Grubb: **The Genetic Markers of Human Immunoglobulins.** 1970. 8 figures. XII, 152 pages.
ISBN 3-540-05211-9

Volume 10
R. J. Lukens: **Chemistry of Fungicidal Action.** 1971. 8 figures, XIII, 136 pages. ISBN 3-540-05405-7

Volume 11
P. Reeves: **The Bacteriocins.** 1972. 9 figures. XI, 142 pages. ISBN 3-540-05735-8

Volume 12
T. Ando, M. Yamasaki, K. Suzuki: **Protamines.** Isolation, Characterization, Structure and Function. 1973. 24 figures, 17 tables. IX, 114 pages. ISBN 3-540-06221-1

Volume 13
P. Jollès, A. Paraf: **Chemical and Biological Basis of Adjuvants.** 1973. 24 figures, 41 tables. VIII, 153 pages.
ISBN 3-540-06308-0

Volume 14
Micromethods in Molecular Biology. Editor: V. Neuhoff With contributions by numerous experts. 1973. 275 figures. (2 in color), 23 tables. XV, 428 pages.
ISBN 3-540-06319-6

Volume 15
M. Weissbluth: **Hemoglobin.** Cooperativity and Electronic Properties. 1974. 50 figures. VIII, 175 pages.
ISBN 3-540-06582-2

Volume 16
S. Shulman: **Tissue Specifitiy and Autoimmunitiy.** 1974. 32 figures. XI, 196 pages. ISBN 3-540-06563-6

Volume 17
Y. A. Vinnikov: **Sensory Reception.** Cytology, Molecular Mechanisms and Evolution. Translated from the Russian by W. L. Gray, B. M. Crook. 1974. 124 figures (173 separate illustrations). XI, 392 pages.
ISBN 3-540-06674-8

Volume 18
H. Kersten, W. Kersten: **Inhibitors of Nucleic Acid Synthesis.** Biophysical and Biochemical Aspects. 1974. 73 figures. IX, 184 pages. ISBN 3-540-06825-2

Volume 19
M. B. Mathews: **Connective Tissue.** Macromolecular Structure and Evolution. 1975. 31 figures. XII, 318 pages. ISBN 3-540-07068-0

Volume 20
M. A. Lauffer: **Entropy-Driven Processes in Biology.** Polymerization of Tobacco Mosaic Virus Protein and Similar Reactions. 1975. 90 figures. X, 264 pages.
ISBN 3-540-06933-X

Volume 21
R. C. Burns, R. W. F. Hardy: **Nitrogen Fixation in Bacteria and Higher Plants.** 1975. 27 figures. X, 189 pages. ISBN 3-540-07192-X

Volume 22
H. J. Fromm: **Initial Rate Enzyme Kinetics.** 1975. 88 figures, 19 tables. X, 321 pages. ISBN 3-540-07375-2

Volume 23
M. Luckner, L. Nover, H. Böhm: **Secondary Metabolism and Cell Differentiation.** 1977. 52 figures, 7 tables. VI, 130 pages. ISBN 3-540-08081-3

Volume 24
Chemical Relaxation in Molecular Biology. Editors: I. Pecht, R. Rigler. 1977. 141 figures, 50 tables. XVI, 418 pages. ISBN 3-540-08173-9

Volume 25
Advanced Methods in Protein Sequence Determination. Editor: S. B. Needleman. 1977. 97 figures, 25 tables. XII, 189 pages. ISBN 3-540-08368-5

Volume 26
A. S. Brill: **Transition Metals in Biochemistry.** 1977. 49 figures, 18 tables. VIII, 186 pages.
ISBN 3-540-08291-3

Volume 27
Effects of Ionizing Radiation on DNA. Physical, Chemical and Biological Aspects. Editors: A. J. Bertinchamps (Coordinating Editor), J. Hüttermann, W. Köhnlein, R. Téoule. 1978. 74 figures, 48 tables. XXII, 383 pages.
ISBN 3-540-08542-4

Volume 28
A. Levitzki: **Quantitative Aspects of Allosteric Mechanisms.** 1978. 13 figures, 2 tables. VIII, 106 pages.
ISBN 3-540-08696-X

Volume 29
E. Heinz: **Mechanics and Energetics of Biological Transport.** 1978. 35 figures, 3 tables. XV, 159 pages.
ISBN 3-540-08905-5

Volume 30
D. Vázquez: **Inhibitors of Protein Biosynthesis.** 1979. 61 figures, 13 tables. X, 312 pages. ISBN 3-540-09188-2

Immobilized Enzymes I

1978. 48 figures, 14 tables. VII, 177 pages
(Advances in Biochemical Engineering, Volume 10)
ISBN 3-540-08975-6

Contents

W. H. Pitcher, Jr.: Design and Operation of Immobilized Enzyme Reactors. – *S. A. Barker, P. J. Somers:* Biotechnology of Immobilized Multienzyme Systems. – *R. A. Messing:* Carriers for Immobilized Biological Active Systems. – *P. Brodelius:* Industrial Applications of Immobilized Biocatalysts. – *B. Solomon:* Starch Hydrolysis by Immobilized Enzymes. Industrial Applications.

Springer-Verlag
Berlin
Heidelberg
New York

Structure and Bonding

Editors: J. D. Dunitz, P. Hemmerich, R. H. Holm, J. A. Ibers, C. K. Jørgensen, J. B. Neilands, D. Reinen, R. J. P. Williams

Volume 20

Biochemistry

1974. 57 figures. IV, 167 pages.
ISBN 3-540-07053-2

Contents

A. S. Mildvan, C. M. Grisham: The Role of Divalent Cations in the Mechanism of Enzyme Catalyzed Phosphoryl and Nucleotidyl Transfer Reactions. – *H. P. C. Hogenkamp, G. N. Sando:* The Enzymatic Reduction of Ribonucleotides. – *W. T. Oosterhuis:* The Electronic State of Iron in Some Natural Iron Compounds: Determinations by Mössbauer and ESR Spectroscopy. – *A. Trautwein:* Mössbauer Spectroscopy on Heme Proteins.

Volume 23

Biochemistry

1975. 50 figures. IV, 193 pages
ISBN 3-540-07332-9

Contents

J. A. Fee: Copper Proteins – Systems Containing the "Blue" Copper Center. – *M. F. Dunn:* Mechanism of Zinc Ion Catalysis in Small Molecules and Enzymes. – *W. Schneider:* Kinetics and Mechanism of Metalloporphyrin Formation. – *M. Orchin, D. M. Bollinger:* Hydrogen-Deuterium Exchange in Aromatic Compounds.

Volume 29

Biochemistry

1976. 51 figures. 48 tables. IV, 219 pages
ISBN 3-540-07886-X

Contents

W. G. Zumft: The Molecular Basis of Biological Dinitrogen Fixation. – *J. J. R. Fraústo da Silva, R. J. P. Williams:* The Uptake of Elements by Biological Systems. – *A. M. Cheh, J. B. Neilands:* The δ-Aminoevulinate Dehydratases: Molecular and Environmental Properties. – *P. J. Sadler:* The Biological Chemistry of Gold. A Metallo-Drug and Heavy-Atom with Variable Valency.